Electromagnetic Modeling and Simulation

Electromagnetic Modeling and Simulation

Editor:

Olga Moreira, Ph.D.

ArclerPress

www.arclerpress.com

Electromagnetic Modeling and Simulation

Editor: Olga Moreira, Ph.D.

708 3rd Avenue, 6th Floor
New York
NY 10017
United States of America
www.arclerpress.com

ISBN: 978-1-68094-396-2

Library of Congress Control Number: 2016951057

About the Editor

Olga Moreira, Ph.D.

Olga Moreira obtained her Ph.D. in Astrophysics from the University of Liege (Belgium) in 2010, her BSc. in Physics and Applied Mathematics from the University of Porto (Portugal). Her post-graduate travels and international collaborations with the European Space Agency (ESA) and European Southern Observatory (ESO) led to great personal and professional growth as a scientist. Currently, she is working as an independent researcher, technical writer, and editor in the fields of Mathematics, Physics, Astronomy and Astrophysics.

Contents

List of Contributors

Guy A. E. Vandenbosch
Katholieke Universiteit Leuven, Belgium

Yu Shao
Wireless Innovation Lab of Yunnan University, School of Information Science and Engineering, Kunming, Yunnan 650091, China

Jing Jing Yang
Wireless Innovation Lab of Yunnan University, School of Information Science and Engineering, Kunming, Yunnan 650091, China

Ming Huang
Wireless Innovation Lab of Yunnan University, School of Information Science and Engineering, Kunming, Yunnan 650091, China

Marek Augustyniak
Gdansk University of Technology, 80-233 Gda´nsk, Poland
DES ART Ltd, 81-969 Gdynia, Poland

Zbigniew Usarek
Gdansk University of Technology, 80-233 Gda´nsk, Poland

Yi Cheng Zhong
Fundamental Science on Extreme High Frequency Laboratory, School of Electronic Engineering, University of Electronic Science and Technology of China, Chengdu 611731, China

Yu Jian Cheng
Fundamental Science on Extreme High Frequency Laboratory, School of Electronic Engineering, University of Electronic Science and Technology of China, Chengdu 611731, China

Yong Fan
Fundamental Science on Extreme High Frequency Laboratory, School of Electronic Engineering, University of Electronic Science and Technology of China, Chengdu 611731, China

Tomasz Karpisz[1]
[1]Institute of Radioelectronics and Multimedia Technology, Warsaw University of Technology, Nowowiejska 15/19, 00-665 Warsaw, Poland
[2]Institute of Electronic Materials Technology, 01-919 Warsaw, Poland
[3]Information Optics Group, Faculty of Physics, Warsaw University, 02-093 Warsaw, Poland

Bartlomiej Salski
Institute of Radioelectronics and Multimedia Technology, Warsaw University of Technology, Nowowiejska 15/19, 00-665 Warsaw, Poland

Ryszard Buczynski
Institute of Electronic Materials Technology, 01-919 Warsaw, Poland
Information Optics Group, Faculty of Physics, Warsaw University, 02-093 Warsaw, Poland

Pawel Kopyt
Institute of Radioelectronics and Multimedia Technology, Warsaw University of Technology, Nowowiejska 15/19, 00-665 Warsaw, Poland

Adam Pacewicz
Institute of Radioelectronics and Multimedia Technology, Warsaw University of Technology, Nowowiejska 15/19, 00-665 Warsaw, Poland

Zhengwei Hao
Electrical and Computer Engineering Department, Northeastern University, Boston, MA 02115, USA

Soheil Saadat
Electrical and Computer Engineering Department, Northeastern University, Boston, MA 02115, USA

Hossein Mosallaei
Electrical and Computer Engineering Department, Northeastern University, Boston, MA 02115, USA

Christof Kaufmann
Fachbereich Elektrotechnik und Informatik, Hochschule Bochum, Bochum, 44801, Germany.

Michael Günther
Lehrstuhl für Angewandte Mathematik/Numerische Analysis, Bergische Universität Wuppertal, Wuppertal, 42119, Germany

Daniel Klagges
Technology Development and Qualification - Simulation, Leopold Kostal GmbH & Co. KG, Lüdenscheid, 58513, Germany

Michael Knorrenschild
Fachbereich Elektrotechnik und Informatik, Hochschule Bochum, Bochum, 44801, Germany

Matthias Richwin
Technology Development and Qualification - Simulation, Leopold Kostal GmbH & Co. KG, Lüdenscheid, 58513, Germany

Sebastian Schöps
Graduate School of Computational Engineering, Technische Universität Darmstadt, Darmstadt, 64293, Germany

E Jan W ter Maten
Lehrstuhl für Angewandte Mathematik/Numerische Analysis, Bergische Universität Wuppertal, Wuppertal, 42119, Germany
Department of Mathematics & Computer Science, TU Eindhoven, PostBox 513, MB Eindhoven, 5600, The Netherlands

Ara Bissal
ABB AB Corporate Research, Forskargränd 7, Västerås 72226, Sweden
Department of Electromagnetic Engineering, Royal Institute of Technology (KTH), Teknikringen 33, Stockholm SE-100 44, Sweden

Anders Eriksson
Department of Mechanics, Royal Institute of Technology (KTH), Osquars backe 18, Stockholm SE-100 44, Sweden

Jesper Magnusson
Department of Electromagnetic Engineering, Royal Institute of Technology (KTH), Teknikringen 33, Stockholm SE-100 44, Sweden

Göran Engdahl
Department of Electromagnetic Engineering, Royal Institute of Technology (KTH), Teknikringen 33, Stockholm SE-100 44, Sweden

Jonathan N. Bringuier
Department of Electrical Engineering, The Pennsylvania State University, University Park, PA 16802, USA

Raj Mittra
Department of Electrical Engineering, The Pennsylvania State University, University Park, PA 16802, USA
King Fahd University of Petroleum and Minerals, Dhahran 31932, Saudi Arabia

Felix Mihai
Department of Mathematical Sciences, College of Science, George Mason University, 4400 University Drive, MS 3F2, Fairfax, VA 22030, USA

Inja Youn
Department of Mathematical Sciences, College of Science, George Mason University, 4400 University Drive, MS 3F2, Fairfax, VA 22030, USA

Igor Griva
Department of Mathematical Sciences, College of Science, George Mason University, 4400 University Drive, MS 3F2, Fairfax, VA 22030, USA

Padmanabhan Seshaiyer
Department of Mathematical Sciences, College of Science, George Mason University, 4400 University Drive, MS 3F2, Fairfax, VA 22030, USA

Chenguo Yao
State Key Laboratory of Power Transmission Equipment & System Security and New Technology, Chongqing University, Chongqing 400044, China

Pan Chen
State Key Laboratory of Power Transmission Equipment & System Security and New Technology, Chongqing University, Chongqing 400044, China
Jiang Bei Power Supply Bureau, Chongqing 401147, China

Congjian Huang
State Key Laboratory of Power Transmission Equipment & System Security and New Technology, Chongqing University, Chongqing 400044, China

Yu Chen
State Key Laboratory of Power Transmission Equipment & System Security and New Technology, Chongqing University, Chongqing 400044, China

Panpan Qiao
State Key Laboratory of Power Transmission Equipment & System Security and New Technology, Chongqing University, Chongqing 400044, China

Song Xue
College of Mechatronics Engineering and Automation, National University of Defense Technology, Changsha 410073, China

Ning He
College of Mechatronics Engineering and Automation, National University of Defense Technology, Changsha 410073, China

Zhiqiang Long
College of Mechatronics Engineering and Automation, National University of Defense Technology, Changsha 410073, China

Jun Xu
Department of Mechanical Engineering, Massachusetts Institute of Technology, 77 Massachusetts Avenue, Cambridge, MA 02139-4307.

Xu Jiang
Department of Mechanical Engineering, Massachusetts Institute of Technology, 77 Massachusetts Avenue, Cambridge, MA 02139-4307

Nicholas Fang
Department of Mechanical Engineering, Massachusetts Institute of Technology, 77 Massachusetts Avenue, Cambridge, MA 02139-4307

Elodie Georget
Aix-Marseille Université, CNRS, Centrale Marseille- Institut Fresnel, Campus universitaire de Saint-Jérôme, 13013 Marseille, France

Redha Abdeddaim
Aix-Marseille Université, CNRS, Centrale Marseille- Institut Fresnel, Campus universitaire de Saint-Jérôme, 13013 Marseille, France

Jean-Michel Geffrin
Aix-Marseille Université, CNRS, Centrale Marseille- Institut Fresnel, Campus universitaire de Saint-Jérôme, 13013 Marseille, France

Mohamed Farhat
Division of Computer, Electrical, and Mathematical Sciences and Engineering, King Abdullah University of Science and Technology (KAUST), Thuwal 23955-6900, Saudi Arabia

Pierre Sabouroux
Aix-Marseille Université, CNRS, Centrale Marseille- Institut Fresnel, Campus universitaire de Saint-Jérôme, 13013 Marseille, France

Stefan Enoch
Aix-Marseille Université, CNRS, Centrale Marseille- Institut Fresnel, Campus universitaire de Saint-Jérôme, 13013 Marseille, France

Sébastien Guenneau
Aix-Marseille Université, CNRS, Centrale Marseille- Institut Fresnel, Campus universitaire de Saint-Jérôme, 13013 Marseille, France

Yinghui Guo
Center for Information Photonics & Communications, School of Information Science & Technology, Southwest Jiaotong University, Chengdu, Sichuan, 610031, China

Lianshan Yan
Center for Information Photonics & Communications, School of Information Science & Technology, Southwest Jiaotong University, Chengdu, Sichuan, 610031, China

Wei Pan
Center for Information Photonics & Communications, School of Information Science & Technology, Southwest Jiaotong University, Chengdu, Sichuan, 610031, China

Liyang Shao
Center for Information Photonics & Communications, School of Information Science & Technology, Southwest Jiaotong University, Chengdu, Sichuan, 610031, China

Bo Zhao
Institute of Target Characteristics and Identification and the State Key Laboratory of Millimeter Waves, Department of Radio Engineering, Southeast University, Nanjing 210096, China

Tie Jun Cui
Institute of Target Characteristics and Identification and the State Key Laboratory of Millimeter Waves, Department of Radio Engineering, Southeast University, Nanjing 210096, China

Erik D. Sapper
Department of Coatings and Polymeric Materials, North Dakota State University, Fargo, ND 58102, USA

Brian R. Hinderliter
Department of Mechanical and Nuclear Engineering, Virginia Commonwealth University, Richmond, VA 23284, USA

Preface

Maxwell's equations form the basis of understanding of problem and solutions involving the interaction of electromagnetic waves with matter. There are two different approaches: analytical formulation (the model is derived from Maxwell's equations under given boundary conditions and medium parameters) and direct numerical simulation methods. Real-world problems such as electromagnetic scattering, electromagnetic radiation, modeling of waveguides can be too complex to solve analytically for a multitude of geometries found in actual devices. Fortunately, the rapidly growing computer capacity and calculation speeds made it possible for numerical simulation methods to overcome this inability. Numerical techniques and methods can now solve a wide range of electromagnetic problems by deriving closed-form solutions of Maxwell's equations under numerous constitutive equations of media and boundary conditions (e.g. by computing the electric and magnetic fields across the given domain, power flow direction, waveguide normal modes, media-generated wave dispersion and scattering).

This book is a collection of the contemporary research articles focused on topics of electromagnetic modeling and simulation methods such as the well-known Finite Element Model (FEM), Finite-difference time-domain (FDTD), Finite integration technique (FIT) and Method of Moments (MoM) applied to different media and boundary conditions. For instance, the first chapters include a review these techniques and tools applied to the study of plasmonic structures, graphene modeling, and electromagnetic nondestructive testing. The following chapters are focused on FEM and FDTD based methods applied to the analysis of photonic fibers, and metamaterials integrated with active and tunable circuit elements. The remaining chapters are focused on electromagnetic modeling of specific systems and technologies. Some examples are the modeling of high voltage direct current in multi-terminal grids, inductive charging systems in wirelessly charge electric vehicles, on-body wireless transmission in Body-Area-Networks (BANs) and more.

Editor
Olga Moreira, Ph.D.

INTRODUCTION

An electromagnet is a magnet that is created using an electric current created by electricity. Since electricity can be turned on and off, so can an electromagnet. It can even be weakened or strengthened by decreasing or increasing the current.

Electromagnetism, science of charge and of the forces and fields associated with charge. Electricity and magnetism are two aspects of electromagnetism.

Electricity and magnetism were long thought to be separate forces. It was not until the 19th century that they were finally treated as interrelated phenomena. In 1905 Albert Einstein's special theory of relativity established beyond a doubt that both are aspects of one common phenomenon. At a practical level, however, electric and magnetic forces behave quite differently and are described by different equations. Electric forces are produced by electric charges either at rest or in motion. Magnetic forces, on the other hand, are produced only by moving charges and act solely on charges in motion.

Electric phenomena occur even in neutral matter because the forces act on the individual charged constituents. The electric force, in particular, is responsible for most of the physical and chemical properties of atoms and molecules. It is enormously strong compared with gravity. For example, the absence of only one electron out of every billion molecules in two 70-kilogram (154-pound) persons standing two metres (two yards) apart would repel them with a 30,000-ton force. On a more familiar scale, electric phenomena are responsible for the lightning and thunder accompanying certain storms.

Electric and magnetic forces can be detected in regions called electric and magnetic fields. These fields are fundamental in nature and can exist in space far from the charge or current that generated them. Remarkably, electric fields can produce magnetic fields and vice versa, independent of any external charge. A changing magnetic field produces an electric field, as the English physicist Michael Faraday discovered in work that forms the basis of electric power generation. Conversely, a changing electric field produces a magnetic field, as the Scottish physicist James Clerk Maxwell deduced. The mathematical equations formulated by Maxwell incorporated light and wave phenomena into electromagnetism. He showed that electric and magnetic fields travel together through space as waves of electromagnetic radiation, with the changing fields mutually sustaining each other. Examples of electromagnetic waves traveling through space independent of matter are radio and television waves, microwaves, infrared rays, visible light, ultraviolet light, X-rays, and gamma rays. All of these waves travel at the same speed—namely, the velocity of light (roughly 300,000 kilometres, or 186,000 miles, per second). They differ from each other only in the frequency at which their electric and magnetic fields oscillate.

Maxwell's equations still provide a complete and elegant description of electromagnetism down to, but not including, the subatomic scale. The interpretation of his work, however, was broadened in the 20th century. Einstein's special relativity theory merged electric and magnetic fields into one common field and limited the velocity of all matter to the velocity of electromagnetic radiation. During the late 1960s, physicists discovered that other forces in nature have fields with a mathematical structure similar to that of the electromagnetic field. These other forces are the nuclear force, responsible for the energy released in nuclear fusion, and the weak force, observed in the radioactive decay of unstable atomic nuclei. In particular, the weak and electromagnetic forces have been combined into a common force called the electroweak force. The goal of many physicists to unite all of the fundamental forces, including gravity, into one grand unified theory has not been attained to date.

An important aspect of electromagnetism is the science of electricity, which is concerned with the behavior of aggregates of charge, including the distribution of charge within matter and the motion of charge from place to place. Different types of materials are classified as either conductors or insulators on the basis of whether charges can move freely through their constituent matter. Electric current is the measure of the flow of charges; the laws governing currents in matter are important in technology, particularly in the production, distribution, and control of energy.

The concept of voltage, like those of charge and current, is fundamental to the science of electricity. Voltage is a measure of the propensity of charge to flow from one place to another; positive charges generally tend to move from a region of high voltage to a region of lower voltage. A common problem in electricity is determining the relationship between voltage and current or charge in a given physical situation.

Everyday modern life is pervaded by electromagnetic phenomena. When a light bulb is switched on, a current flows through a thin filament in the bulb; the current heats the filament to such a high temperature that it glows, illuminating its surroundings. Electric clocks and connections link simple devices of this kind into complex systems such as traffic lights that are timed and synchronized with the speed of vehicular flow. Radio and television sets receive information carried by electromagnetic waves traveling through space at the speed of light. To start an automobile, currents in an electric starter motor generate magnetic fields that rotate the motor shaft and drive engine pistons to compress an explosive mixture of gasoline and air; the spark initiating the combustion is an electric discharge, which makes up a momentary current flow.

Coulomb's Law

Many of these devices and phenomena are complex, but they derive from the same fundamental laws of electromagnetism. One of the most important of these is Coulomb's law, which describes the electric force between charged objects. Formulated by the 18th-century French physicist Charles-Augustin de Coulomb, it is analogous to Newton's law for the gravitational force. Both gravitational and electric forces decrease with the square of the distance between the objects, and both forces act along a line between them. In Coulomb's law, however, the magnitude and sign of the electric force are determined by the charge, rather than the mass, of an object. Thus, charge determines how electromagnetism influences the motion of charged objects. (Charge is a basic property of matter. Every constituent of matter has an electric charge with a value that can be positive, negative, or zero. For example, electrons are negatively charged, and atomic nuclei are positively charged. Most bulk matter has an equal amount of positive and negative charge and thus has zero net charge.)

According to Coulomb, the electric force for charges at rest has the following properties:

(1) Like charges repel each other; unlike charges attract. Thus, two negative charges repel one another, while a positive charge attracts a negative charge.

(2) The attraction or repulsion acts along the line between the two charges.

(3) The size of the force varies inversely as the square of the distance between the two charges. Therefore, if the distance between the two charges is doubled, the attraction or repulsion becomes weaker, decreasing to one-fourth of the original value. If the charges come 10 times closer, the size of the force increases by a factor of 100.

(4) The size of the force is proportional to the value of each charge. The unit used to measure charge is the coulomb (C). If there were two positive charges, one of 0.1 coulomb and the second of 0.2 coulomb, they would repel each other with a force that depends on the product 0.2×0.1. If each of the charges were reduced by one-half, the repulsion would be reduced to one-quarter of its former value.

Static cling is a practical example of the Coulomb force. In static cling, garments made of synthetic material collect a charge, especially in dry winter air. A plastic or rubber comb passed quickly through hair also becomes charged and will pick up bits of paper. The synthetic fabric and the comb are insulators; charge on these objects cannot move easily from one part of the object to another.

Principle of Charge Conservation

Coulomb's law, the principle of charge conservation is a fundamental law of nature. According to this principle, the charge of an isolated system cannot change. If an additional positively charged particle appears within a system, a particle with a negative charge of the same magnitude will be created at the same time; thus, the principle of conservation of charge is maintained. In nature, a pair of oppositely charged particles is created when high-energy radiation interacts with matter; an electron and a positron are created in a process known as pair production.

The smallest subdivision of the amount of charge that a particle can have is the charge of one proton, $+1.602 \times 10^{-19}$ coulomb. The electron has a charge of the same magnitude but opposite sign—*i.e.,* -1.602×10^{-19} coulomb. An ordinary flashlight battery delivers a current that provides a total charge flow of approximately 5,000 coulomb, which corresponds to more than 10^{22} electrons, before it is exhausted.

Electric current is a measure of the flow of charge, as, for example, charge flowing through a wire. The size of the current is measured in amperes and symbolized by i. An ampere of current represents the passage of one coulomb of charge per second, or 6.2 billion billion electrons (6.2×10^{18} electrons) per second. A current is positive when it is in the direction of the flow of positive charges; its direction is opposite to the flow of negative charges.

Electric Fields and Forces

The force and conservation laws are only two aspects of electromagnetism, however. Electric and magnetic forces are caused by electromagnetic fields. The term field denotes a property of space, so that the field quantity has a numerical value at each point of space. These values may also vary with time. The value of the electric or magnetic field is a vector—i.e., a quantity having both magnitude and direction. The value of the electric field at a point in space, for example, equals the force that would be exerted on a unit charge at that position in space.

Every charged object sets up an electric field in the surrounding space. A second charge "feels" the presence of this field. The second charge is either attracted toward the initial charge or repelled from it, depending on the signs of the charges. Of course, since the second charge also has an electric field, the first charge feels its presence and is either attracted or repelled by the second charge, too.

The electric field from a charge is directed away from the charge when the charge is positive and toward the charge when it is negative. The electric field from a charge at rest is shown in Figure 1 for various locations in space. The arrows point in the direction of the electric field, and the length of the arrows indicates the strength of the field at the midpoint of the arrows.

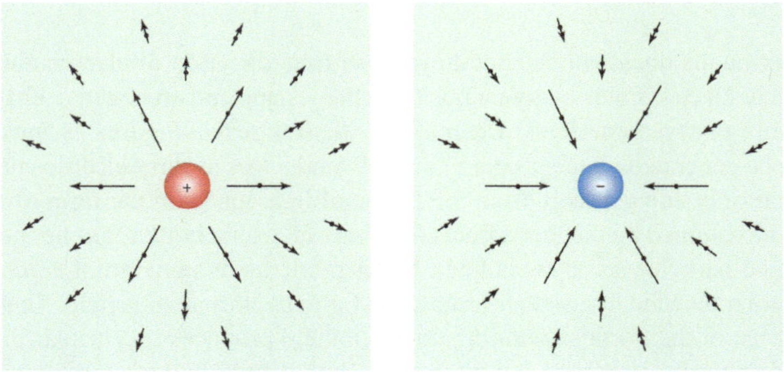

Figure 1: *Electric fields.* (Left) Field of a positive electric charge; (right) field of a negative electric charge.

If a positive charge were placed in the electric field, it would feel a force in the direction of the field. A negative charge would feel a force in the direction opposite the direction of the field.

In calculations, it is often more convenient to deal directly with the electric field than with the charges; frequently, more is known about the field than about the distribution of charges in space. For example, the distribution of charges in conductors is generally unknown because the charges move freely within the conductor. In static situations, however, the electric field in a conductor in equilibrium has a definite value, zero, because any force on the charges inside the conductor redistributes them until the field vanishes. The unit of electric field is newtons per coulomb, or volts per metre.

The electric potential is another useful field. It provides an alternative to the electric field in electrostatics problems. The potential is easier to use, however, because it is a single number, a scalar, instead of a vector. The difference in potential between two places measures the degree to which charges are influenced to move from one place to another. If the potential is the same at two places (i.e., if the places have the same voltage), charges will not be influenced to move from one place to the other. The potential on an object or at some point in space is measured in volts; it equals the electrostatic energy that a unit charge would have at that position. In a typical 12-volt car battery, the battery terminal that is marked with a + sign is at a potential 12 volts greater than the potential of the terminal marked with the − sign. When a wire, such as the filament of a car headlight, is connected between the + and the − terminals of the battery, charges move through the filament as an electric current and heat the filament; the hot filament radiates light.

Magnetic Fields and Forces

The magnetic force influences only those charges that are already in motion. It is transmitted by the magnetic field. Both magnetic fields and magnetic forces are more complicated than electric fields and electric forces. The magnetic field does not point along the direction of the source of the field; instead, it points in a perpendicular direction. In addition, the magnetic force acts in a direction that is perpendicular to the direction of the field. In comparison, both the electric force and the electric field point directly toward or away from the charge.

The present discussion will deal with simple situations in which the magnetic field is produced by a current of charge in a wire. Certain materials, such as copper, silver, and aluminum, are conductors that allow charge to flow freely from place to place. If an external influence establishes a current in a conductor, the current generates a magnetic field. For a long straight wire, the magnetic field has a direction that encircles the wire on a plane perpendicular to the wire. The strength of the magnetic field decreases with distance from the wire. The arrows in Figure 2 represent the size and direction of the magnetic field for a current moving in the direction indicated. Figure 2A shows an end view with the current coming toward the reader, while Figure 2B provides a three-dimensional view of the magnetic field at one position along the wire.

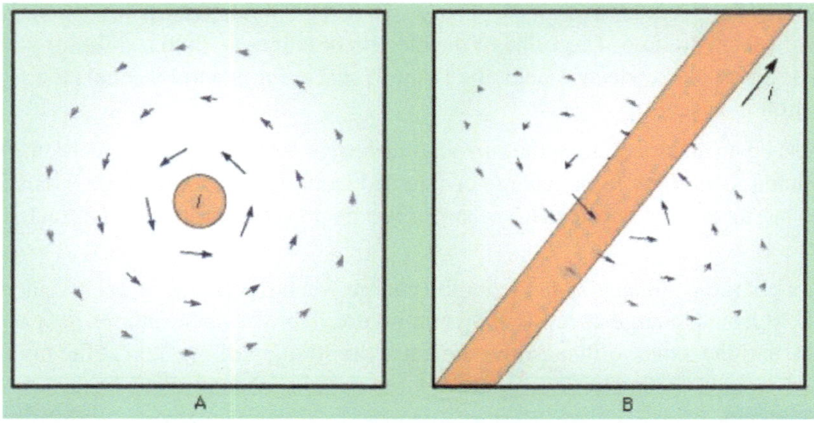

In subsequent figures, continuous lines will be used to represent the direction of electric and magnetic fields. These lines emphasize the important fact that electric fields begin on positive charges and end on negative charges, while magnetic fields do not have beginnings or ends and close on themselves. The magnetic field shown in Figure 2 is unusually simple. Highly complex and useful magnetic fields can be generated by the proper choice of conductors to carry electric currents. Under development are thermonuclear fusion reactors for obtaining energy from the fusion of light nuclei in the form of very hot plasmas of hydrogen isotopes. The plasmas have to be confined by magnetic fields (dubbed "magnetic bottles") as no material container can withstand such high temperatures. Charged particles are also confined by magnetic fields in nature. Large numbers of charged particles, mostly protons and electrons, are trapped in huge bands around the Earth by its magnetic field. These bands are known as the Van Allen radiation belts. Disturbance of the Earth's confining magnetic field produces spectacular displays, the so-called northern lights, in which trapped charged particles are freed and crash through the atmosphere to Earth.

Interaction of a Magnetic Field with a Charge

How does the magnetic field interact with a charged object? If the charge is at rest, there is no interaction. If the charge moves, however, it is subjected to a force, the size of which increases in direct proportion with the velocity of the charge. The force has a direction that is perpendicular both to the direction of motion of the charge and to the direction of the magnetic field. There are two possible precisely opposite directions for such a force for a given direction of motion. This apparent ambiguity is resolved by the fact that one of the two directions applies to the force on a moving positive charge while the other direction applies to the force on a moving negative charge. Figure 3 illustrates the directions of the magnetic force on positive charges and on negative charges as they move in a magnetic field that is perpendicular to the motion.

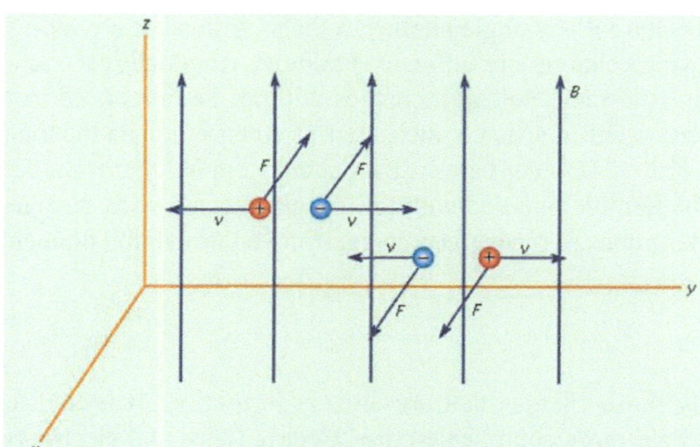

Depending on the initial orientation of the particle velocity to the magnetic field, charges having a constant speed in a uniform magnetic field will follow a circular or helical path.

Electric currents in wires are not the only source of magnetic fields. Naturally occurring minerals exhibit magnetic properties and have magnetic fields. These magnetic fields result from the motion of electrons in the atoms of the material. They also result from a property of electrons called the magnetic dipole moment, which is related to the intrinsic spin of individual electrons. In most materials, little or no field is observed outside the matter because of the random orientation of the various constituent atoms. In some materials such as iron, however, atoms within certain distances tend to become aligned in one particular direction.

Magnets have numerous applications, ranging from use as toys and paper holders on home refrigerators to essential components in electric generators and machines that can accelerate particles to speeds approaching that of light. The practical application of magnetism in technology is greatly enhanced by using iron and other ferromagnetic materials with electric currents in devices like motors. These materials amplify the magnetic field produced by the currents and thereby create more powerful fields.

While electric and magnetic effects are well separated in many phenomena and applications, they are coupled closely together when there are rapid time fluctuations. Faraday's law of induction describes how a time-varying magnetic field produces an electric field (see below Faraday's law of induction). Important practical applications include the electric generator and transformer. In a generator, the physical motion of a magnetic field produces electricity for power. In a transformer, electric power is converted from one voltage level to another by the magnetic field of one circuit inducing an electric current in another circuit.

The existence of electromagnetic waves depends on the interaction between electric and magnetic fields. Maxwell postulated that a time-varying electric field produces a magnetic field. His theory predicted the existence of electromagnetic waves in which each time-varying field produces the other field. For example, radio waves are generated by electronic circuits known as oscillators that cause rapidly oscillating currents to flow in antennas; the rapidly varying magnetic field has an associated varying electric field. The result is the emission of radio waves into space (see electromagnetic radiation: Generation of electromagnetic radiation).

Many electromagnetic devices can be described by circuits consisting of conductors and other elements. These circuits may operate with a steady flow of current, as in a flashlight, or with time-varying currents. Important elements in circuits include sources of power called electromotive forces; resistors, which control the flow of current for a given voltage; capacitors, which store charge and energy temporarily; and inductors, which also store electrical energy for a limited time. Circuits with these elements can be described entirely with algebra. (For more complicated circuit elements such as transistors, see semiconductor device and integrated circuit).

Two mathematical quantities associated with vector fields, like the electric field E and the magnetic field B, are useful for describing electromagnetic phenomena. They are the flux of such a field through a surface and the line integral of the field along a path. The flux of a field through a surface measures how much of the field penetrates through the surface; for every small section of the surface, the flux is proportional to the area of that section and depends also on the relative orientation of the section and the field. The line integral of a field along a path measures the degree to which the field is aligned with the path; for every small section of path, it is proportional to the length of that section and is also dependent on the alignment of the field with that section of path. When the field is perpendicular to the path, there is no contribution to the line integral. The fluxes of E and B through a surface and the line integrals of these fields along a path play an important role in electromagnetic theory. As examples, the flux of the electric field E through a closed surface measures the amount of charge contained within the surface; the flux of the magnetic field B through a closed surface is always zero because there are no magnetic monopoles (magnetic charges consisting of a single pole) to act as sources of the magnetic field in the way that charge is a source of the electric field.

Electromagnetic Simulation

Using approximate computations to Maxwell's equations and Faraday's law, electromagnetic simulations are models of electromagnetics and their effects on the environment and physical structures about them. An electromagnetic simulation can be used to aim a satellite antenna in the right direction for maximum channels and clarity and judge its performance or to determine wave propagation when not in free space. These simulations can aid in the efficient design of computer chips and point to how to improve performance in major electronics by locating the incompatibilities of components within them. Electromagnetic radiation being picked up and scattered and then absorbed by small particles is used in simulations for science projects at the European Organization for Nuclear Research (CERN) laboratories for their particle accelerators projects. Electromagnetic simulation programs are also being used as tools in college physics labs to teach more effectively as students receive hands-on experience with solving problems using them.

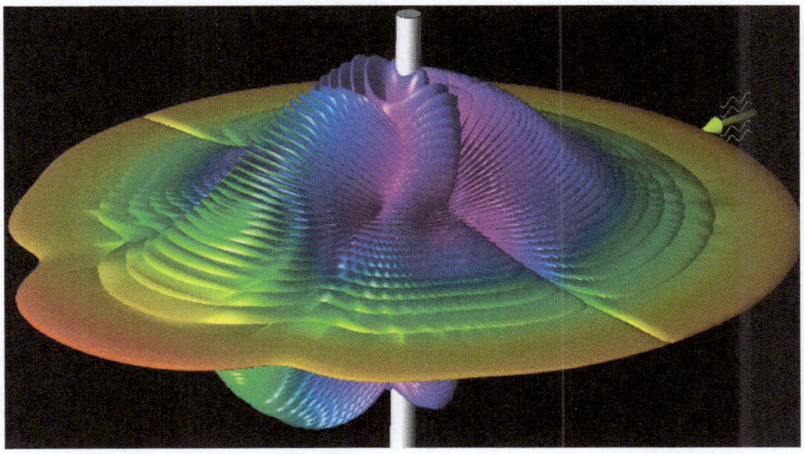

Solving Maxwell's equations at each point in an orthogonal or non-orthogonal grid is one of the avenues to use grids to discretize space by creating a topological survey of the space. Solving these equations in an electromagnetic simulation often reveals problems in computer memory and power as they usually can only be done on supercomputers by time-stepping for each instant of time throughout a whole domain, solving the Maxwell equations as they go or split-stepping using time iterations and fast Fourier transforms. In fluid mechanics, the boundary method or "method of moments" (MoM) can be applied to solve engineering problems, acoustics, and electromagnetics. This focuses calculations only on the border areas of a space rather than the volume values at every time step of the entire space.

A kitchen microwave is analogous to what is known as a Faraday cage, which illustrates how an electromagnetic simulation model might be useful in electromagnetic protection. Electrical currents can be blocked by metal walls or other such shielding devices whereas magnetic currents merely can be moved around the obstruction. In Faraday's cage, when the walls of the cage are grounded, an electric current's path is disturbed by electrons acting as electric charge carriers in a mesh pattern and compensate for the field; this cause the electric current to dissipate. Just as the mesh screen on the front of a microwave door blocks microwaves from escaping the device because microwaves are larger than the tiny holes in the mesh, an electromagnetic mesh simulation can design good protective shielding from electrical currents.

An electromagnetic simulation method that solves Maxwell equations by cycling through an electric field for one instant and then cycling through a magnetic field for the next instant and repeatedly alternating over and over again is known as the finite-difference time-domain (FDTD) method for producing simulations. EM wave interaction with material structures engineering problems have been solved by this method more than any other in the US since around 1990. It is used in solving for radar signature technologies, wireless technologies and biomedical imaging, just to name a few of its applicable uses.

Wave modeling for electromagnetic simulation and analysis of circuits can be performed using the partial element equivalent circuit (PEEC) three-dimensional (3-D) full-wave modeling method. Integral equations are interpreted as Kirchhoff's voltage law and, using PEEC, are applied to a PEEC cell which gives the 3-D geometries solution of a complete circuit, allowing additional circuits to be piggy-backed onto the direct current design. Using models such as this in electromagnetic simulation saves time and money costs in the manufacture of integrated circuits.

College physics departments are starting to make use of video games designed to give students lessons via electromagnetic simulation to visually depict to the students the phenomena of physics representations. This can help students gain a better grasp of concepts and afford their brains experiences that reveal to them weaknesses in their own understanding and the steps to take to strengthen these. Both students and instructors alike have found that both faster and more in-depth learning can be facilitated using real-world examples of physics concept solving by means of electromagnetic simulation software.

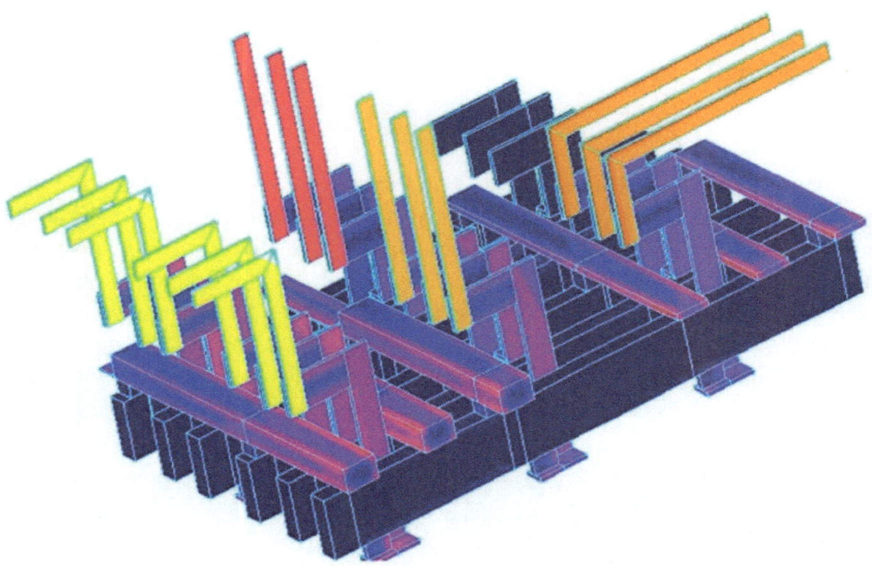

CHAPTER 1 COMPUTATIONAL ELECTROMAGNETICS IN PLASMONICS

Guy A. E. Vandenbosch

INTRODUCTION

In electromagnetics, numerical techniques have been essential in the development of new technology in the last two decades. The rapidly growing computer capacity and calculation speeds make accurate solutions of very complex problems feasible. This has been especially true in the design of microwave and millimeter wave components and antennas. Whereas 30 years ago, the design of an antenna was based on simple analytical models, or trial and error strategies, nowadays, simulations seem to be as crucial to the design as real measurements.

The situation is quite different in plasmonics. Plasmonics is a quite novel research field and the application of computational electromagnetics in plasmonics can be categorized as "very recent". There are many challenges that still need to be faced and "missing links" that have to be solved.

The plasmonic structures targeted are structures in the order of magnitude of a wavelength at plasmonic frequencies, i.e. at near IR and optical frequencies, and beyond. Although this frequency range is totally different from the traditional range where computational tools have been developed, i.e. the microwave range, in most cases no special modeling techniques have to be used. By far most plasmonic topologies reported in literature have been analyzed / designed with the well-known numerical techniques implemented within in-house developed or commercial software packages. This means that in this chapter the major numerical techniques can be overviewed in a general sense, referring to standard literature. These techniques will not be derived or explained here in full detail. Instead, this chapter focuses on those aspects that come into the picture when the structure is plasmonic.

After the section on techniques and tools, this chapter will focus on the performance of these techniques and tools for plasmonic structures. This is done through an overview of benchmarks available in literature and by considering a few thoroughly analyzed structures. Missing links will be pointed out and suggestions will be given for the future.

FUNDAMENTAL PHYSICAL MODELING DIFFERENCES

This section discusses the differences between classical topologies handled with computational electromagnetics and plasmonic topologies. Classical means structures at much lower frequencies, for example at microwave frequencies, where computational electromagnetics is a well-developed mature field.

Frequency

First of all, it is essential to point out that the interaction between light and plasmonic structures in the frequency bands considered can still be analyzed with a high degree of accuracy using classical electromagnetic theory. Although the frequency is orders of magnitude higher in plasmonics compared to microwaves, Mawell's laws are the same. They are linear, and thus scalable. As long as quantum effects do not have to be taken into account, which is still the case for the plasmonic applications considered, since the structures are not that small [1], there is no fundamental problem. The underlying formulation of Maxwell's equations can remain unaffected.

Citation: Guy A. E. Vandenbosch (2012). Computational Electromagnetics in Plasmonics, Plasmonics - Principles and Applications, Dr. Ki Young Kim (Ed.), InTech, DOI: 10.5772/51166.

The fact that at this small scale, no quantum effects have to be taken into account is really a crucial observation. It means that the concept of a "scatterer", a device able to be excited by electromagnetic waves rather than particles, still works. Basically, the coupling between an electromagnetic (light-) wave and a plasmonic scatterer is thus the same as it is at microwave frequencies, and can be studied in the same way.

Volumetric Currents

A consequence of the high frequencies is the small scale: the elementary building blocks of the topologies considered are at nanoscale. There are two important issues related to this. First, nanoscale fabrication technology of today is only able to generate 3D type structures (i.e. volumes). Thin 2D sheets, where the thickness of the metal is orders of magnitude smaller than the transversal dimensions of the pattern, as commonly used at microwave frequencies (for example a conducting strip or patch), are not possible. Second, for nanostructures operating at plasmonic frequencies the skin depth may be comparable with the structural dimensions, so that currents do flow over the complete volume and the device has to be described with volumetric currents. A surface current description is not sufficient.

Material Characteristics

At microwave frequencies, in most cases the material properties are constant over the frequency bands considered. Also, apart from the losses, most metals behave more or less in the same way, i.e. as good conductors. At IR and optical frequencies however, most metals have very dispersive properties. Permittivities and conductivities may vary orders of magnitude over these frequency bands. The real part of the permittivity may even be negative. In Fig. 1, the permittivity of Cu and Ag are depicted, both real and imaginary part, which corresponds to conductivity. These data were obtained through experimental ellipsometry. Note that the difference between the two metals is enormous. It is evident that this strong variation has a serious impact on the behavior of a device over the frequency band. The variability of material characteristics thus has to be taken into account fully into the modeling.

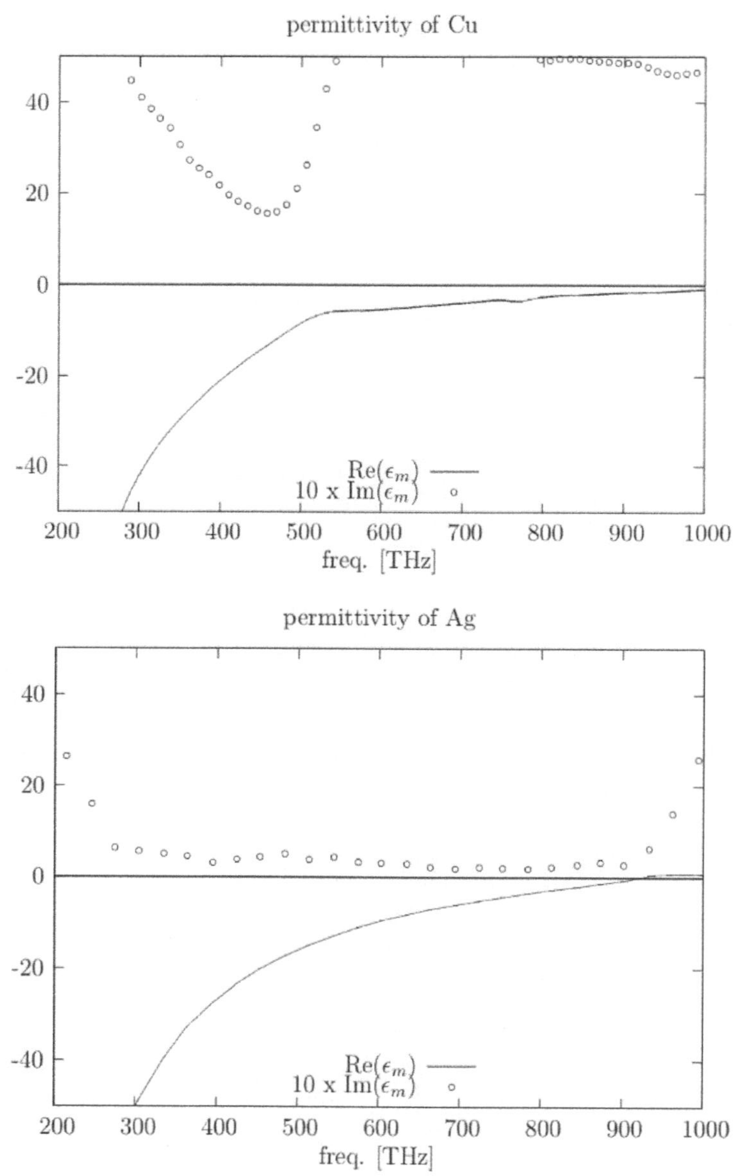

Figure 1. Top: permittivity of Cu; bottom: permittivity of Ag.

MODELING TECHNIQUES

In this section the full wave modeling techniques are introduced and categorized on the basis of their solution method: Finite Elements (FE), Finite Differences in the Time Domain (FDTD), Finite Integration Technique (FIT), and Integral Equations (IE) solved by the Method of Moments (MoM). Based on their theoretical specificities, the application of each method in the case of plasmonics is discussed.

The cradle of computational electromagnetics can be found in the microwave research community. Since this community traditionally is dealing with structures in the order of wavelengths, right from its beginning days, it had no choice than to try to rigorously solve Maxwell's equations. Most standard reference works on full-wave computational electromagnetics by consequence can be found within this community. A history and a comprehensive overview of the different numerical techniques and of their application in computational electromagnetics (CEM) may be found in [2]. Recent work on the last developments in CEM [3] concentrates on the two main approaches of differential and integral methods. A Good review and perspectives concerning the relationship between differential and integral equations (IE) modeling is recommended in a review paper by Miller [4]. A thorough discussion on the different techniques with clarifying examples is also given in [5].

Differential Equation Techniques

In time domain, Maxwell's equations are

$$\nabla \times \mathbf{E} = -\frac{d\mathbf{B}}{dt} \quad (1)$$

$$\nabla \times \mathbf{H} = \frac{d\mathbf{D}}{dt} + \mathbf{J} \quad (2)$$

with E and B the electric field and the magnetic induction, respectively, H and D the magnetic field and the electric induction, respectively, and J the electric current flowing. In free space, and in a homogeneous, isotropic, time-invariant, linear medium (most materials behave like this in the microwave frequency range), the following relations hold

$$\mathbf{D} = \varepsilon \mathbf{E} \quad (3)$$

$$\mathbf{B} = \mu \mathbf{H} \quad (4)$$

They are called the constitutive relations, with ε and μ the permittivity and permeability of the medium surrounding the observation point considered. It is crucial to point out that in plasmonics in general (3) and (4) cannot be used. For example, the very dispersive properties of metals at optical frequencies (and beyond) prohibit this for these materials.

In time domain, the time variation has to be determined. In frequency domain, it is assumed that all field quantities are varying in a sinusoidal way. This means that the variation with time is known, and only the variation of the fields in space has to be determined. Using complex notation for the frequency domain with a exp(jωt) time dependency, Maxwell's equations become

$$\nabla \times \mathbf{E} = -j\omega \mathbf{B} \quad (5)$$

$$\nabla \times (\mu^{-1}\mathbf{B}) = j\omega\varepsilon \mathbf{E} + \mathbf{J} \quad (6)$$

with ω=2π·frequency the pulsation. Note that in (5) and (6) ε and μ are used. In general, here they are depending on frequency, and in this way dispersion is fully taken into account (see also section 2.3). In the field of plasmonics, this is an important difference between computational tools in time and in frequency domain. Computational tools in frequency domain may use directly the well-known concepts of permittivity and permeability, albeit that they become frequency dependent. Computational tools in time domain cannot use these concepts directly. As will be shown later, this complicates things.

From a mathematical perspective, Maxwell's equations are differential equations relating vector fields to each other. Differential equation methods in Computational Electromagnetics are methods that directly consider Maxwell's equations (or the Helmholtz wave equations derived from them), with little analytical preprocessing. Basically, these differential equations, which are valid in any point of space and time, are solved by approximating them by difference equations, which are valid in a discrete set of points in space and time. This is done by chopping up space (and time if time domain is considered) in little pieces in which the field variation has a pre-described profile. This reduces the problem of fields varying over space (and time) to a discrete (matrix) problem that can be handled on a computer. The differences between the several differential equation methods are related to the different ways in which space (and time) can be chopped up.

Since the number of unknowns is proportional to the volume and the resolution considered, differential equation methods are particularly suitable for modeling small full three-dimensional volumes that have complex geometrical details, for example smaller closed-region problems involving inhomogeneous media [6]. Intrinsically, differential equations are less suited for open problems. The reason is that in principle they require a discretization of the entire space under consideration. This space is limited in case of closed problems, but corresponds to infinite space in case of open problems. In practice, this problem is solved by the introduction of techniques like Absorbing Boundary Conditions, and Perfectly Matched Layers (PML) [7]. They mimic the wave arriving at the boundary as travelling further to infinity. The quality of these truncating techniques nowadays is very high so that, in practice, the intrinsic problem with open structures has been overcome, albeit it in an approximate numerical way.

The most popular differential equation-based methods are the Finite Element Method (FEM), for example utilized in Ansoft's HFSS software package, and the Finite-Difference Time Domain method (FDTD), which is employed for example by CST's Time Domain transient solver (in the particular case of Cartesian grids), and by Lumerical.

The Finite Element Method (Fem) [7], [8]

FEM is a method based on solving partial differential equations. It is most commonly formulated based on a variational expression. It subdivides space in elements, for example tetrahedra. Fields inside these elements are expressed in terms of a number of basic functions, for example polynomials. These expressions are inserted into the functional of the equations, and the variation of the functional is made zero. This yields a matrix eigenvalue equation whose solution yields the fields at the nodes. FEM gives rise to a very sparse matrix equation, which can be solved using dedicated matrix algebra technology, leading to very fast solution times, considering the huge number of unknowns.

Its first formulations were developed as matrix methods for structural mechanics. This lead to the idea to approximate solids and Courant (1942) introduced an assembly of triangular elements and the minimum of potential energy to torsion problems [9]. The first paper on the application of FEM to electrical problems appeared in 1968 [10]. An extensive review on the history of FEM in electromagnetics was published in an issue of the Antennas and Propagation Magazine [11]. FEM normally is formulated in the frequency domain, i.e. for time-harmonic problems. This means that, as for IE-MoM, the solution has to be calculated for every frequency of interest.

Numerous references can be given developing, explaining, and using FEM. A good book to start with is [8]. A software tool using FEM and very widely spread is Ansoft HFSS.

The Finite-Difference Time-Domain Technique (Fdtd) [12], [13], [14]

The nature of Maxwell's differential equations is that the time derivative of the H-field is dependent on the curl of the E-field, and the time derivative of the H-field is dependent on the curl of the E-field. These basic properties result in the core FDTD time-stepping relation that, at any point in space, an updated value of an E/H-field in time is dependent on the stored value of the E/H-field and the numerical curl of the local distribution of the H/E-field in space. The numerical translation into a time-stepping algorithm was introduced by Yee in 1966. Indeed, swapping between E-field and H-field updates allows to define a marching-on-in-time process wherein sampled fields of the continuous electromagnetic waves under consideration are used. These waves can be seen to propagate in the Yee lattice, a numerical three-dimensional space lattice comprised of a multiplicity of Yee cells, see Fig. 2. More specifically, Yee proposed a leapfrog scheme for marching-on in time wherein the E-field and H-field updates are staggered so that E-field updates are observed midway during each time-step between successive H-field updates, and vice versa. A huge advantage of FDTD is that this explicit time-stepping scheme avoids the need to solve simultaneous equations, so no matrix inversions are necessary. It also yields dissipation-free numerical wave propagation. Negative is that this scheme results in an upper bound on the time-step to ensure numerical stability. This means that simulations may require many thousands of time-steps for completion. The use of the Yee lattice has proven to be very robust in numerical calculations.

FDTD is extremely versatile since the interaction of an electromagnetic wave with matter can be mapped into the space lattice by assigning appropriate values of permittivity to each electric field component, and permeability to each magnetic field component. This can be done without seriously compromising the speed of the method.

The fact that time is observed directly, and the versatility of the method, make it probably the most efficient technique for complex 3D transient problems. An implementation of the FDTD technique dedicated to plasmonics and photonics can be found in the commercial tool Lumerical [15].

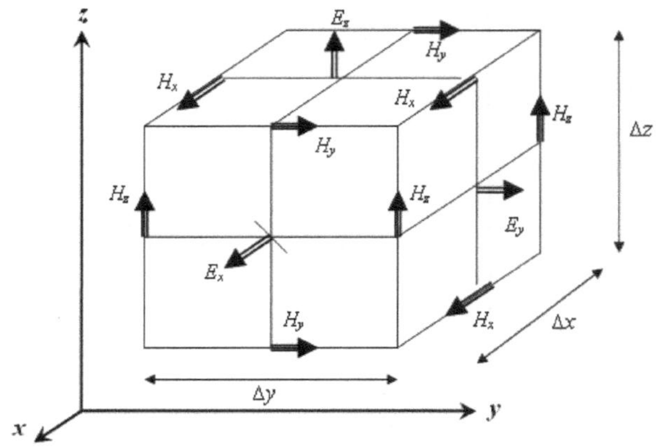

Figure 2. Standard Cartesian Yee cell used in the FDTD technique.

The Finite Integration Technique (FIT) [16]

The Finite Integration Technique was introduced by Weiland in 1977. The word integration does not imply any relation with integral equations. FIT first describes Maxwell's equations on a grid space. The matrix equations for the electromagnetic integral quantities obtained by FIT possess some inherent properties of Maxwell's equations, for example with respect to charge and energy conservation. This makes them very attractive from a theoretical point of view. FIT can be formulated on different kinds of grids, e.g. Cartesian or general non-orthogonal ones, which is a clear advantage. In the time-domain, the resulting discrete grid equations of FIT are, at least in "some" cases, identical to the discrete equations derived with the classical Finite-Difference Time-Domain (FDTD) method. In contrast to FIT, which is applied to the integral form of the field equations, FDTD (as a subset of the finite integration method) is applied to the differential form of the governing Maxwell curl equations. Theoretical links between FDTD and the FIT approach realized in CST may be found in [17]. A comparative study on Time Domain methods was presented in [18]. In some sense, FIT can thus be considered as a powerful generalization of the FDTD technique. A software tool using FIT and very widely spread is CST Microwave Studio.

Integral Equation Techniques (IE)

Integral equation methods make use of Maxwell's equations in integral equation form to formulate the electromagnetic problem in terms of unknown currents flowing on the object to be described. These currents are induced by a field incident on the object. This incident field can be a real incident field, travelling in space, or a bounded wave feeding the object for example via a transmission line. An integral equation solution is fundamentally based on the combination of two equations.

In the case of Electric Field Integral Equations (the case of prime importance to the plasmonics community), the solution is based on considering the electric field. The first equation is

$$\mathbf{E}^{sca}(\mathbf{r}) = \int_{V'} \mathbf{G}(\mathbf{r},\mathbf{r}')\cdot \mathbf{J}^{ind}(\mathbf{r}')dV' \tag{7}$$

which gives the scattered electric field E^{sca} generated in an arbitrary observation point r in space in terms of an induced volumetric electric current distribution J^{ind} flowing within the volume V' of the object considered. The kernel G(r, r') is a so-called dyadic Green's function, which is a tensor, and which relates a Dirac impulse type current flowing in r' to the electric field it generates. The main strength of the integral equation technique is that in many cases the dyadic Green's function can be calculated either directly analytically, such as in homogeneous media, or as some kind of inverse Fourier transform of an analytic function, as is the case for example in layered media. It is crucial to understand that the dyadic Green's function is actually a solution of Maxwell's equations and thus rigorously takes into account the background medium, for example a stack of dielectric layers. This means that unknown currents only have to be assumed on the objects embedded within this background medium. This results in an enormous reduction of unknowns compared to differential equation techniques, which have to model these dielectric layers in just the same way as the objects embedded within them. Basically, the problem formulation automatically covers the entire surrounding space without making any fundamental approximations. As a consequence, the corresponding solution is automatically valid in every point of the background medium. Far field radiation phenomena, surface waves in layered structures, etc., that are vital for efficient and accurate analysis, are analytically included in the solution.

In the case of plasmonics, the second equation is

$$\begin{aligned}\mathbf{J}^{ind}(\mathbf{r}') &= j\omega\,(\varepsilon(\mathbf{r}')-\varepsilon_0)\,\mathbf{E}^{ind}(\mathbf{r}')\\ &= j\omega\,(\varepsilon(\mathbf{r}')-\varepsilon_0)\,(\mathbf{E}^{sca}(\mathbf{r}')+\mathbf{E}^{sca}(\mathbf{r}'))\end{aligned} \tag{8}$$

It expresses the boundary condition which has to be enforced within the object under consideration. It is a relation between the total field, which is the sum of the incident and the scattered field, and the volumetric electric current, which can be considered as partially being a conduction current (due to the imaginary part of the permittivity), and a polarization current (due to the real part of the permittivity).

Combining (7) and (8) yields the equation from which the currents can be solved. This can be done by chopping up the object considered (only the object, and thus not entire space !!!) in little pieces in which the current variation has a pre-described profile (a so-called basis function). This reduces the problem of currents varying over the object to a discrete (matrix) problem that can be handled on a computer. IE-MoM gives rise to a dense matrix equation, which can be solved using standard matrix algebra technology.

Next to the reduced number of unknowns, there are other theoretical advantages linked to the IE-MoM method. The second advantage is that, if properly formulated, IE-MoM *is variationally stable* since most of the output parameters are expressed in integral form over the equivalent currents. This means that even if the calculated currents differ considerably from the exact solution, integral parameters over both currents may remain very similar. Further, this will be illustrated by showing that even with a rather rough mesh high quality physical results may be obtained. A third advantage is that MoM does not heavily suffer from field singularities for example near sharp edges, since they are analytically incorporated inside the Green's functions. For differential equation methods special care (= a fine mesh) should be taken to describe correctly these field singularities.

The main disadvantage of IE-MoM is that, although there are many efforts in that direction, there is still a lack of matrix solvers operating on dense matrices which are comparable in efficiency to the solvers used in the differential equation techniques (mainly FEM), which yield sparse matrices. This has precluded the use of the very flexible volumetric IE-MoM technique for modeling complex structures on widespread computer systems.

IE-MoM is normally applied in the frequency domain, i.e. for time-harmonic problems. This means that the solution has to be determined at each frequency of interest.

Much more details can be found in the classic and basic Method of Moments (MoM) book [19]. There are many variants of the method. In general, boundary equations can be enforced also at the boundaries of volumes (utilizing Surface Integral Equations (SIE) [20]), next to inside the entire volumes themselves(applying Volume Integral Equations (VIE) [21] at the inside of the components, as described above). Also, the integral equations can be written down in different forms (dyadic form, mixed-potential form, hybrid forms, etc. [22]), which give rise to specific implementations. Further, the first theoretical developments in the field of computational plasmonics are appearing in literature [23].

In the following sections, several aspects linked with the use of integral equation techniques for plasmonic structures are discussed.

3D Volumetric Current

The fact that a 3D volumetric current has to be described, as mentioned already, invokes the need for either a real volumetric MoM implementation or the more common surface approach, where the plasmonic component is described by applying the equivalence principle at its surface. The surface approach has already been used, for example in [24]. The volumetric approach as described in [21], has very recently been introduced in the plasmonics community [25]. For realistic structures, where volumes can be embedded within a layer structure, it offers a very flexible solution technique.

Material Characteristics

Although strongly varying material characteristics may seem trivial to implement in an integral equation scheme, in some cases more severe consequences occur. For example, some Green's function calculation schemes are based on the fact that the material properties do not change over the frequency band of interest. Any particular implementation of IE-MoM needs this essential relaxation into its formalism, which possibly needs to be adapted.

An advantage of a frequency domain technique like IE-MoM compared to a time domain technique like FDTD is that the ellipsometric measurement data can be used directly in the tool, without any further fitting. FDTD techniques developed for the optical range tend to fit the dielectric response of the metal to the experimentally determined dielectric permittivity using a Drude model [26] or more sophisticated models (for example in Lumerical). However, this may create problems (see further).

Occurrence of Layer Structures

As explained above, IE technique solvers are formulated making use of Green's functions. These Green's functions can be formulated for multi-layered structures, such that the background medium of the structure may consist of an arbitrary number of horizontal, infinitely stretched, dielectric and metallic layers, which are taken into account analytically. This environment is particularly interesting for plasmonic structures, because they are traditionally deposited on a flat layered dielectric substrate, for example glass. This glass can be contained in the background environment. The only remaining components are local scattering objects with medium or small dimensions compared to the wavelength, such as the dipole in section 5.2.1, but also dots, rods, monomers, dimmers, rings, discs, etc.. With volumetric integral equations these components are replaced by equivalent volume currents, which appear as the primary unknowns in the resulting integral equations. Since the substrates used in normal circumstances are huge compared to the wavelength, edge effects are extremely small and can be neglected. It may thus be concluded that plasmonic structures have specific features which make a IE-MoM solution quite attractive.

SOFTWARE TOOLS

In the following sections, several solvers are briefly described. Several commercial solvers and one academic solver are considered. It has to be emphasized that the author does not claim that this overview is complete. Since the widespread use of computational electromagnetics in plasmonics is quite recent, it is highly probably that there are more solvers that the author is not aware of.

Commercial Software Tools

HFSS [27]: FEM

Since it was one of the first tools in the market, and also due to its generality and flexibility, HFFS is one of the tools heavily used in industrial microwave and millimeter wave design environments. The purpose of HFSS is to extract parasitic parameters (S, Y, Z), visualize 3D electromagnetic fields (near- and far-field), and generate SPICE models, all based on a 3D FEM solution of the electromagnetic topology under consideration. This software is extremely popular and is used for all kinds of purposes. Numerous results for plasmonic topologies can be easily found by googling the words HFSS and plasmonics.

The first step in a new simulation project is to define the geometry of the structure. The geometry is modeled in the GUI or can be imported from another program (AutoCAD, STEP, …). Then the user continues by defining material properties, boundary conditions, and excitations to the different domains and surfaces of the geometry. If desired one can control the meshing process manually. Finally, the solution parameters are defined. The most important are the solution frequency, the order of the base functions and the matrix solver type. The direct matrix solver is default, but an iterative solver can be used as well. Very useful is the automatic adaptive mesh generation and refinement, which in many cases frees the designer of worrying about which mesh/grid to choose. First a solution for an initial mesh is determined and the solution is assessed. If the solution does not qualify, the mesh is refined and a new solution is computed. This procedure is repeated until one of the exit criteria is fulfilled. It is important to note that HFSS offers no curved elements for a better approximation of curved objects. This means that the mesh along a curved surface has to be chosen very fine. This can be achieved with the 'surface approximation' parameter. Finally the solution, i.e. electric and magnetic fields, currents, and S-parameters can be visualized in 1D, 2D and 3D. All solutions can be exported as files. An 'Optimetrics' toolbox is offered for optimizations, parameter, sensitivity and statistical analysis.

Comsol Multiphysics [28]: FEM

Comsol Multiphysics uses the finite element method to solve partial differential equations in 2D and 3D. It is a multiphysics code, which means that it can handle not only electromagnetics, but also acoustics, mechanics, fluid dynamics, heat transfer, etc.. The geometry, material parameters and boundary conditions have to be set up in the graphical user interface (GUI). Then, after defining boundaries and domains the mesh is generated. This includes the selection of the order of the basis functions and the curved mesh elements (order higher than one means curved element). They are used for a better approximation of curved boundaries. The meshing can be steered by specifying a number of nodes and their distribution on each edge of the model. Comsol Multiphysics offers a wide variety of matrix solvers to solve the system of equations. There are direct solvers, more suitable for smaller problems, and iterative solvers suitable for larger problems. Comsol Multiphysics can generate 2D and 3D plots. Quantities derived from the electromagnetic field, like Poynting vector, energy, and energy loss, are available as predefined variables. Comsol Multiphysics does not allow parameter sweeps or optimization. COMSOL is very easily combined with MATLAB. A simulation set up with the GUI can be saved as a Matlab script, which can be easily modified. The Matlab optimization toolbox can be used to optimize a topology.

JCMsuite [29]: FEM

JCMsuite is a software package based on FEM. It contains dedicated tools for certain simulation problems appearing in nano-optics and plasmonics. It incorporates scattering tools, propagation mode tools, and resonance mode tools.The time-harmonic problems can be formulated in 1D, 2D and 3D.The topologies can be isolated or may occur in periodic patterns, or a mixture of both. Interesting is that dedicated tools for problems posed on cylindrically symmetric geometries are available. The electromagnetic fields are discretized withhigher order edge elements. JCMsuite contains an automatic mesh generator (with adaptive features), goal-oriented errorestimators for adaptive grid refinement, domain-decomposition techniques and fast solvers.

CST [30]: FIT

CST Microwave Studio (CST MWS) is based on the finite integration technique (FIT). It allows to choose the time domain as well as the frequency domain approach. Despite the presence of transient, eigenmode, and frequency domain solvers within CST MWS, the transient solver is considered as the flag ship module. The Time Domain Solver calculates the broadband behavior of electromagnetic devices in one simulation run with an arbitrarily fine frequency resolution. The modeling of curved structures using the Perfect Boundary

Approximation® technique and the modeling of thin perfectly electric conducting sheets with the Thin Sheet Technique® tries to cope with the typical difficulties inherent to FDTD methods for classical structures. Several mesh types can be chosen. The automatic mesh generator detects the important points inside the structure (fixpoints) and locates mesh nodes there. The user can manually add fixpoints on a structure, as well as fully control the number of mesh lines in each coordinate with regards to the specified wavelength. Energy based adaptation of the mesh allows to refine it in a predefined number of passes, providing a mesh refinement of sophisticated design features for the price of longer overall simulation times. Although the FIT in principle can handle material parameters changing over the dielectric volumes defined, this is not implemented yet.

CST, as a general purpose software package being a real competitor for HFSS in the traditional microwave field, has gained a lot of popularity in the last few years. Also for the analysis and design of plasmonic structures, more and more results obtained with CST can be found in literature (just google the words CST and plasmonics). A problem sometimes observed with CST is a ripple in the frequency response in case the tool settings are not appropriate. This is due to the fact that the flagship of CST is inherently a time domain solver.

Lumerical [15]: FDTD

Lumerical is the leading software tool in the plasmonics and photonics community. It is based on the FDTD algorithm, and can be used for 2D and 3D topologies. The topology has to be generated in the GUI or can be imported with GDSII/SEM files. Basic shapes include triangles, rectangular blocks, cylinders, conic surfaces, polygons, rings, user-defined (parametric) surfaces, spheres and pyramids. Several boundary conditions can be used: absorbing (PML), periodic, Bloch, symmetric, asymmetric, and metal boundaries. A non-uniform mesh can be used and automesh algorithms are provided. Since it is a dedicated time domain solver, it has a sophisticated library of Lorentz, Drude, Debye models for the material parameters. Excitation of the structure is possible with waveguide sources, dipoles, plane waves, focused beams and diffraction-limited spots. A scripting language is available to customize simulation. Data can be exported to Matlab or in ASCII.

Non-Commercial Software Tools

MAGMAS 3D [31]: IE-MOM

MAGMAS 3D is the IE-MoM code developed at the Katholieke Universiteit Leuven, Belgium. It was originally developed in cooperation with the European Space Agency for planar and quasi-planar antenna and scattering structures embedded in a multilayered dielectric background medium and operating in the microwave frequency range. Starting from the topology considered, frequency band needed, and type of excitation, it calculates the network, radiation, and scattering characteristics of the structure under consideration. It is based on a full-wave Mixed-Potential formulation of Electric Field Integral Equations (MP-EFIE), originally applied only to 2D surface currents [22], [32]. New theoretical techniques were developed and implemented within the framework: the Expansion Wave Concept (EWC) [33], [34], the Dipole Modeling Technique (DMT) [35], special de-embedding procedures [36], etc..In 2007, it was extended with the capability to handle volumetric 3D currents with the VIE technique introduced in [21]. This was crucial in view of its application in plasmonics. In 2009, this led to the first verified simulated results in this field [25]. Now, it is extensively used for plasmonics. The MAGMAS mesh is based on a combination of rectangular and triangular mesh cells.Full mesh control is available in manual meshing mode. A Graphical User Interface is available.

To the best knowledge of the author, to date (June 2012), it is the only IE-MoM framework able to handle arbitrary plasmonic structures embedded in a multilayered environment, with the very general and flexible VIE technique. No similar commercial solvers exist, in contrast to the situation at microwave frequencies, where a multitude of IE solvers can be found.

Other. Non-commercial Tools Found in Literature

At this moment plasmonics is a mainly experimentally driven research field. Whereas for example in the field of traditional antenna research, numerous papers can be found on modeling as such, this is much less the case in plasmonics, especially in the high impact journals. Most papers are concerned with the description of the physical phenomena occurring and occasionally just mention the numerical tool used.

Most researchers in the plasmonics field use the commercial tools available, and as far as I can see, the main one is Lumerical. In the publications that do treat the numerical modeling of plasmonic structures as such, mainly the FDTD or the FEM technique is applied [37], [38]. Very few papers consider the IE-MoM technique, and if they do, it is the Surface IE technique [39], [40], [41]. In [42] Chremmos uses a magnetic type scalar integral equation to describe surface plasmon scattering by rectangular dielectric channel discontinuities. Even the use of the magnetic current formalism to describe holes, classical at microwave frequencies, already has been used in plasmonics [43]. However, these dedicated developments cannot be categorized under the title „analysis framework" in the sense that they do not allow to handle a wide range of different topologies.

BENCHMARKING

Nowadays, physicists and engineers rely heavily on highly specialized full wave electromagnetic field solvers to analyze, develop, and optimize their designs. Computer-aided analysis and optimization have replaced the design process of iterative experimental modifications of the initial design. It is evident that the underlying solution method for a software tool may significantly influence the efficiency and accuracy by which certain structure types are analyzed. Nevertheless, the commercial focus increasingly switches from such key theoretical considerations to improvements in the area of layout tools and system-level design tools. Therefore, users may get the wrong impression that a given solver is automatically suited to solve any kind of problem with arbitrary precision. This is of course not true.

This section verifies the plausibility of such expectations by presenting a benchmark study for a few plasmonic structures. The study focuses on the capabilities and limitations of the applied EM modeling techniques that usually remain hidden for the user.

Benchmarking in Literature

In literature, not many comparisons between solvers in the field of plasmonics can be found. In [44] Hoffmann et al. consider a single plasmonic topology, a pair of Au spheres, and analyze this structure with several electromagnetic field solvers: COMSOL Multiphysics (FEM), JCMsuite (FEM), HFSS (FEM), and CST Microwave Studio (the FEM solver available in the Studio is used, not the flagship FDTD solver). The output parameter considered is the electric field strength in between the two spheres. Note that all solvers tested are based on the Finite Elements technique in the frequency domain, so it can be expected that the main issues with these solvers for this topology are the same. Very interesting is the fact that in the paper itself CST is categorized as "inaccurate". However, afterwards this was corrected. It seems that wrong material settings were used, and after correction accurate results were obtained. This issue led to the fact that this benchmark example is now presented as a reference example on the CST website.

In [45], several numerical methods are tested for 2D plasmonic nanowire structures: not only the Finite Element Method (FEM) and the Finite Difference Time-Domain (FDTD) technique, but also less „commercial" methods like the Multiple Multipole Program (MMP), the Method of Auxiliary Sources (MAS), and the Mesh-less Boundary Integral Equation (BIE) method are tested. By comparing the results, several conclusions can be drawn about their applicability and accuracy for plasmonic topologies. Differential techniques like FEM and FDTD can reach a high level of accuracy only with a high discretization. In 2D, this is readily affordable on present-day computer systems. There, these techniques have a clear advantage in terms of speed, matrix size, and accuracy. In 3D, due to the rocketing size of the problem, this may not always be that obvious. The advantages may thus disappear in a full 3D analysis of geometrically complicated structures. This happens because matrices become denser or more ill-conditioned. The main conclusion of the paper is that the most efficient method depends on the problem dimension and complexity. This is a similar conclusion as was reached in [46] within the context of the analysis and design of planar antennas.

Comparison between Differential and Integral Equation Techniques

In this section, the differential and integral equation techniques are compared. This is done by choosing a representative solver from each category and using it in the analysis of a basic plasmonic topology. In the category differential techniques, Lumerical is chosen, as it is the most widespread commercial solver in the plasmonic community. Since, as far as we know, there are no commercial integral equation solvers in the plasmonics area, MAGMAS 3D is chosen in this category, the in-house developed solver at Katholieke Universiteit Leuven. The basic topology selected is the plasmonic dipole. It is a structure that can function as a scatterer, but also as a real nano-antenna. The main result of this section is that it draws conclusions on different computational aspects involved in the modeling of nanostructures with the two solvers.

Plasmonic Dipole

The plasmonic dipole is depicted in Fig. 3. It is a structure consisting of two nanorods with a gap in between. When the gap is "shortcircuited", it functions as a single rod, when the gap is open, it functions as a device generating an enhanced electric field there. When it is connected to other nanocircuits, it may function as a nano-antenna, both in transmit (Fig. 3b), and in receive (Fig. 3c). The two rods may be fabricated from metals like Au, Ag, Cu, Al, Cr, etc.. In this section the width W and height H of the dipole are kept constant at 40 nm. This value is well-chosen, since it is a value which can be fabricated with sufficient accuracy using present-day nanofabrication technology. The gap width can have different values, depending on the case considered. Note however that 10 nm is about the minimum that can be fabricated with reasonable accuracy nowadays.

Quality of the Input

It cannot be over-emphasized that Computational Electromagnetics solvers in general produce correct results consistently for a multitude of different structures only if two conditions are met:

1. The user has sufficient general background in the field, and a more specific knowledge about the solution technique implemented in the solver that he is using. If not, there is a huge danger that the solver is not used in a proper way. For example, although this is not necessary to "operate" the solver, if the user has no basic knowledge and does not realize how the discretization / meshing scheme works, it is impossible to understand and assess the effect of a proper meshing. In the majority of the cases, there will be no problem with that, and automatic meshing schemes will be able to produce good results. However, in unusual and/or challenging cases, often met in scientific research, this issue will be a crucial factor.
2. The user needs to have sufficient knowledge about the actual tool that he is using and its peculiarities, more specific about the way a certain problem has to be imported into the solver. A very good illustration of this has already been addressed in section 5.1 where CST was actually not properly used and produced incorrect results that were published in open literature. Afterwards, this was corrected and the problem disappeared. However, following the proper line of reasoning is not always straightforward. Even experienced researchers sometimes easily can make mistakes. It is the opinion of the author that this is happening too much, even in a lot of peer-reviewed scientific papers.

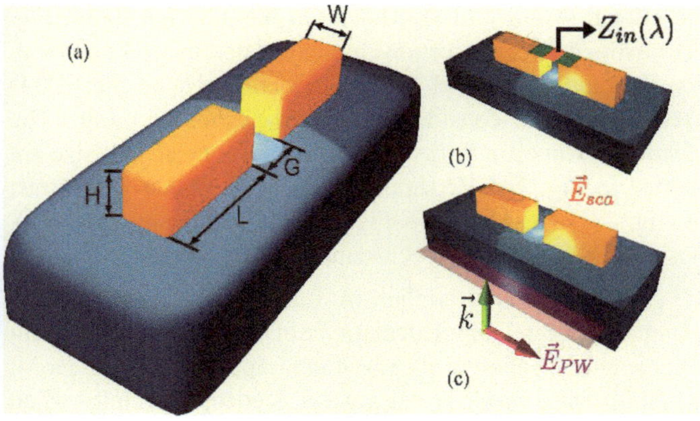

Figure 3. The dipole model studied. a) W and H are set equal to 40 nm and the gap G can vary. b) The dipole as transmitting antenna with a model for the feeding structure located in the gap. c) The dipole as receiving antenna excited by a plane wave.

Modeling of the Feed

The first parameter studied is the impedance that is seen when the nanodevice operates as a transmitting or receiving antenna. In Fig. 4, the impedance simulated with both MAGMAS and Lumerical is given for an Al dipole of 200 nm length and with a gap of 10 nm. The input impedance is calculated as $Z = V/I$, where Z, V and I are the input impedance, the voltage over the gap G, and the source current, respectively. The exciting source is located in the middle of the gap. MAGMAS uses a physical current filament of finite width *w*. Lumerical does not have a built-in physically feasible current source model. It uses a (less realistic) pure electrical dipole source, which means that the current *I* has to be evaluated "manually" by the user based on the magnetic field distribution and Ampere's law. This also means that the impedance will be somewhat depending on the resolution chosen during meshing. It is seen that there is an excellent agreement in the trend of the curves, i.e. the change of input impedance with frequency for different widths of the source current filament in MAGMAS (5 nm and 13.33 nm) is the same as the one in Lumerical. However, in general there is an off-set. Only for a specific width of the current filament in MAGMAS, nl. 5 nm, the two excitation mechanisms become almost completely equivalent and even the offset between the two solvers disappears. A similar agreement was also observed for both gold and silver dipoles.

The main conclusion of this section is that there is still an issue with the modeling of (localized) feeds, especially in Lumerical. The reason is that the plasmonics community is used to consider objects as scatterers, excited by an incident wave. In this case, the models implemented are satisfactory. However, in view of designing nanocircuits, localized feeds and impedances derived from them become important. The basic rule is that the feeding model has to correspond as closely as possible to the actual feed topology that is going to be used later in practice. In this respect, the localized feeding models available in Lumerical today are unsatisfactory.

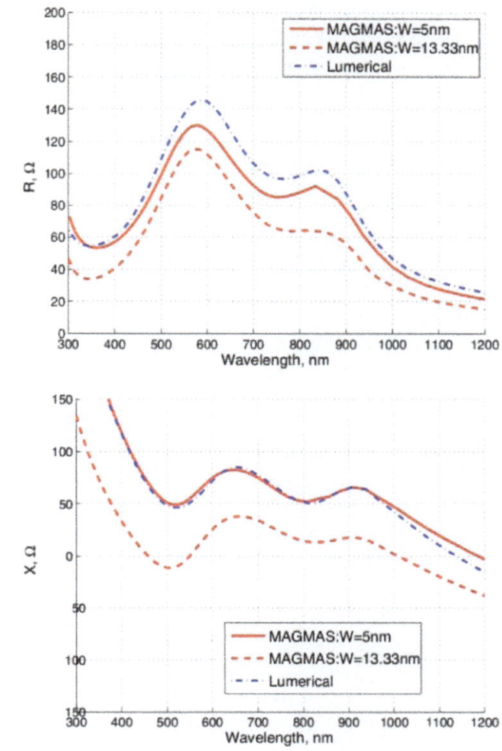

Figure 4. Comparison of impedances obtained with MAGMAS and Lumerical for an aluminum dipole on a glass substrate, L = 200 nm and gap = 10 nm. Two widths *w* are used in MAGMAS for the source current filament. R and X stand for real and imaginary part.

Modeling of Losses

The importance of taking into account losses is obvious. Metals at plasmonic frequencies may be very lossy and neglecting this would result in totally erroneous results. However, there is a difference between time domain and frequency domain solvers. Since Lumerical is an FDTD based tool, it works in time domain. This means that it cannot work directly with (complex) permittivities and permeabilities. The material data have to be transformed from functions of frequency into functions of time. This is a very difficult process, involving a convolution, and is subject to various constraints. Local curve fitting of permittivities is very difficult to implement in a time domain solver. Lumerical fits the sampled measurement data with its own multi-coefficient material model, which "provides a superior fit compared to the standard material models (such as Lorentz, Debye, Drude, etc)". However, this produces artefacts, as is illustrated in Fig. 5. There, the relative error is given between the measured imaginary part of the permittivity (which represents losses) and the value obtained for this imaginary part after fitting, for three metals. It is seen that for gold and aluminum, the error is in the order of 20 %, which is already quite elevated. However, in the case of silver, the error reaches a full 100 %, which is unacceptable. In some cases, it will result in a totally wrong prediction for example of the radiation efficiency of nano-antennas [47]. If one forces the material characteristics to be exactly the same in the two solvers, for example by using the interpolated values also in MAGMAS, instead of the real measured ones, the two solvers produce very similar results, see Fig. 6.

The main conclusion is that a highly accurate prediction of losses in Lumerical is still an issue. The material models have to be further refined, especially for silver.

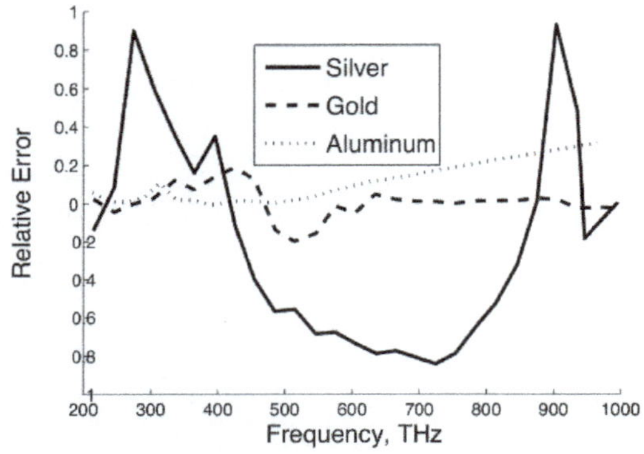

Figure 5. Relative error between interpolated and measured imaginary part of permittivity for silver, gold and aluminum.

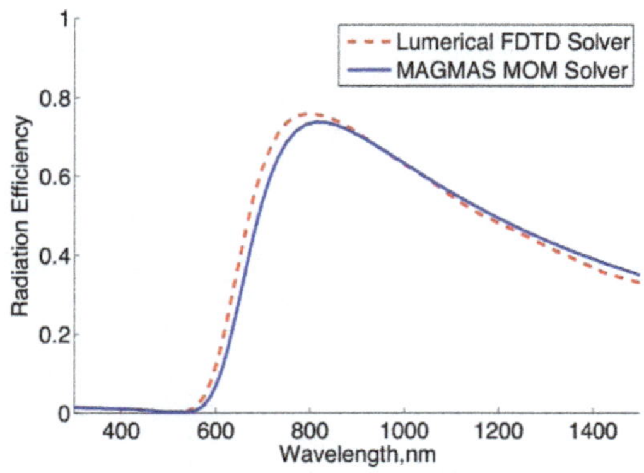

Figure 6. Comparison between FDTD and IE-MoM for a gold dipole in free space, L = 250 nm and the gap is 10 nm. Identical material parameters are used in both solvers.

Meshing

It is well known that the mesh quality and resolution are key factors in the accuracy of any solver. The electromagnetic coupling between nearby segments may differ considerably due to the specific meshing used, especially in parts of the structure where rapid topological changes occur. The question is whether the meshes used in the solvers are adequate. This issue was investigated by performinga convergence study in terms of the resolution of the meshes [48].

The extinction cross section of a gold monomer (no gap, or in other words G = 0 nm) calculated using MAGMAS 3D with different meshes is plotted in Fig. 7(a). The analysis of these data demonstrates clearly the stability of IE-MoM. The results obtained even with a very rough mesh 6x1x1 (41.7 *nm* x 40*nm* x 40 *nm*) provide already a very good estimation of the antenna resonance properties. The differences in extinction cross section calculated with the rough and fine meshes are almost negligibly small. This result is obtained thanks to the variational stability of IE-MoM. On the adjacent figure Fig. 7(b) extinction cross sections are calculated using Lumerical. As expected, in general the results obtained with the FDTD method depend considerably on the chosen mesh. The calculated wavelengths as a function of the mesh cell size are plotted in Fig. 7(c). It should be also noted that in contrast to Fig. 7(a) (IE-MoM) in Fig. 7(b) (FDTD) not only peak positions but also their levels depend clearly on the mesh.

The conclusion is that in Lumerical a fine mesh is mandatory for reliable calculations of the monomer resonant wavelength. Thanks to its variational approach, in IE-MoM, a highly dense mesh is not always needed, especially when scattering problems are studied.

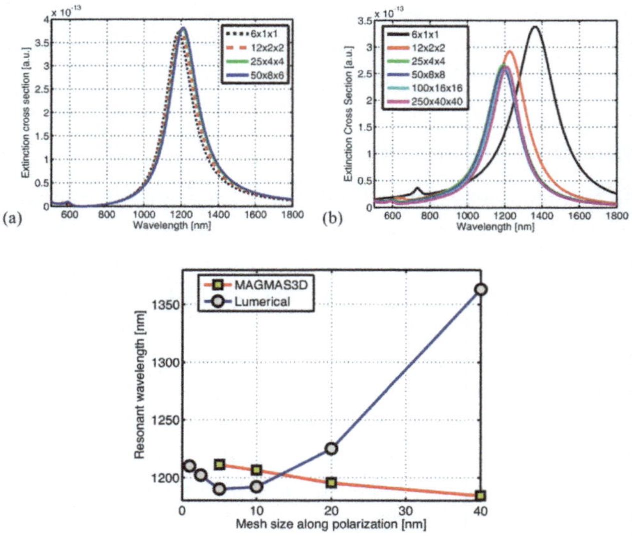

Figure 7.a and b: extinction cross section of a 250 nm x 40 nm x 40 nm gold monomer on a substrate with n=1.5 (a. MAGMAS 3D, b. Lumerical), c. convergence performance of both solvers.

Calculation Speed

Information on the calculation times for the convergence study of the previous section is given in Table 1. It has to be emphasized that calculation times cannot be directly compared. The computers used were different, and the FDTD tool used 10 processors in parallel. Further, since it is a time domain technique, a calculation time per frequency point cannot be given for FDTD. Nevertheless, combining the data in this table with the data concerning the convergence, given in Fig. 7, it is easily seen that IE-MoM outperforms FDTD for this particular case.

Plasmonic Nanojets

In this section it is illustrated what can be reached with computational electromagnetics in the field of plasmonics. The results in this section are extracted from the paper [49], where full details can be found.

Table 1. Calculation times as a function of the mesh for the gold monomer.

IE-MoM		FDTD	
Mesh	Calculation time per frequency point (s)	Mesh	Total calculation time (s) (10 processors in parallel)
6 x 6 x 1	5.4	6 x 1 x 1	38
12 x 2 x 2	10.4	12 x 2 x 2	55
25 x 4 x 4	33.6	25 x 4 x 4	225
50 x 8 x 6	238	50 x 8 x 8	672
		100 x 16 x 16	5893
		250 x 40 x 40	20100

Similarly to a pebble hitting a water surface, the energy deposition by femtosecond laser pulses on a gold surface can produce local melting and back-jet. Contrary to water though, a gold surface can be nanopatterned and the excitation of surface plasmon resonances leads to the appearance of hotspots, literally. This was explicitly proven for the first time in a nanopatterned gold surface, composed of a G shaped periodic structure.It was observed there that laser-induced melting and back-jet occur precisely in the plasmonic hotspots. The aid of computational electromagnetics to rigorously analyze the structure by predicting the hotspots was crucial in explaining this phenomenon. Much more details can be found in [49]. It is clear that this type of insight into the basic interaction mechanisms of light with matter at the nanoscale opens up new possibilities for applications.

Another example of what can be reached by applying computational electromagnetics in plasmonics can be found in [50].

Figure 8. Illuminating the sample with femtosecond laser pulses produces a polarizationdependentpattern of nanobumps with sharp tips on the sample surface [49].Left: horizontally polarized light, right: vertically polarized light.

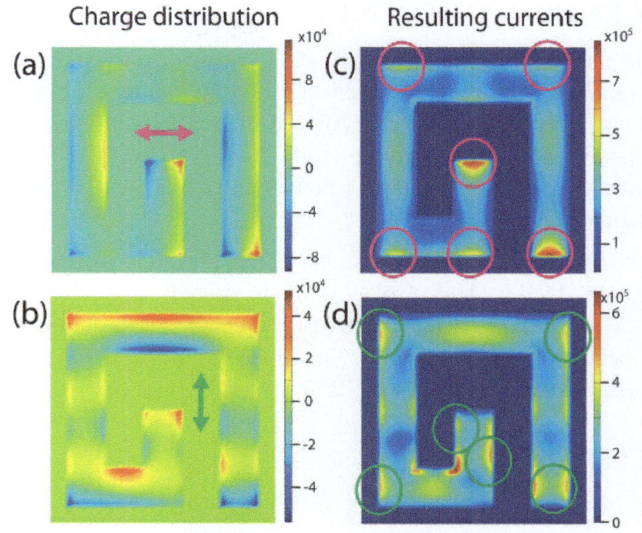

Figure 9. The pattern of melted nanobumps matches that of the plasmonic currents, where Ohmic losses locally increase the temperature [49]. The electric field of light incident on the nanostructures causes the surface charges to oscillate in response to the direction of light polarization: horizontal, in (a), and vertical, in (b). The locations of the resulting electric currents are shown in (c) and (d) respectively. The current maxima are indicated with colored circles matching the pattern of nanobumps in Fig. 8.

CONCLUSIONS

In this chapter, abrief introductory overview has been given on the use of existing computational techniques and software tools for the analysis and design of plasmonic structures. This field is booming, and many modeling techniques developed at lower frequencies, i.e. in the microwave range, are now being transferred to the plasmonics community, of course with the necessary adaptations. Although the differential equation techniques are the most widespread, both in scientific literature, and in the commercial scene, it is proven that integral equation techniques are a valid alternative. In some cases they are even superior. The author sees a lot of opportunities in this area.

In most cases, the tools are used to analyze structures. Few papers really make dedicated designs. However, this will change in the future. Inevitably, the plasmonics community will follow a similar patch as the traditional antenna community. Whereas 30 years ago, the design of an antenna was based on the accumulated expertise and know-how of many years, nowadays most antennas are designed almost in a single pass by experienced designers using commercial tools. This is possible since the tools in this field have sufficient accuracy and matureness in order to be able to do that. It is my belief that as soon as plasmonics will make the unavoidable shift from the physics to the engineering community, this last community will aim at conceiving applications which will involve designs heavily based on computational electromagnetics.

I would like to end this chapter with the following guideline. The use of two different solvers, based on different theoretical methods (integral and differential) may provide an excellent means to characterize the quality of simulation results. If the two results are in good agreement, it is highly likely that the results are correct. If the two results are in disagreement, a deeper investigation of the structure and its modeling is absolutely necessary.

ACKNOWLEDGEMENT

The author gratefully acknowledges the following persons: Zhongkun Ma for the impedance and radiation efficiency comparison between MAGMAS and Lumerical, Dr. Vladimir Volskifor the convergence study with MAGMAS and Lumerical, and V. K. Valev for making available the plasmonic nanojet figures.

We also would like to express our gratitude to the KU Leuven Methusalem project and its beneficiary Prof. V. V. Moshchalkov, whose activities created a real stimulus to become personally active in this field, and the fund for scientific research Flanders (FWO-V) for the financial support.

REFERENCES

1. I. Ahmed, E. H. Khoo, E. Li, R. Mittra, ". A. hybrid, for. approach, coupled. solving, Maxwell, equations. Schrödinger, in. arising, simulation. the, nano-devices". I. E. E. E. of, Antennas, Propagation. Wireless, Vol. Letters, 9149172010
2. D. B. Davidson,", A. review, important. of, developments. recent, full-wave. C. E. M. in, R. F. for, engineering,". I. E. E. E. microwave, 3rd Int, Conf. Comp. Electromagnetics and Its Applications, pp. PS/1PS/4, Nov. 2004
3. C. W. Townbridge, J. K. Sykulski,", Sykulski,"Some Key Developments in Computational Electromagnetics and Their Attribution," IEEE Antennas and Propag. Magazine, 426503508Apr. 2006
4. E. K. Miller,", A. Selective, of. Survey, Electromagnetics,". I. E. E. E. Computational, Antennas. Trans, vol. Propag, 9Sept. 198812811305
5. F. Peterson, S. L. Ray, and R. Mittra, "Computational methods for electromagnetics", IEEE Press- Oxford University Press, 1998.
6. P. Awadhiya, Barba, L. Kempel, ". Finite-element, programming. method, easy???". I. E. E. E. made, Antennas, Magazine. Propagation, vol, 7379Aug. 2003
7. J. M. Jin, The Finite Element Method in Electromagnetics, second edition, John Willey & Sons, Inc., New York, 2002.
8. J. L. Volakis, A. Chatterjee, and L. C. Kempel, Finite element method for electromagnetics, IEEE Press, Oxford University Press, 1997.
9. R. L. Courant, ". Variational, for. methods, solution. the, problems. of, equilibrium. of, Bulletin. vibration,", the. of, Mathematical. American, . Society, pp, 1231943
10. S. Ahmed, ". Finite-element, for. method, problems,". waveguide, Letters. Electronics, vol, 18387389Sept. 1968
11. R. Coccioli, T. Itoh, G. Pelosi, P. P. Silvester, ". Finite-Element, in. Methods, A. Microwaves, Bibliography,". I. E. E. E. Selected, Antennas, Propag, Magazine, 3863448Dec. 1996
12. K. S. Yee, ". Numerical, of. Solution, Boundary. Initial, Problems. Value, Maxwell's. Involving, in. Equations, Media,". I. E. E. E. Isotropic, on. Trans, Propag. Antennas, 1. vol, pp, 3023071966
13. Taflove, Computational electrodynamics: the finite difference time domain method, Artech House, 1997.
14. D. M. Sullivan, ". Electromagnetic, using. simulation, F. D. T. D. the, Wiley. method"-I, E. E. E. Press, 2000978-0-78034-747-2
15. www.lumerical.com
16. T. Weiland, ". A. discretization, for. method, solution. the, Maxwell's. of, for. equations, fields". six-component, Electronics, A. E. Ü. Communications, 3. vol, No, 31977116120
17. Bossavit, L. Kettunen,", Kettunen,"Yee-like schemes on a tetrahedral mesh, with diagonal lumping," Int. J. Numer. Model., 421291421999
18. M. Celuch-Marcysiak, W. K. Gwarek, ". Comparative, of. study, time-domain. the, for. methods, computer. the, analysis. aided, microwave. of, Int. circuits,", Conf. on Comp. in Electromagnetics., 3034Nov. 1991
19. R. F. Harrington, "Field Computation by Moment Methods", New York: Macmillan, 1968.
20. T. K. Sarkar, E. Arvan, S. Ponnapalli, ". Electromagnetic, from. scattering, bodies". I. E. E. E. dielectric, Trans, Antennas Propag., 37673676May 1989
21. Y. Schols, G. A. E. Vandenbosch, ". Separation, horizontal. of, dependencies. vertical, a. in, integral. surface/volume, approach. equation, model. to, . quasi, D. structures, multilayered. in, ". I. E. E. E. media, Trans, Antennas Propagat., 55410861094April 2007
22. M. Vrancken, G. A. E. Vandenbosch, dyadic-mixed-potential. Hybrid, spectral-space. combined, integral-equation. domain, of. analysis, quasi-, D. structures, stratified. in, I. E. E. E. media, Microwave. Trans, Tech. Theory, 5. vol, pp, 2162252003
23. X. Zheng, V. K. Valev, N. Verellen, Y. Jeyaram, A. V. Silhanek, V. Metlushko, M. Ameloot, G. A. E. Vandenbosch, V. V. Moshchalkov, ". Volumetric, of. Method, Moments, multi-level. conceptual, blocks. building, nanotopologies". for, Journal. Photonics, Vol, 1267282Jan. 2012
24. J. Smajic, C. Hafner, L. Raguin, K. Tavzarashvili, M. Mishrickey, ". Comparison, Numerical. of, for. Methods, Analysis. the, Plasmonic. of, Journal. Structures", Computational. of, Nanoscience. Theoretical, vol, 37637742009
25. G. A. E. Vandenbosch, V. Volski, N. Verellen, and V. V. Moshchalkov, "On the use of the method of moments in plasmonic applications", Radio Science, 46, RS0E02, doi:10.1029/2010RS004582, 21 May 2011.
26. R. Qiang, R. L. Chen, J. Chen, ". Modeling, properties. electrical, gold. of, at. films, frequency. infrared, F. D. T. D. using, International. method", of. Journal, Infrared, Waves. Millimeter, vol, 8126312702004
27. www.ansoft.com
28. www.comsol.com
29. www.jcmwave.com
30. CST GmbH- Computer Simulation Technology, CST Microwave Studiouser manual, 2006http://www.cst.com

31. G. A. E. Vandenbosch, M. A. G. M. A. S. . D. , http://www.esat.kuleuven.be/telemic/antennas/magmas/
32. G. A. E. Vandenbosch, A. R. Van de Capelle, integral. Mixed-potential, formulation. expression, the. of, field. electric, a. in, dielectric. stratified, to. medium-application, case. the, a. of, current. probe, I. E. E. E. source, Trans, Antennas Propagat., 40806817July 1992
33. F. J. Demuynck, G. A. E. Vandenbosch, A. R. Van de Capelle, expansion. The, concept. wave, I. part, calculation. efficient, spatial. of, functions. Green's, a. in, dielectric. stratified, I. E. E. E. medium, Trans, Antennas Propagat., 46397406Mar. 1998
34. G. A. E. Vandenbosch, F. J. Demuynck, expansion. The, concept. wave, I. I. a. part, way. new, model. to, coupling. mutual, microstrip. in, I. E. E. E. antennas, Trans, Antennas Propagat., 46407413March 1998
35. B.L.A.Van Thielen and G. A. E. Vandenbosch, Method for the calculation of mutual coupling between discontinuities in Planar Circuits, IEEE Trans. Microwave Theory and Techniques, 50155164January 2002
36. E. A. Soliman, G. A. E. Vandenbosch, E. Beyne, R. P. Mertens, Characterization. Multimodal, Planar. of, Structures. I. E. E. E. Microwave, Trans, Microwave Structures, IEEE Trans. Microwave Theory and Techniques, 521175182January 2004
37. R. Kappeler, D. Erni, X. , D. Cui, L. Novotny, ". Field, of. computations, antennas". optical, of. Journal, Computational, Nanoscience. Theoretical, vol, 36866912007
38. R. Qiang, R. L. Chen, J. Chen, ". Modeling, properties. electrical, gold. of, at. films, frequency. infrared, F. D. T. D. using, International. method", of. Journal, Infrared, Waves. Millimeter, vol, 8126312702004
39. B. Gallinet, O. J. Martin, ". Scattering, plasmonic. on, arrays. nanostructures, with. a. modeled, integral. surface, Photonics. formulation", Nanostructures-Fundamentals, vol. Applications, 42782842009
40. A. M. Kern, O. J. F. Martin, ". Surface, formulation. integral, . D. for, of. simulations, plasmonic, permittivity. high, J. nanostructures", Opt, Soc. Am. A, 26485942010
41. B. Gallinet, A. M. Kern, O. J. F. Martin, ". Accurate, modeling. versatile, electromagnetic. of, on. scattering, nanostructures. periodic, a. with, integral. surface, J. approach", Opt, Soc. Am. A, 271022612271 2010
42. I. Chremmos, ". Magnetic, integral. field, analysis. equation, surface. of, scattering. plasmon, rectangular. by, channel. dielectric, J. discontinuities", Opt, Soc. Am. A Opt. Image Sci. Vis., 27185942010
43. J. Alegret, P. Johansson, M. Kall, ". Green's, calculations. tensor, Plasmon. of, of. resonances, holes. single, pairs. hole, thin. in, films". gold, Journal. New, Physics. of, vol, 102008
44. J. Hoffmann, C. Hafner, P. Leidenberger, J. Hesselbarth, S. Burger, of. Comparison, field. electromagnetic, for. solvers, . D. the, of. analysis, nano. plasmonic, Proc. S. P. I. antennas, SPIE 739073900J73900J2009
45. J. Smajic, C. Hafner, L. Raguin, K. Tavzarashvili, M. Mishrickey, ". Comparison, Numerical. of, for. Methods, Analysis. the, Plasmonic. of, Journal. Structures", Computational. of, Nanoscience. Theoretical, vol, 37637742009
46. Y. Vasylchenko, W. Schols, De Raedt, G. A. E. Vandenbosch, ". Quality, of. assessment, techniques. computational, tools. software, planar. for, analysis". I. E. E. E. antenna, Propagat. Antennas, Magazine, Feb. 2009
47. G. A. E. Vandenbosch, Z. ". Upper, for. bounds, solar. the, harvesting. energy, of. efficiency, Nano. nano-antennas", Vol. Energy, 3494502May 2012
48. F. Pelayo, G. De Arquer, V. Volski, N. Verellen, G. A. E. Vandenbosch, V. V. Moshchalkov, ". Engineering, input. the, of. impedance, nano. optical, antennas. dipole, geometry. materials, effect". I. E. E. E. excitation, Trans, Antennas Propagat., 59931443153Sep. 2011
49. V. K. Valev, D. Denitza, X. Zheng, A. Kuznetsov, B. Chichkov, G. Tsutsumanova, E. Osley, V. Petkov, B. De Clercq, A. V. Silhanek, Y. Jeyaram, V. Volskiy, P. A. Warburton, G. A. E. Vandenbosch, S. Russev, O. A. Aktsipetrov, M. Ameloot, V. V. Moshchalkov, T. Verbiest, ". Plasmon-enhanced, laser. sub-wavelength, plasmonic. ablation, Advanced. nanojets", Vol. Materials, 10OP29OP35March 2012
50. V. K. Valev, A. V. Silhanek, B. De Clercq, W. Gillijns, Y. Jeyaram, X. Zheng, V. Volskiy, O. A. Aktsipetrov, G. A. E. Vandenbosch, M. Ameloot, V. V. Moshchalkov, T. Verbiest, ". U-shaped, for. switches, information. optical, at. processing, nanoscale". the, Vo. Small, 1825732576Sep. 2011

CHAPTER 2

COMPUTATIONAL ELECTROMAGNETIC METHODS FOR GRAPHENE MODELING

Yu Shao, Jing Jing Yang, and Ming Huang

INTRODUCTION

Graphene consists of a monolayer of carbon atoms arranged on a two-dimensional honeycomb lattice which is made up of hexagons. Since its first discovery in 2004, graphene has attracted tremendous research interest in various fields due to its distinctive properties [1–3]. Graphene is a rising star not only in material science and condensed matter physics, but also in the electronic and photonic device communities [4].

Graphene has unique electronic band structure, and the electrons in it behave as massless Dirac fermion [5,6]. Graphene acquires a pronounced electric field effect which means carrier concentration can be tuned by electrostatic gating. It has been shown that, in bilayer graphene, the electronic gap between conduction bands and the valence can be tuned between zero and midinfrared energies, which makes bilayer graphene the only known semiconductor with a tunable energy gap [7, 8]. Zhang et al. [9] presented a widely tunable electronic bandgap in electrically gated graphene, and they realized a gate-controlled, continuously tunable bandgap of up to 250 meV. Large-area graphene films of the order of centimeters on copper substrates were realized by chemical vapor deposition using methane [10], which facilitated the fabrication of graphene transistors. A good review of graphene transistors in both logic and radiofrequency applications is provided in [11].

Citation: Yu Shao, Jing Jing Yang, and Ming Huang, "Computational Electromagnetic Methods for Graphene Modeling," International Journal of Antennas and Propagation, vol. 2016, Article ID 7478621, 9 pages, 2016. doi:10.1155/2016/7478621.

Besides electron-device application, graphene has also been recognized as a novel optical material for photonic application. Compared to traditional metal, graphene exhibits several favorable properties. Particularly, surface plasmon polaritons (SPPs), which are electromagnetic waves coupled charge excitations on the surface of a conductor, can be excited in graphene. The plasmons in graphene are tightly confined and the volumes of SPPs in graphene can be 10^6 times smaller than those in free space [12]. This property leads to strong light-matter interaction in graphene [13]. Additionally, the dielectric properties of graphene can be electrically or chemically tuned by changing the charge carrier density and Fermi energy [13–15]. Brar et al. [12] created and probed plasmons in graphene with $\lambda_p \le \lambda_0/100$ and resonant energies as high as 310 meV for 15 nm nanoresonators. Fei et al. [16] showed common graphenes $/SiO_2/Si$ back-gated structure support propagating surface plasmons and they altered both the amplitude and the wavelength of these plasmons by varying the gate voltage. The graphene-based plasmonics may enable the novel optic devices working in different frequency ranges—from terahertz to the visible range—with extremely high speed, low driving voltage, low power consumption, and compact sizes [17]. Vakil and Engheta in [18] showed that graphene can be made into a one-atom-thick platform for infrared metamaterials and transformation optical devices by manipulating spatially inhomogeneous and nonuniform conductivity patterns. Ju et al. [19] used metamaterial made up of periodic graphene microribbon arrays for terahertz plasmon excitations and demonstrated that the plasmon resonance can be tuned over a broad terahertz frequency range by adjusting microribbon width and electrostatic doping.

Our group has been dedicated to the study of molecular sensors for a long time. Francescato et al. [20] presented a platform for broadband molecular spectroscopy based on the propagation of strongly confined antibonding plasmons supported by graphene sandwiches. This novel scheme measures the absorption spectrum of the molecule and shows a broadband capability and high sensitivity. Recently, Yang et al. [21] proposed a cylindrical graphene plasmon waveguide and investigated its application in molecular sensing.

The terahertz antenna based on graphene has promising application in wireless communications in nanosystems. The tunable property of graphene facilitates the designing of reconfigurable antenna. Huang et al. [22] presented a beam reconfigurable antenna based on a switchable high-impedance surface using single-layer graphene. The graphene-based antennas which have broad wavelength tuning range are proposed in [23, 24].

In this paper, the electromagnetic simulation of graphene is introduced and some widely used computational electromagnetic methods for graphene modeling are reviewed. The advantages and drawbacks of each method are discussed and numerical examples of these methods are given to illustrate their performance and application.

ELECTROMAGNETIC SIMULATION OF GRAPHENE

To investigate the electromagnetic property of graphene, Maxwell equations need to be solved to simulate the wave propagation in graphene. In most cases, Maxwell equations have no analytical solutions except for a few simple canonical problems. Therefore, numerical methods are crucial to understand the wave guiding and scattering by graphene. There are three popular computational electromagnetic methods to simulate graphene: finite-difference time-domain method (FDTD), finite element method (FEM), and method of moment (MoM). Each method has its own advantages and disadvantages depending on the specific problem. Many commercial software packages based on these methods are available, such as FDTD Solutions, CST, HFSS, COMSOL, IE3D, and FEKO. All the software packages can be used to model graphene; however, none of these are developed specifically for graphene which makes them less efficient in dealing with this monoatomic layer device.

Graphene is modeled as a thin surface with complex conductivity which can be tuned by external electric field bias or chemical doping. The surface conductivity of graphene is commonly described by Kubo formula [32]:

$$\cdot \int_0^\infty \varepsilon \left(\frac{\partial f_d(\varepsilon)}{\partial \varepsilon} - \frac{\partial f_d(-\varepsilon)}{\partial \varepsilon} \right) d\varepsilon - \int_0^\infty \frac{f_d(-\varepsilon) - f_d(\varepsilon)}{(\omega - j2\Gamma)^2 - 4(\varepsilon/\hbar)^2} d\varepsilon \Bigg], \tag{1}$$

where Γ is the phenomenological scattering rate, e is the electron charge, ε is the energy state, and $\hbar$ denotes the reduced Plank constant. The first term of (1) is corresponding to the intraband contribution while the second term is due to the interband contribution. That is,

$$\sigma_g = \sigma_{\text{intra}} + \sigma_{\text{inter}}. \tag{2}$$

The intraband term in (2) can be evaluated by

$$\sigma_{\text{intra}} = \frac{e^2 k_B T}{\pi \hbar^2 (2\Gamma + j\omega)} \left[\frac{\mu_c}{k_B T} + 2 \ln \left(e^{-\mu_c/k_B T} + 1 \right) \right] \tag{3}$$

and the interband term in (2) is approximated by

$$\sigma_{\text{inter}} = -\frac{je^2}{4\pi\hbar} \ln \left(\frac{2|\mu_c| - (\omega - j2\Gamma)\hbar}{2|\mu_c| + (\omega - j2\Gamma)\hbar} \right) \quad (k_B T \ll |\mu_c|), \tag{4}$$

where μ_c is the chemical potential, T is the temperature, and k_B is the Boltzmann constant.

In electromagnetic simulation, the 2D graphene surface is usually approximated by a 3D dielectric slab whose 3D conductivity is evaluated by the following equation:

$$\sigma_{3D} = \frac{\sigma_g}{d}, \tag{5}$$

where d is the thickness of the graphene layer. The accuracy of 3D dielectric slab model degrades as d increases. It can be shown that, using the 2D conductivity, the SPP wavelength is [25]

$$\lambda_{\text{SPP}} = \lambda_0 \left(1 - \left(\frac{2}{\eta_0 \sigma} \right)^2 \right)^{-0.5}. \tag{6}$$

The 3D dielectric slab can support even and odd modes. The odd mode has the wavelength of

$$\lambda_{\text{odd}} = 2\pi \left(-\frac{2}{d} \coth^{-1} \varepsilon_{3D} \right)^{-1} \tag{7}$$

and (6) can be approximated by (7) only if the following three conditions are satisfied [25]:

$$\frac{d}{\lambda_{\text{SPP}}} \ll 1, \quad |\sigma| \ll \frac{2}{\eta_0}, \quad \left| \frac{\sigma}{d} \right| > 2\omega\varepsilon_0. \tag{8}$$

Figure 1 shows the relative error of using the dielectric slab model for graphene with respect to normalized d and σ; more details can be found in [25].

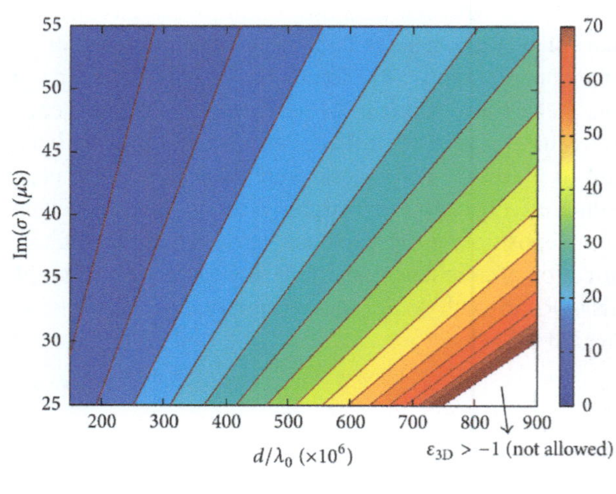

Figure 1: Relative error of the dielectric slab model with respect to normalized d and σ [25].

FINITE-DIFFERENCE TIME-DOMAIN METHOD

FDTD is time-domain method which has specific advantage in transient problem and wideband problem. The broadband results can be obtained in one simulation. Compared to frequency-domain method, it takes less computing memory and simulation time. Additionally, FDTD is an iterative scheme which eliminates solving large linear system; hence, it is simple, robust, and easy to implement.

There are three general approaches to model graphene in FDTD: (I) using standard FDTD method with very fine mesh within the graphene sheet; (II) using the subcell FDTD method; (III) using surface resistive boundary condition.

In the first approach, graphene is modeled as thin layer with finite thickness and the surface conductivity is transformed to volume permittivity. According to the Courant-Friedrichs-Lewy (CFL) stability condition of FDTD, fine mesh leads to small time step, so this approach is both memory- and time-consuming which reduces its efficiency. To alleviate the time step constraint, some unconditionally stable FDTD methods are developed for graphene simulation. In [33], the locally one-dimensional (LOD) FDTD is applied for the efficient simulation of graphene device. They demonstrated that the LOD-FDTD can be 60% faster than standard FDTD with reasonable accuracy. Wang et al. [26] proposed an unconditionally stable one-step leapfrog alternating-direction-implicit (ADI) FDTD to study the SPPs in graphene structure. In Figure 2, they showed the SPPs propagating along the spiral waveguide. The one-step ADI-FDTD method gives accurate result which is comparable to conventional FDTD. They also showed that the conventional FDTD took 1.7×10^5 s for this problem, while ADI-FDTD only took 0.8×10^5 s with CFLN = 10.

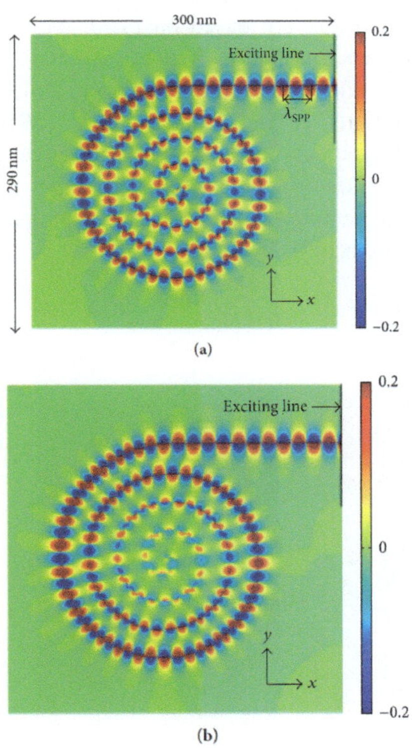

Figure 2: The normalized E_y component of guided SPPs along the spiral graphene waveguide at time t=1.9258ps calculated by (a) conventional FDTD and (b) the proposed improved ADI-FDTD with CFLN = 10 [26].

In the second approach, graphene is treated as thin layer occupying a fraction of FDTD cell [27]. The Yee cell for the subcell FDTD method is shown in Figure 3, where E_z component is split into $E_{z,i}$ and $E_{z,o}$ in cells occupying graphene [27]. However, this scheme is complex in mathematics and needs special type of PML to model infinitely thin sheets [34].

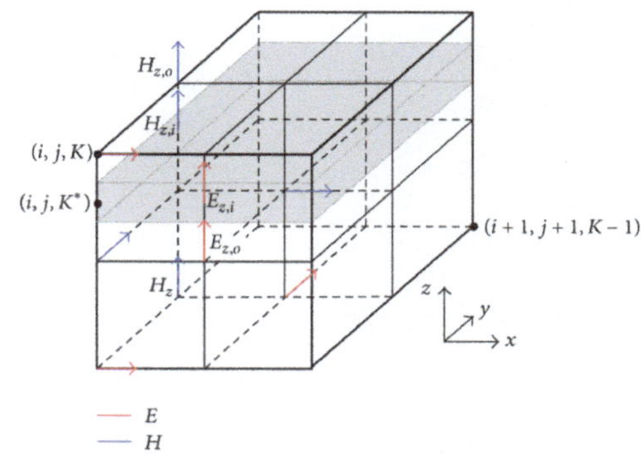

Figure 3: Modified Yee cell for subcell FDTD method at the location of graphene (shaded area) [27].

In the third approach, graphene is modeled as a zero-thickness conductive sheet over which the fields satisfy the surface boundary condition. Nayyeri et al. [28, 35] presented this scheme in detail. As shown in Figure 4(a), a conductive sheet locates at $K + 1/2$. The tangential component of H field is discontinuous over the sheet, so H_x is split into ${}^1H_x$ and ${}^2H_x$, which satisfy the following discretized Faraday law:

$$\mu_1 \frac{{}^1H_x^{n+1/2} - {}^1H_x^{n-1/2}}{\Delta t} = \frac{E_y^n(K+1/2) - E_y^n(K)}{\Delta x/2}, \tag{9a}$$

$$\mu_1 \frac{{}^2H_x^{n+1/2} - {}^2H_x^{n-1/2}}{\Delta t} = \frac{E_y^n(K+1) - E_y^n(K+1/2)}{\Delta x/2}. \tag{9b}$$

The surface boundary condition is

$${}^2H_x - {}^1H_x = \sigma_s E_y \tag{10}$$

from which $E_y^n(K + 1/2)$ can be written as

$$E_y^n\left(K + \frac{1}{2}\right) = \frac{1}{2\sigma_s}\left[\left({}^2H_x^{n+1/2} + {}^2H_x^{n-1/2}\right) - \left({}^1H_x^{n+1/2} + {}^1H_x^{n-1/2}\right)\right]. \tag{11}$$

Substituting (11) into (9a) and (9b), ${}^1H_x$ and ${}^2H_x$ can be derived as

$${}^1H_x^{n+1/2} = \frac{1}{1 - c_1 c_2}\left(F_1^n + c_1 F_2^n\right), \tag{12a}$$

$${}^2H_x^{n+1/2} = \frac{1}{1 - c_1 c_2}\left(F_2^n + c_2 F_1^n\right). \tag{12b}$$

The expressions of c_1, c_2, F_1^n, and F_2^n are shown in [28]. Figure 4(b) is a 3D FDTD cell containing conductive sheet at $K + 1/2$. In this case, tangential components of magnetic field (H_x, H_y) and normal component of electric field (E_z) are discontinuous over the sheet, so they are split into two parts on the two sides of the surface. Their updating equations can be derived in a similar way to 1D case and are presented in [28].

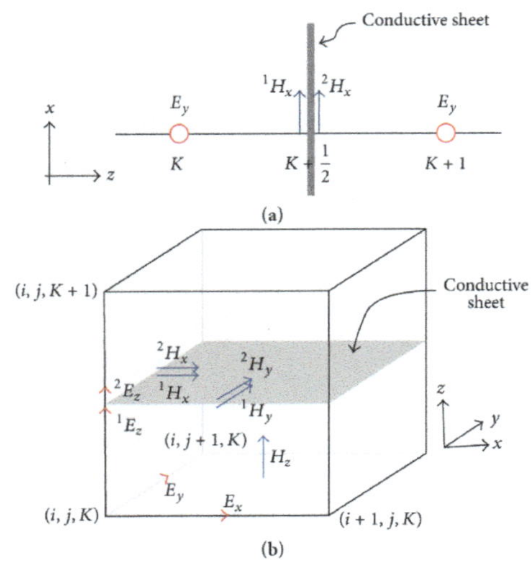

Figure 4: (a) 1D FDTD cell including a conductive sheet at grid $K + 1/2$. (b) 3D FDTD cell with a conductive sheet at grid $K + 1/2$. [28].

The surface boundary condition approach has been validated [28]. Figure 5(a) shows a 2D problem in which an infinite line source radiates above an infinite graphene sheet. The pattern of E_z at the wavelength of 100 μm is shown in Figure 5(b). A strong agreement between the proposed FDTD result and that derived from semianalytic method is obtained. In addition, unlike the subcell FDTD method, the classical PML is applicable to the surface boundary condition scheme.

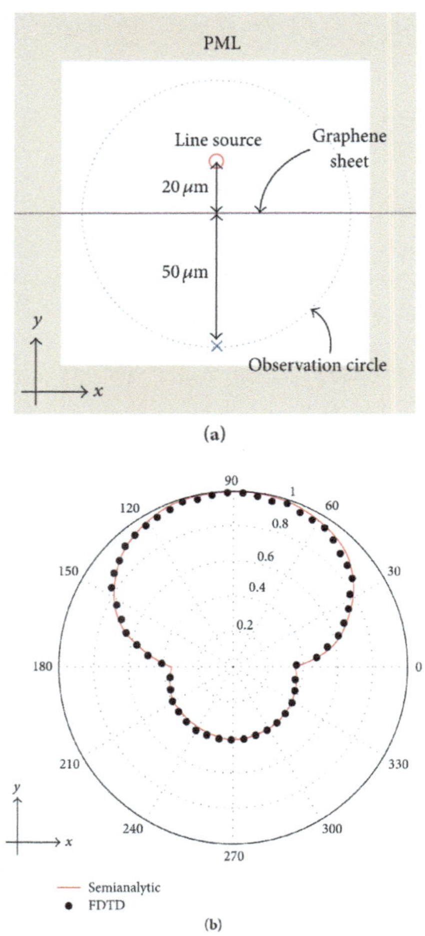

Figure 5: (a) Line source scattering by a graphene sheet. (b) Normalized pattern of E_z at the wavelength of $\lambda_0 = 100\,\mu m$ [28].

In time-domain simulation, the conductivities of graphene shown in (2)–(4) need to be converted from frequency domain into time domain. The intraband conductivity, which has a Drude-like expression, can be easily converted into time domain. However, the interband conductivity has a complex logarithmic form which needs a vector fitting technique. A summation of partial fractions in terms of complex conjugate pole-residues is applied to approximate the conductivity. The detailed description of the fitting technique can be found in [35–37].

FINITE ELEMENT METHOD

FEM is a full-wave numerical technique for electromagnetic boundary-value problems. The basic principle of this method is to discretize the whole computing domain with a finite number of subdomains in which the unknown function is expanded by simple interpolation functions with unknown coefficients. Then, a system of algebraic equations of these unknown coefficients is obtained by using Ritz variational or Galerkin's method. Finally, by solving this linear system, the approximated solution of the entire domain can be obtained.

In FDTD, the computing space is discretized by orthogonal grid; this staircase approximation will reduce the modeling fidelity when it comes to complex geometry, while in FEM, where the triangles or tetrahedral elements are applied, arbitrary geometries can be modeled accurately. As a result, FEM has advantage in complex and inhomogeneous problems. Furthermore, FEM is a frequency domain method which makes it efficient in dealing with narrow-band problems.

Brar et al. [12] solved Maxwell's equation by FEM and modeled graphene as a thin sheet with 0.1 nm thickness; the results suggested that graphene can increase light-matter interactions at infrared energies. Software packages such as HFSS and COMSOL are based on FEM, and they are used widely in graphene simulation. Recently, we used COMSOL to investigate the transmission properties of a cylindrical graphene plasmon waveguide [21]. Tamagnone et al. [29] simulated a reconfigurable graphene antenna with HFSS; the structure of the antenna is shown in Figure 6(a); the input impedance of the antenna can be tuned by changing the chemical potential as shown in Figure 6(b).

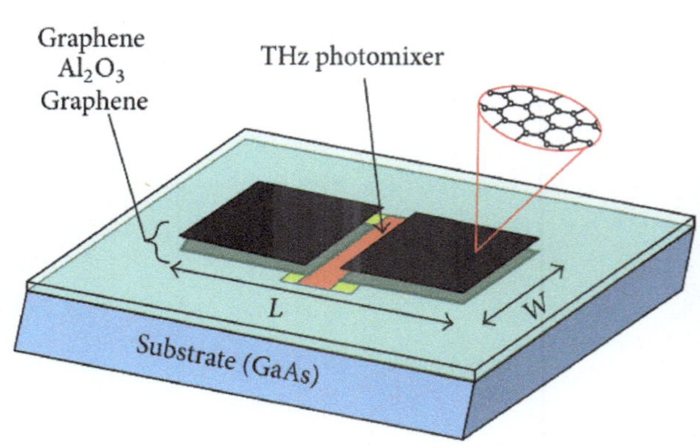

(a)

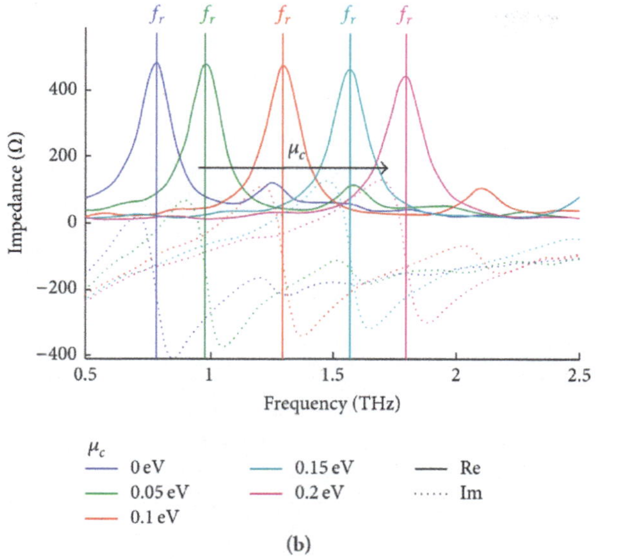

Figure 6: (a) Structure of a reconfigurable graphene antenna. (b) Input impedance of the antenna with respect to chemical potential [29].

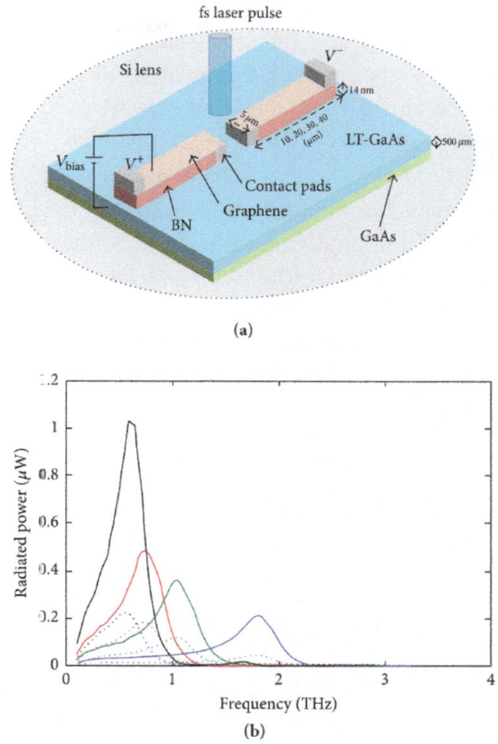

Figure 7: (a) Structure of graphene RF plasmonic antenna fed with a photoconductive material. (b) Radiated power with respect to frequency for graphene antennas with lengths 10 μm (blue), 20 μm (green), 30 μm (red), and 40 μm (black) [30].

METHOD OF MOMENTS

In contrast to FDTD and FEM which solve differential equations, MoM is a technique used to solve electromagnetic surface or volume integral equations in the frequency domain. In MoM, the quantities of interest are not the fields but the electromagnetic sources (surface or volume current), so only the surface of the geometry needs to be discretized. The surface current is discretized into wire segments and/or surface patches. A linear system can be constructed by the method of weighted residuals and the results of the linear equations give the surface current. The far-field result can be derived from the surface current by Green's function. Because it only requires calculating the boundary values instead of the values throughout the space, MoM is highly efficient for electrically large objects and is widely used in solving radiation and scattering problems. However, when applied to complex inhomogeneous cases, it will be very computationally expensive and less efficient.

The analytical expressions of dyadic Green's function for graphene are derived in [38]. Shapoval et al. [39] proposed integral equations based on surface-impedance boundary condition to analyze plane wave scattering and absorption by graphene-strip gratings. The method of moments for graphene nanoribbons was developed in [40–44], in which the issue of nonlocality of graphene conductivity was taken into account. Nonlocal effect arises from spatial dispersion of graphene which is nonnegligible when dealing with slow modes supported by graphene nanoribbons. The spatially dispersive intraband conductivity tensor was derived in [41].

Software packages such as IE3D and FEKO are based on MoM. IE3D is applied to simulate the microwave propagation in a coplanar waveguide over graphene from 40 MHz to 110 GHz [45]. Cabellos-Aparicio et al. [30] used FEKO to study the radiated power of a graphene plasmonic antenna fed by photoconductive source. The antenna structure is shown in Figure 7(a) and the radiated power with respect to frequency is shown in Figure 7(b). The detailed parameters of the photoconductive antenna can be found in [30].

DISCONTINUOUS GALERKIN TIME-DOMAIN METHOD

The graphene involved problems are often multiscale. Graphene is monoatomic and its thickness is much smaller than wavelength, so it is an electrically fine structure. In contrast, the substrate belongs to electrically coarse structure because its dimension is much greater than wavelength. As we have mentioned before, time-domain methods have the advantage that the broadband characterization can be obtained with only a single simulation. However, FDTD method has a serious efficiency problem in multiscale simulation because it requires high discretization density to model electrically fine structure due to its cartesian grid. The finite element time-domain (FETD) method is capable of modeling complex and fine structures and achieving high-order accuracy with high-order basis functions. The major drawback is a global linear system of equations that needs to be solved at each time step. The multiscale problems usually contain a large number of unknowns; FETD will be very computationally expensive in this case. Discontinuous Galerkin time-domain (DGTD) method is promising in multiscale problems [46]. DGTD allows for domain decomposition. A multiscale structure is divided into several subdomains and each subdomain can be discretized separately. All operations in DGTD are local, so large global matrix is split into several smaller matrices. Unlike FETD, the matrices of DGTD are inverted and stored before time marching, and different time integration scheme can be used in different subdomain. Additionally, DGTD is naturally adapted to parallel computing. Recently, DGTD has been used in nanophotonics field and considered as a viable alternative to the well-established FDTD and FETD methods [47].

Li et al. [31] proposed DGTD method with resistive boundary condition for the electromagnetic analyzing of graphene. They also applied this method to the magnetized graphene from microwave to THz range, where the anisotropic and disperse surface conductivity is involved [48]. Figure 8(a) shows the reflection Γ_R, transmission Γ_T, and absorption Γ_A coefficients of an infinitely large graphene sheet under the illumination of a normally incident plane wave; the results of DGTD agree very well with the theoretical data. Figure 8(b) is the normalized extinction cross section of a freestanding graphene patch; good consistency is achieved between DGTD and integral equation method. More numerical examples can be found in [31].

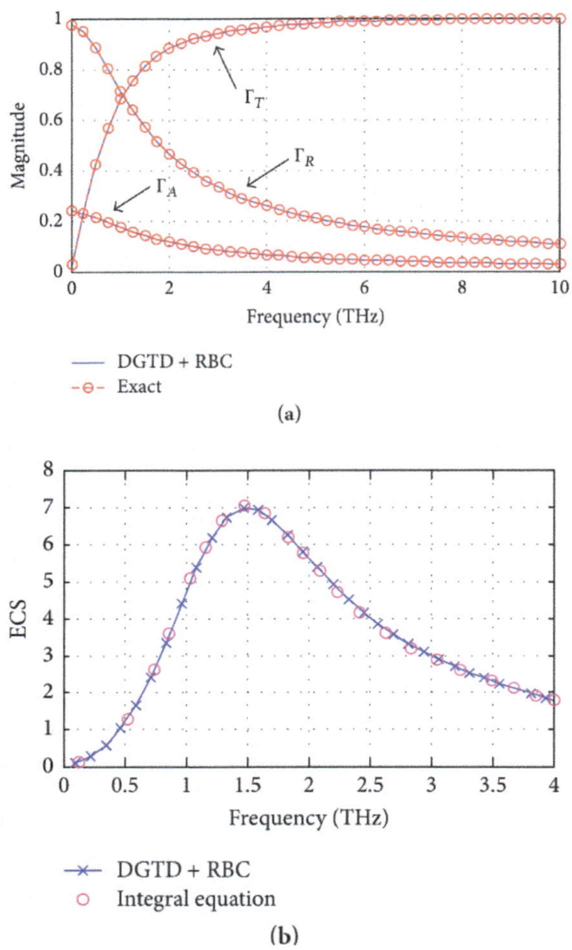

Figure 8: (a) The magnitude of reflection Γ_R, transmission Γ_T, and absorption coefficients Γ_A calculated by DGFD as well as the theoretical value. (b) Normalized extinction cross section of a graphene patch calculated by DGTD and integral equation method [31].

CONCLUSION

Due to its intriguing properties, graphene is used more and more widely in electronic and photonic community. This paper gives a brief review of application of graphene, as well as the computational methods which can be used to investigate its electromagnetic properties. Although a number of famous commercial software packages are available for graphene simulation, none of them are developed specifically for graphene; therefore, they suffer from inefficiency. Each computational method has its own advantages and drawbacks; one should choose the appropriate method according to the specific problem. FDTD is simple and easy to implement; however, it is less efficient in modeling fine structure. FEM is flexible in geometry modeling but solving large global system makes it computationally expensive. DGTD is very suitable for multiscale problem and we believe it will be more and more widely used in modeling and designing graphene-assisted device.

ACKNOWLEDGMENTS

This work was supported by the National Natural Science Foundation of China (Grants nos. 61161007, 61261002, 61461052, and 11564044), the Specialized Research Fund for the Doctoral Program of Higher Education (Grants nos. 20135301110003 and 20125301120009), China Postdoctoral Science Foundation (Grants nos. 2013M531989 and 2014T70890), and the Key Program of Natural Science of Yunnan Province (Grants nos. 2013FA006 and 2015FA015).

REFERENCES

1. K. S. Novoselov, A. K. Geim, S. V. Morozov et al., "Electric field effect in atomically thin carbon films,"Science, vol. 306, no. 5696, pp. 666–669, 2004.
2. A. K. Geim and K. S. Novoselov, "The rise of graphene," Nature Materials, vol. 6, no. 3, pp. 183–191, 2007.
3. C. Lee, X. Wei, J. W. Kysar, and J. Hone, "Measurement of the elastic properties and intrinsic strength of monolayer graphene," Science, vol. 321, no. 5887, pp. 385–388, 2008.
4. P. Avouris, "Graphene: electronic and photonic properties and devices," Nano Letters, vol. 10, no. 11, pp. 4285–4294, 2010.
5. K. S. Novoselov, A. K. Geim, S. V. Morozov et al., "Two-dimensional gas of massless Dirac fermions in graphene," Nature, vol. 438, no. 7065, pp. 197–200, 2005.
6. A. H. Castro Neto, F. Guinea, N. M. R. Peres, K. S. Novoselov, and A. K. Geim, "The electronic properties of graphene," Reviews of Modern Physics, vol. 81, no. 1, pp. 109–162, 2009.
7. T. Ohta, A. Bostwick, T. Seyller, K. Horn, and E. Rotenberg, "Controlling the electronic structure of bilayer graphene," Science, vol. 313, no. 5789, pp. 951–954, 2006.
8. E. V. Castro, K. S. Novoselov, S. V. Morozov et al., "Biased bilayer graphene: semiconductor with a gap tunable by the electric field effect," Physical Review Letters, vol. 99, no. 21, Article ID 216802, 2007.
9. Y. Zhang, T.-T. Tang, C. Girit et al., "Direct observation of a widely tunable bandgap in bilayer graphene," Nature, vol. 459, no. 7248, pp. 820–823, 2009.
10. X. Li, W. Cai, J. An et al., "Large-area synthesis of high-quality and uniform graphene films on copper foils," Science, vol. 324, no. 5932, pp. 1312–1314, 2009.
11. F. Schwierz, "Graphene transistors," Nature Nanotechnology, vol. 5, no. 7, pp. 487–496, 2010.
12. V. W. Brar, M. S. Jang, M. Sherrott, J. J. Lopez, and H. A. Atwater, "Highly confined tunable mid-infrared

plasmonics in graphene nanoresonators," Nano Letters, vol. 13, no. 6, pp. 2541–2547, 2013.

13. F. H. L. Koppens, D. E. Chang, and F. J. G. de Abajo, "Graphene plasmonics: a platform for strong light–matter interactions," Nano Letters, vol. 11, no. 8, pp. 3370–3377, 2011.

14. J. Chen, M. Badioli, P. A. Gonzalez, et al., "Optical nano-imaging of gate-tunable graphene plasmons,"Nature, vol. 487, pp. 77–81, 2012.

15. M. Jablan, H. Buljan, and M. Soljačić, "Plasmonics in graphene at infrared frequencies," Physical Review B, vol. 80, no. 24, Article ID 245435, 2009.

16. Z. Fei, A. S. Rodin, G. O. Andreev et al., "Gate-tuning of graphene plasmons revealed by infrared nano-imaging," Nature, vol. 486, no. 7405, pp. 82–85, 2012.

17. A. N. Grigorenko, M. Polini, and K. S. Novoselov, "Graphene plasmonics," Nature Photonics, vol. 6, no. 11, pp. 749–758, 2012.

18. A. Vakil and N. Engheta, "Transformation optics using graphene," Science, vol. 332, no. 6035, pp. 1291–1294, 2011.

19. L. Ju, B. Geng, J. Horng et al., "Graphene plasmonics for tunable terahertz metamaterials," Nature Nanotechnology, vol. 6, no. 10, pp. 630–634, 2010.

20. Y. Francescato, V. Giannini, J. Yang, M. Huang, and S. A. Maier, "Graphene sandwiches as a platform for broadband molecular spectroscopy," ACS Photonics, vol. 1, no. 5, pp. 437–443, 2014.

21. J. Yang, J. Yang, W. Deng, F. Mao, and M. Huang, "Transmission properties and molecular sensing application of CGPW," Optics Express, vol. 23, no. 25, pp. 32289–32299, 2015. ·

22. Y. Huang, L.-S. Wu, M. Tang, and J. Mao, "Design of a beam reconfigurable thz antenna with graphene-based switchable high-impedance surface," IEEE Transactions on Nanotechnology, vol. 11, no. 4, pp. 836–842, 2012.

23. I. Llatser, C. Kremers, A. Cabellos-Aparicio, J. M. Jornet, E. Alarcón, and D. N. Chigrin, "Graphene-based nano-patch antenna for terahertz radiation," Photonics and Nanostructures—Fundamentals and Applications, vol. 10, no. 4, pp. 353–358, 2012.

24. Y. Yao, M. A. Kats, P. Genevet et al., "Broad electrical tuning of graphene-loaded plasmonic antennas,"Nano Letters, vol. 13, no. 3, pp. 1257–1264, 2013.

25. E. Forati, G. W. Hanson, A. B. Yakovlev, and A. Alù, "Planar hyperlens based on a modulated graphene monolayer," Physical Review B, vol. 89, no. 8, Article ID 081410, 2014.

26. X.-H. Wang, W.-Y. Yin, and Z. Chen, "Broadband modeling surface plasmon polaritons in optically pumped and curved graphene structures with an improved leapfrog ADI-FDTD method," Optics Communications, vol. 334, pp. 152–155, 2015.

27. G. D. Bouzianas, N. V. Kantartzis, C. S. Antonopoulos, and T. D. Tsiboukis, "Optimal modeling of infinite graphene sheets via a class of generalized FDTD schemes," IEEE Transactions on Magnetics, vol. 48, no. 2, pp. 379–382, 2012.

28. V. Nayyeri, M. Soleimani, and O. M. Ramahi, "Modeling graphene in the finite-difference time-domain method using a surface boundary condition," IEEE Transactions on Antennas and Propagation, vol. 61, no. 8, pp. 4176–4182, 2013.View at MathSciNet ·

29. M. Tamagnone, J. S. G. Diaz, J. R. Mosig, and J. P. Carrier, "Reconfigurable terahertz plasmonic antenna concept using a graphene stack," Applied Physics Letters, vol. 101, no. 21, Article ID 214102, 2012.

30. A. Cabellos-Aparicio, I. Llatser, E. Alarcón, A. Hsu, and T. Palacios, "Use of terahertz photoconductive sources to characterize tunable graphene RF plasmonic antennas," IEEE Transactions on Nanotechnology, vol. 14, no. 2, pp. 390–396, 2015.

31. P. Li, L. J. Jiang, and H. Bağcı, "A resistive boundary condition enhanced DGTD scheme for the transient analysis of graphene," IEEE Transactions on Antennas and Propagation, vol. 63, no. 7, pp. 3065–3076, 2015. View at MathSciNet ·

32. G. W. Hanson, "Dyadic green›s functions for an anisotropic, non-local model of biased graphene,"IEEE Transactions on Antennas and Propagation, vol. 56, no. 3, pp. 747–757, 2008.

33. I. Ahmed, E. H. Khoo, and E. Li, "Efficient modeling and simulation of graphene devices with the LOD-FDTD method," IEEE Microwave and Wireless Components Letters, vol. 23, no. 6, pp. 306–308, 2013.

34. X. Yu and C. D. Sarris, "A perfectly matched layer for subcell FDTD and applications to the modeling of graphene structures," IEEE Antennas and Wireless Propagation Letters, vol. 11, pp. 1080–1083, 2012. · ·

35. V. Nayyeri, M. Soleimani, and O. M. Ramahi, "Wideband modeling of graphene using the finite-difference time-domain method," IEEE Transactions on Antennas and Propagation, vol. 61, no. 12, pp. 6107–6114, 2013.View at MathSciNet ·

36. H. Lin, M. F. Pantoja, L. D. Angulo, J. Alvarez, R. G. Martin, and S. G. Garcia, "FDTD modeling of graphene devices using complex conjugate dispersion material model," IEEE Microwave and Wireless Components Letters, vol. 22, no. 12, pp. 612–614, 2012. · ·

37. D.-W. Wang, W.-S. Zhao, X.-Q. Gu, W. Chen, and W.-Y. Yin, "Wideband modeling of graphene-based structures at different temperatures using hybrid FDTD method," IEEE Transactions on Nanotechnology, vol. 14, no. 2, pp. 250–258, 2015.

38. A. Y. Nikitin, F. J. Garcia-Vidal, and L. Martin-Moreno, "Analytical expressions for the electromagnetic dyadic green›s function in graphene and thin layers," IEEE Journal of Selected Topics in Quantum Electronics, vol. 19, no. 3, Article ID 4600611, 2013.

39. O. V. Shapoval, J. S. Gomez-Diaz, J. Perruisseau-Carrier, J. R. Mosig, and A. I. Nosich, "Integral equation analysis of plane wave scattering by coplanar graphene-strip gratings in the thz range," IEEE Transactions on Terahertz Science and Technology, vol. 3, no. 5, pp. 666–674, 2013.

40. R. Araneo, G. Lovat, and P. Burghignoli, "Graphene nanostrip lines: dispersion and attenuation analysis," in Proceedings of the 16th Workshop on Signal and Power Integrity (SPI ‹12), pp. 75–78, IEEE, Sorrento, Italy, May 2012.

41. G. Lovat, G. W. Hanson, R. Araneo, and P. Burghignoli, "Semiclassical spatially dispersive intraband conductivity tensor and quantum capacitance of graphene," Physical Review B, vol. 87, no. 11, Article ID 115429, 2013.

42. P. Burghignoli, R. Araneo, G. Lovat, and G. Hanson, "Space-domain method of moments for graphene nanoribbons," in Proceedings of the 8th European Conference on Antennas and Propagation (EuCAP ‹14), pp. 666–669, IEEE, The Hague, The Netherlands, April 2014.

43. R. Araneo, P. Burghignoli, G. Lovat, and G. W. Hanson, "Modal propagation and crosstalk analysis in coupled graphene nanoribbons," IEEE Transactions on Electromagnetic Compatibility, vol. 57, no. 4, pp. 726–733, 2015.

44. A. Fallahi, T. Low, M. Tamagnone, and J. Perruisseau-Carrier, "Nonlocal electromagnetic response of graphene nanostructures," Physical Review B, vol. 91, Article ID 121405, 2015.

45. M. Dragoman, D. Neculoiu, A. Cismaru et al., "Coplanar waveguide on graphene in the range 40 MHz-110 GHz," Applied Physics Letters, vol. 99, no. 3, Article ID 033112, 2011.

46. J. Chen and Q. H. Liu, "Discontinuous Galerkin Time-Domain methods for multiscale electromagnetic simulations: a review," Proceedings of the IEEE, vol. 101, no. 2, pp. 242–254, 2013.

47. S. Descombes, C. Durochat, S. Lanteri, L. Moya, C. Scheid, and J. Viquerat, "Recent advances on a DGTD method for time-domain electromagnetics," Photonics and Nanostructures—Fundamentals and Applications, vol. 11, no. 4, pp. 291–302, 2013.

48. P. Li and L. J. Jiang, "Modeling of magnetized graphene from microwave to THz range by DGTD with a scalar RBC and an ADE," IEEE Transactions on Antennas and Propagation, vol. 63, no. 10, pp. 4458–4467, 2015.

CHAPTER 3

FINITE ELEMENT METHOD APPLIED IN ELECTROMAGNETIC NDTE

Marek Augustyniak and Zbigniew Usarek

Keywords: Finite element method, Electromagnetic non-destructive testing, MFLEddy current testingEMATs

INTRODUCTION

Non-destructive testing and evaluation (NDT, NDE or NDTE) attracts lasting attention driven by demands of reliability and economic service of engineering structures. Contemporary engineering—and NDT development in particular—becomes increasingly associated with numerical modelling [1]. Among available modelling approaches, finite element analysis (FEA) has taken the lead in both academic and commercial applications. It is much more versatile than any analytical model. As compared to two major concurrent numerical approaches, i.e. boundary element method (BEM) and finite difference method (FDM), it is more intuitive, subject to less fundamental limitations (e.g. concerning unstructured mesh, nonlinearities or couplings) and is promptly available in several computer tools provided with a comfortable graphical user interface (GUI) and exhaustive user manuals.

Finite element analysis can complement and partially replace experimental ENDT for reasons listed below:

- the simulation allows for generating scenarios with a full control over all variables and phenomena
- it is impractical and in some cases unfeasible to measure electromagnetic parameters (magnetic induction, current density) inside a solid specimen [2], whereas these can be easily retrieved from FEA
- the measurement of a detailed distribution of the magnetic and/or electric field around an engineering object is time-consuming and requires painstaking data processing; by contrast, FEA software directly generates the resulting contour plots
- FEA tends to be more economic than experiment, especially when generating an array of results for subsequent inverse problem solution

Table 1 presents authors' attempt to arrange the techniques by the frequency range and the level of complexity. The latter variable corresponds to both the underlying physics and the practical difficulties in obtaining reliable results by either experiment or simulation. Any ENDT method, including those located at the bottom of the classification, require careful calibration and interpretation. Therefore, the table contains only subjective indications. For example, remote field eddy current testing (RFECT) actually is superior in terms of complexity than the static MFL technique.

Table 1 A classification of ENDT methods by the range of frequency and relative complexity; numbers in parentheses indicate the relevant paragraph in the review

Complexity ↑	Static	1 - 1000 Hz	kHz - MHz
	passive MFL (5.3)	MBN and MAE (2.3)	Electromagnetic waves (5.1)
	MFL w. velocity effect (2.2)	Pulsed MFL (2.3)	EMAT (4)
	Static MFL (2.1)	RFECT (3)	ECT (3)

Frequency of excitation

The choice of ENDT technique influences the strategy of finite element simulation. The key options to be selected in an electromagnetic numerical analysis are summarized below:

Citation: Marek Augustyniak and Zbigniew Usarek, Finite Element Method Applied in Electromagnetic NDTE, DOI: 10.1007/s10921-016-0356-6.

- time regime: static/harmonic/transient
- coupling (multiphysics): none/weak/strong
- boundary conditions: flux-parallel, flux-normal, fixed degree-of-freedom (DOF), coupling of DOFs
- nonlinearity: none/nonlinear B(H)/material anisotropy/ velocity effects
- element formulation (magnetic scalar or vector potential, edge-flux formulation)
- dimensionality (2D/2D-axisymmetric/3D)
- software (commercial/academic)

Simulation options listed above are described in more details in [6].

There are several commercial electromagnetic FEA software brands on the market, including MagNet, COMSOL, JSOL, MAXWELL, ANSYS Multiphysics, OPERA, FLUX and others. Their basic common functionality is computation of magnetostatics or electrostatics. Most codes can handle as well harmonic or transient problems involving eddy currents. Some tools are remarkable for the implementation of advanced functions, such as the magnetic hysteresis loop or a robust solution of a moving conductor induction. However, the market evolves rapidly, and the functionality of different tools tends to converge.

Important contributions of numerical modelling to ENDT are reported in three major groups of publications. Firstly, some monographs are available on the application of finite element method in electromagnetics [7–9]. In some of the books the electromagnetic NDT is the major topic [10, 11]. Secondly, there are regular journals devoted to progress in NDT (incl. NDE, NDT&E, RNDE, "Insight" and others) containing both numerical and experimental developments in the field. Finally, an eminent dissemination role is played by proceedings of major NDT-related conferences (ISEM, ENDE, ECNDT), where simulation and modelling tends to be a full-fledged topic. For the sake of example, several recent papers from the ENDE Proceedings [12–17] have been summarised further on in this review.

Bibliographic reviews on FEA applications in various engineering disciplines were systematically published by Mackerle (e.g. [3, 4]), including an exhaustive bibliography on finite element modelling in NDT [5], encompassing the time span between 1976 and 1997 (plus an addendum reaching 2003). In the domain of electrical, magnetic an electromagnetic methods, that review focuses on ECT and the potential drop technique. Our review is complementary to the valuable Mackerle's work, offering up-to-date references and a discussion of specific features of modelling ENDT phenomena.

The following chapters present a review of finite element simulations applied in ENDT, followed by a discussion inspired by own experience in the field. In each chapter one ENDT method is briefly introduced, and some technical aspects of its finite element solution are given. Representative papers in the field are mentioned starting from the 1970ties. Alternative numerical methods and less typical ENDT applications are occasionally invoked. Although the Authors intended to provide a possibly comprehensive and balanced summary of the subject, this review remains a very individual and subjective insight into the vast amount of the published literature.

FEA IN MAGNETIC FLUX LEAKAGE NDT

Static MFL

Static MFL methodology involves magnetizing a portion of a structure and recording the flux at the surface, in order to detect its anomalous spatial distribution. Usually a local magnetization close to saturation is required, because a leakage flux amplitude is generally proportional to the magnetization level. However, too high level of magnetization may lead to decrease a signal-to-noise ratio. The reason is an offset introduced by a background component of the signal. Most common sources of a magnetizing field, electromagnets or yokes with permanent magnets are used.

To design and optimize any MFL system a thorough understanding of magnetic circuit is required. The magnetostatic FEM solver is an efficient tool in MFL-related design and analysis [18]. The FEA solution of a MFL problem requires either a single nonlinear run (static analysis with B(H) curves) or a series of solutions at consecutive time points (transient analysis). The modelling can be 2D or 3D. The material data consists of magnetization curves defined up to saturation ($\mu R \sim 1.0$)($\mu R \sim 1.0$) and possibly electrical conductivities in the transient problem. The smoothness of B(H) curves can be essential for obtaining convergent solution [19]. The mesh density may be uniform, and the standard recommendation holds as to using the hexahedral element shapes whenever it is possible. The resulting vector fields (primary: B—the magnetic induction, and secondary: J—the current density) are vector real values. The numerical methods which cannot deal with nonlinear problems (e.g. BEM, FDM) are not suitable for MFL simulation [20]. Consequently, FEM is preferable in solving the equations governing electromagnetic field in MFL system.

One of the first numerical studies on the MFL defectoscopy including finite element calculation was published by Hwang and Lord in 1975 [21]. They predicted the magnetic field distribution around a rectangular slot on the surface of a circular ferromagnetic bar. The stress-flux correlations were numerically studied as early as in 1987 by Atherton and Czura [22]. They attributed the measured flux leakage over pipeline corrosion pits to stress-induced permeability changes and provided a 2D FEA to solve the inverse problem, i.e. to determine, how significant permeability change takes place in specimen.

The article [23] includes a study, in which apparently minor modifications of a model (inclusion of the detector assembly) produced dramatic qualitative changes of the results. It was demonstrated by 2D FEA, that anomalous leakage fluxes in the vicinity of pipe defects were non-linearly dependent on a defect depth. Further on, the constant H-excitation in a model gave significantly different results from a more realistic constant-B source.

In Atherton [24] a systematic study was performed towards the inverse problem solution of MFL over far side pipe grooves. The author demonstrated a strong linear correlation between anomalous radial flux densities calculated with 2D FEA and those obtained experimentally from 3D synthetic corrosion pits.

The following paragraphs summarise selected results of more recent research

The magnetostatic NDT was modelled with FEA by Al-Naemi, Hall and Moses [18]. The 2D (qualitative) vs 3D (quantitative) models generated and computed in MagNet software were compared against an experiment. The study reproduced the typical stray field behaviour over a defect in a flat plate. Although material data description is missing in the paper, one may presume that linear magnetic permeability was defined.

In Katoh et al. [25], the yoke-magnetization in MFL-testing was modelled with 2D static FEM modelling, using a self-developed software. The simulation was relatively basic, and did not aim at solving any inverse problem.

Another basic sensitivity analysis of the static magnetic circuit composed of a C-core magnet and the metallic specimen can be found in [26]. Linear magnetic permeability of the material, the core length and the spacing were modelled with COMSOL Multiphysics. Although the Barkhausen noise measurement was presented as the study's context, only the primary variables, i.e. magnetic induction B and field intensity H were computed.

Gaunkar et al. [26] redesigned the C-core magnetising setup with COMSOL FEA software, studying the influence of a core material, pole spacing and pole tip curvature. They employed a DC excitation and did not take saturation effects into account.

Babbar et al. [27] represented both the shape and residual stress influence on MFL patterns, over a slightly stamped (dented) plate. They used MagNet software to perform a static solution with a nonlinear, anisotropic B(H) curves. Interestingly, apart from the electromagnetic analysis, they started with structural FEA, subsequently dividing the plate into 13 discrete regions, having varying magnetic properties correlated with stress state. The MFL signal resulting from different stressed regions were studied separately and in combination with the shape effect. The authors found the excellent agreement between the FEA-derived and experimental MFL pattern.

Some papers with a precisely defined industrial context are invoked in the following paragraphs.

Christen et al. from Switzerland [1] examined the influence of steel wrapping on magneto-inductive testing of the main cables of suspension bridges. They employed MagNet software to create and solve a 3D model of a 2 m-long solid body with nonlinear B(H) characteristics. The set-up optimisation was sought, by studying the effect of a sensor lift-off and a magnetic excitation intensity. Based on the FEA results, a postprocessing procedure was proposed for separating and eliminating disturbance introduced by the steel wrapping.

In Gloria et al. [28] the pipeline corrosion defects were detected with a modified MFL, not requiring the saturation of a material. 2 and 3D numerical models were constructed using Gmsh and GetDP freeware finite element programs, and the geometrical parameters of the set-up were optimised. The simulation was compared favourably against an experiment.

A NDT group from India [29] examined with MFL steam generator tubes of a prototype fast breeder reactor. They created a 3D magnetostatic FEA model in COMSOL. They suggested a parallel use of MFL (for a detection of localised defects) and remote field ECT, sensitive to wall thinning, although no ECT simulation was provided.

A group from Queen's University, Canada [30], created a 3D magnetostatic model in MagNet software, in the context of pipeline MFL NDT. Effects of alignment of nearby pits and stress-induced anisotropy were studied in a non-linear model.

Kikuchi et al. [31] examined the feasibility of NDT evaluation of magnetic properties and hardness of two-layered specimens by magnetostatic single-yoke probes. In spite of using a 2D approach in a simulation, they observed a good correlation between experimental and FEA-derived initial magnetisation curves for layered plates.

Coughlin et al. [32] examined the effect of stress on MFL responses from elongated corrosion pits in pipeline X70 steel excited with radial MFL set-ups. In this case the finite element 3D stress simulation was carried out, but no magnetic FEA.

The magnetic "signature" of cracks has been subject to relatively little research, as compared with the effect of stress or corrosion pits. Gao et al. [15] have published their resultss in ENDE Conf. Procs., claiming the competitiveness of their method of interpreting flux leakage near cracks over the standard 3D MFL approach. The standard 3D MFL, in turn, was numerically studied in [33], and an efficient yet accurate inversion algorithm in 2D was produced.

Some atypical usages of numerical simulation as a supportive tool for MFL are presented below.

Snarskii et al. [34] proposed a method of integral equations for the 3D solution of MFL over a surface defect in a plate of a finite thickness. They demonstrated the same accuracy as in FEA (Opera 3D software), but higher robustness. The demonstrated computation speed-up can be important when generating a large database of direct problem patterns, required for any inverse problem solution.

Mahendran and Philip [35] discussed a new magnetic emulsion to enhance visual magnetic NDT. Moreover, they included a comprehensive review of analytical methods of reproducing the field distribution, and pointed to some FEA-based studies, including that by Katoh et al.

Mukhopadhyay and Srivastava [36] tested efficiency of a discrete wavelet transform for a noise reduction MFL signal from a series of defects in a buried pipeline, and claimed the superiority of the method over its alternatives.

The observed trends and perspectives of further research are described in the last chapter of this preview.

MFL(V): Velocity Effects

High-speed non-destructive inspection systems using MFL method are in great demand in online inspection and defect characterisation, especially in pipeline and rail track maintenance [20]. The main component of these systems is an autonomous mobile device containing magnetic field sources and sensors, usually named pipeline inspection gauge (P.I.G.). Such a NDT set-up aims at deducing the shape and dimensions of a flaw, out of the voltage induced in a pick-up coil moving relatively fast (e.g. 1–10 m/s) along a studied structure.

MFL with the velocity effect shares the fundamental principles with the static MFL. An additional difficulty, in a signal interpretation and modelling, stems from the formation of eddy currents when an excitation source moves over or inside a pipeline. The first typical numerical approach to this problem is quasi-static, i.e. the simulation resembles the magnetostatic one, but an extra movement component is added to the governing equation (e.g. [37]). Another approach consists in representing the movement of the set-up over several time-steps, applying some type of electromagnetic coupling between the objects in the relative movement (e.g. [38]). Both the approaches encounter numerical difficulties (inaccuracy, unconvergence) as the relative velocity and material's magnetic permeability increase.

The start of finite element modelling of MFL with the velocity effect dates back to the 1980s (e.g. Shannon and Jackson [39] or Rodger et al. [40]). The effect of magnetizer velocity on MFL was subsequently simulated in 1992 by Nestleroth and Davis [41] using axisymmetric FEA, followed by works of Katragadda et al. [42, 43].

Park and Park [44] analysed the velocity-induced eddy currents in MFL type ENDT, constructing a 3D finite element model. They designed a compensation scheme of the signal distortion, for velocities of the PIG reaching 5 m/s, coming up with a 'pure' defect-related signal, as if the probe is immobile.

Ireland and Torres [45] presented 2D finite element simulations in which a section of a pipe is magnetized along its circumference under both static and moving tool conditions. They employed the standard quasi-static approach with velocity equation component. They highlighted difficulties associated with maintaining a stable magnetic circuit when the set-up is moving. They noted, that a magnetic field profile is extremely complex under both stationary and dynamic tool conditions, and the repeatability of MFL defect patterns can be poor.

In another work [46] the same authors presented a preliminary research on a circumferential magnetizing set-up for pipeline inspection. They performed a dynamic FE analysis for the tool moving at 4 m/s. Both 2 and 3D models were generated, and the B(H) nonlinearity was taken into account. Several limitations and complexities of the method were indicated, including the difficulty to magnetize a pipe circumferentially and the dependence of the signal on the angular position of the set-up.

Li et al. [20] performed a simulation of MFL in a pipeline, with the set-up travelling at high speed, up to 30 m/s. The 2D model (quasi-static with velocity terms) is created in ANSOFT Maxwell. The authors discussed the signal-to-noise ratio and the optimum configuration of sensors.

Nestleroth and Davis [47] studied pairs of permanent magnets rotating around the central axis of a pipe in proximity of its surface. The generated magnetic field was measurably disturbed over a defect. A finite element simulation, validated with experimental data, allowed for investigating a design space with various parameters such as a geometry, material properties, and excitation frequency.

The mentioned works are a representative, but non-exhaustive selection from amongst numerous studies on MFL with a moving source. New research results regularly emerge (e.g. [48, 49]) and the subject is likely to attract lasting attention from the NDT community.

MFL(f): Frequency Effects

Judging from the number of publications and commercial implementations, the magnetic Barkhausen effect is the most important among micromagnetic phenomena applicable in ENDT. There is a significant amount of experimental articles aiming at elucidating the nature of MBN [50–52]. The micromagnetic modelling approaches to MBN representation were recently reviewed by Zapperi and Durin [53], and some results from stochastic (Monte Carlo) models were published as well [54]. However, relatively little has been done so far to develop macromagnetic or multiscale modelling schemes.

When using a C-core magnet in ENDT, replacing a DC excitation (typical for MFL) with a low-frequency (0–10 Hz) AC excitation allows acquiring additional data from the studied object, however the interpretation of signals is a challenging task. FEA plays an irreplaceable role in a comprehensive description of a time- and space-distribution of the magnetic induction within a studied object (plate, tube), which is impossible with any available analytical or experimental method. The FE solution of a magnetic field distribution inside a bulk ferromagnetic object is complicated by the fact, that excitation frequencies as low as 1 Hz generate a skin effect, which significantly disturbs the static field patterns [55–57]. Additional, special degrees of freedom (D.O.F.s) such as time-integrated electric potential may be necessary. Symmetry planes and initial conditions require special modelling approach. A time-transient scheme is preferred over a harmonic calculation, which usually leads to an incorrect representation of skin-effects due to the assumed B(H) linearity.

Augustyniak et al. [58] showed, that both the magnetic Barkhausen effect and the magnetoacoustic emission in a steel plate magnetized by a C-core electromagnet are significantly disturbed by eddy currents at excitation frequencies as low as 1 Hz. At least qualitative characteristics of these effects can be reproduced with FEA (ANSYS).

Another sensitivity analysis of the static magnetic circuit composed of a C-core magnet and the metallic specimen can be found in [26]. The material linear magnetic permeability, core length, and spacing were modelled with COMSOL Multiphysics. Although the Barkhausen noise measurement was presented as the study's context, only the primary variables, i.e. magnetic induction B and field intensity H were computed.

The Japanese group [59] focused on standardized samples used to assess the average induction within an object magnetized dynamically with a C-core set-up. They built a 3D model of the C-core and plates, made of different steels, and concluded, that the standard shim is applicable only to materials exhibiting a high magnetic permeability. Interestingly, they made use of a pseudo-nonlinear harmonic FEA instead of a fully-nonlinear time-transient solution, and found acceptable comparison against experiment.

A recent work [57] presents a detailed analysis of a time- and space-distribution of the nonlinear magnetic field inside a steel plate magnetized with a double-core electromagnet with a separate control of AC excitation currents on both branches. These results, obtained with transient electromagnetic FEA are complementary to the previous research by Nagata [60], who applied the method of boundary elements for calculation of magnetic fields and eddy currents induced by a pair of orthogonal C-cores.

Another step after having determined the magnetic field time- and space-distribution is reproducing the characteristics of Barkhausen effect or magnetoacoustic emission. In both

ENDT methods there is an interaction of external, macroscopic excitation with microstructure, generating a series of short signals, either magnetic (MBN) or acoustic (MAE).

Spanish team [61] focused on angular anisotropy of MBN in pipes due to hot-rolling. Microscale FEM simulations of the magnetic flux density in an idealized steel sample containing the ferrite matrix and the pearlite bands were performed.

Pulsed MFL belongs to the methods under development. It bears some analogy to the acoustic borehole logging, i.e. the frequency content of the magnetic field resulting from a pulsed excitation carries information about the depth and extent of possible anomaly. Researchers from the University of Huddersfield [62] put forward a pulsed magnetic flux leakage technique (PMFL) for a crack detection and characterization. They indicated the limitations of a DC MFL, and suggested the superiority of PMFL, where the probe is driven with a square waveform and the rich frequency components can provide information from different depths due to the skin effects. They used FEMLAB (a transient, 2D, finite difference scheme) to study effects of surface and sub-surface cracks on the magnetic field.

Finally, an interesting hybrid MFL method called magneto-optic direct alternating imaging (MODAI) was studied by Novotny et al. [63]. They applied 2D harmonic FEA to demonstrate the feasibility of detection of defects by magneto-optic films. The optical effect was produced by either an AC or DC excitation. The authors investigated the AC excitation at 10 kHz and claim the possibility of detecting small cracks, weak magnetic phases in nonmagnetic materials, corrosion pits, etc.

EDDY CURRENT TESTING (ECT)

Eddy current testing (ECT) is one of the most effective techniques for detecting cracks and flaws in conducting materials. In ECT devices, an alternate current flows in an exciting coil placed near a specimen suspected to have a flaw. Induced eddy currents affect a signal detected by the surrounding pick-up coils, influenced by a position, shape, and other characteristics of defects or variations of material properties [64]. The signal is usually analysed in terms of complex impedance plane trajectory. A magnitude and phase of a voltage drop induced in a detection coil is influenced by a variation of geometry and material parameters $(\mu,\sigma)(\mu,\sigma)$ of the studied object. The method applies equally to ferromagnetic and paramagnetic materials, including aluminium alloys. ECT techniques are widely (but not exclusively) used for the characterization of safety-critical components, e.g. employed in aeronautical and nuclear engineering.

A FEA solution of an EC problem requires a single harmonic linear run for each frequency of interest. A modelling can be axisymmetric, 2D or 3D. Material data consists of linear magnetic permeabilities and electrical conductivities of all the regions. The resulting vector fields (primary: J—current density and secondary: B—magnetic induction) are composed of real and imaginary parts, convertible into a magnitude or phase. Unlike in a harmonic structural analysis, resonances do not occur, so sampling of frequency at relatively large intervals is acceptable. A mesh density has to be refined within a sub-surface of an object in order to adequately represent the skin depth effect. An integration of B over the coil's cross-section and a subsequent time derivation produce an acceptable approximation of the coil's complex impedance, provided that the frequencies are limited so that the skin effect in the coil wire can be neglected. More discussion on numerical approaches to eddy currents can be found in [6]. Another valuable reference is the TEAM Workshop Project [65], involving FEA simulation benchmarks of several electromagnetic set-ups, including some EC-related problems.

First attempts to reproduce eddy current characteristics with FEA date back to late 70ties [66]. In the 80ties, intense research was conducted, led by Ida, Lord, Udpa and others [67–70]. Industrial case-studies were developed, e.g. the study by Palanisamy and Lord [71] focusing on a condenser tubing.

Design and validation of a new sensor configuration is one of leading topics in EC NDT and is often supported with FEA [72–74]. Aming at an optimized sensor configuration, Robaina et al. [75] developed a 3-coil set-up with two excitation coils and a middle acquisition coil, serving to determine material properties by means of an eddy-current-related change of a magnetic flux. Apart from the experiment, they produced an ANSYS model of the coils, calculating the electric potential inside the coil wires, the resulting magnetic flux, and finally the eddy current density inside the copper sample. Nonlinearities were neglected because of a relatively low field and current density values.

EC-sensor optimization was as well the main objective of the TEAM problem no. 27 [76], with eddy current NDT applied to some deep flaws. The aim was to optimize dimensions of a coil and an excitation current to have the highest level of a signal as possible. It is interesting to note, that the eddy currents were excited by a pulsed voltage excitation, and not a harmonic steady excitation.

In a study dedicated to ECT defectoscopy in airplane maintenance, Rosell and Persson [77] performed an experimental eddy current inspection of small fatigue cracks in Ti–6AL–4V sample and compare it against a finite element model. They noted, that as static loads were applied across the crack faces, electrical connections arise within the crack, which has a strong, detectable influence on the eddy current signal. The work focused on optimum method of incorporating this electrical contact effect into the simulation.

Apart from defectoscopy, the EC set-ups can be used to evaluate some material properties, including electrical conductivity, magnetic permeability, porosity, and tensile strength. One of basic examples is the TEAM benchmark no 15, concerning a non-destructive evaluation of materials with ECT. In another example, Ma et al. [78] created a FE model in Ansoft Maxwell representing a double-coil eddy current testing set-up over samples of Al foams manufactured by the sintering-dissolution process. Simulations gave a multi-frequency (10 Hz–1 MHz) response of the sensor/material geometry for 22 conductivities in the range of 0.1 to ~~40 MS/m. The response was proportional to a change of mutual inductance of coils. The ultimate goal of both the experiment and the accompanying simulation was a reliable porosity evaluation with eddy currents.

In Augustyniak et al. [79], both a 2 and 3D modelling of a three-coil EC set-up over a boiler tube was presented and compared against an experiment. Standard practical factors—lift-off, wall thickness and tube diameter were studied with

FEA at frequencies up to ~10kHz~10kHz. The proposed amplified differential signal was shown to carry information on the amount of magnetic ferrite phase forming on the tube during service.

Sablik and Augustyniak [80] proposed an application of EC sensors in the context of steel mills. Transient 2D FEA time variation of a magnetic flux calculated in ANSYS was transformed by FFT into a set of harmonics, and their relative amplitudes were compared. Both the computation and experiment show an increase of the amplitude of the third harmonic with an increasing tensile strength of the plate.

Remote field eddy current testing (RFECT) is a major, industrially important variation of the standard eddy-current defectoscopy. While in conventional ECT, an exciter and detector coils are placed at the same location and operate at relatively high frequencies (1 kHz–10 MHz), in the RFECT both coils are several pipe diameters apart with a low frequency of operation from 40 to 160 Hz. The choice of a frequency reduces the skin effect and enables detection of flaws localized at the opposite side of the examined object, usually a pipe. In the 1980ties, RFECT phenomena were already experimentally observed successfully analysed using the finite element method. [81–84]. In Kim et al. [85], RFECT was applied to corroded gas transmission pipelines. Initial FEA studies showed that the rotating magnetic field PIG is sensitive to axially oriented tight cracks. Kasai et al. [86] performed an experimental and FEA 3D study in order to assess detectability of back-side flaws on flat ferromagnetic plates, in the context of corrosion of oil storage tanks. They modified the sensor and examined the signal itself as well as the signal-to-noise ratio. In the context of ENDE conferences, Mihalache et al. [17] developed a multifrequency RFECT algorithm applied specifically to the magnetic steam generation (SG) tubes of a fast breeder reactor (FBR). They determined algorithm parameters using proprietary 3D numerical FEA, and validated the concept with experimental measurements conducted on a small test tank.

FEA was successfully employed in non-standard EC techniques, including pulsed eddy-current defectoscopy [87] and stochastic approaches [88]. A special case was published by Kinoshita [89], who used MAXWELL 3D software to reproduce static and harmonic behaviour of magnetic field. The proposed 'electromagnetic impedance' NDT method was supposed to be able to detect fatigue progress of an aluminium alloy plated with ferromagnetic Ni–CO–P material. A permanent magnet in the simulation had a nonlinear anhysteretic B(H) relationship. The fundamental system relationship is determined between the coil's differential impedance and strains in the sample.

The modelling of pulsed ECT has been extensively covered in recent editions of ENDE conf. proceedings. Three examples have been selected for thes review. Zhang et al. [12] performed the 2D-FEM simulation in ANSYS reaching a successful correlation with an analytical model implemented in MATLAB. The problem at hand concerned pulsed eddy currents produced in a ferromagnetic plate with a flat bottom hole. Miorelli et al. [14] proposed a concept of surrogate models of Pulsed ECT, validating them in 3D using COMSOL. Finally, Xin an Lei [13] addressed a difficult problem of incorporating minor hysteresis loops into the calculation, in the context of measurement of pipeline wll thickness.

As popular and useful as it is, standard FEA is not the only approach to EC numerical prediction. Fetzer et al. [90] put forward a coupled FEM-BEM solution of the TEAM problem no. 8, concerning eddy-current NDT. They argued, that representation of the surrounding air with boundary elements greatly reduced computational effort.

Another alternative approach consisted in coupling FEA with some non-standard algorithms. For example, Ida [91] has recently proposed the coupling of FEM with surface impedance boundary conditions (SIBC), allowing elimination of the mesh in a conductor beyond the skin depth zone, thus increasing the speed of the solution without compromising accuracy. Similarly, Sabbagh and co-workers put forward an eddy-current NDT modelling scheme based on volume-integral equations [92–94]. The approach proved to be very successful in the computation of flaw responses in a number of simple geometries, but exhibited limitations in description of complicated surfaces.

More details on ECT methodologies with some references to numerical modelling can be found in [95].

EMATS (ELECTROMAGNETIC ACOUSTIC TRANSDUCERS)

NDT based on EMATs is remarkable for its capability of detecting defects situated far away from an excitation source. The method can be considered as an extension of the standard ultrasonic defectoscopy particularly well suited for non-contact wave generation. EMATs work on nonmagnetic conducting materials (Lorentz force), or ferromagnetic materials (combination of Lorentz force, magnetostriction and magnetization forces). Optimization of EMATs in ferromagnetic materials is often accomplished using computational simulations that account for all the mentioned three main types of transduction mechanism. However, modelling the magnetization force is the least understood part of EMAT simulation and various authors often use controversial methods that lead to contradictory predictions. The discussion of these controversies has been avoided in this review, therefore only the papers considering the Lorentz force and/or magnetostriction are cited.

Although EMATs are broadband transducers and can function with pulsed excitation, they are often excited with a narrow band tone burst to maximize the signal-to-noise ratio, and the majority of simulations deal with a sinusoidal single-frequency excitation.

FE analysis of EMATs represents elastic wave propagation resulting from local application of magnetic fields. It is usually composed of weakly coupled electromagnetic and structural (ultrasound) analyses. The electromagnetic FEA involved determination of a distribution of a static (bias) magnetic field, followed by a transient or harmonic calculation of time-varying B, eddy currents and Lorentz forces. Explicit solvers (ABAQUS Explicit, LS-DYNA, RADIOSS) are best suited for the structural part.

First numerical representations of EMAT date back to the 70ties, with the works by Thompson [96] and Kawashima [97]. Majority of subsequent works were oriented towards an improvement of set-up's performance by modification of geometrical parameters.

An EMAT was modelled in 2D and experimentally tested by Hao et al. [98] They generated a sequence of finite element calculations for a static magnetic field, a pulsed eddy current field, and resulting transient ultrasonic wave.

An exhaustive work by Mirkhani et al. [2] presented a practical numerical study aiming at enhancing the characteristics of an EMAT set-up (mainly signal-to-noise ratio). The FEA strategy included ANSYS EMAG computation followed by LS-DYNA 2D representation of ultrasonic waves. The pre-processor HyperMesh was applied at the latter stage, with the maximum mesh size equal to $\lambda/15\lambda/15$.

Wang et al. [99] proposed an enhancement to the previous FEA models of an EMAT, taking into account a dynamic magnetic field in a meander coil, coming up with the distribution of eddy currents, static magnetic field, Lorentz forces and resultant elastic waves. The FEA results corroborated with an experiment.

Dhayalan [100] presented a numerical analysis of multimode Lamb waves interacting with artificial defects and compares these calculations with measurements on a thin aluminium plate. They started with a 2D electromagnetic model (COMSOL software), reinserting resultant Lorentz forces into a subsequent ultrasonic computation (ABAQUS Explicit).

Kim et al. [101] employed ANSYS to reproduce the magnetic transducer consisting of a nickel grating, permanent magnets providing static bias magnetic field and a set of coils supplying or sensing a time-varying field. They used a topology optimization scheme over the permeable area of a transducer, in order to maximize both the static bias induction and the time-varying excitation field.

Dutton et al. [102] proposed a novel configuration of permanent magnets in an EMAT, consisting of two square magnets with similar magnetic poles facing each other. Effects of pole separation and lift-off were investigated. Both a simulation (FEMLAB) and an experiment consistently indicated an increase of the achieved magnetic induction in the region of interest by a factor of 1.8, as compared with a conventional one-magnet solution.

Huang et al. [103] applied ANSOFT FEA software to study optimum lift-offs in an EMAT system producing US waves at 250 kHz, designed so that only a magnetostrictive contribution took place in the specimen. They concluded, that the lift-off of a receiver could be very small, but the lift-off of a transmitter should be around 2 mm to ensure the stability of the system.

Zhou et al. [16] developed a FEM-BEM hybrid model to study the electromagneto-mechanical coupling influence on the simulated EMAT signals, considering the presence of the magnetostrictive force as well as the Lorentz force.

In spite of extensive published research, some fundamental controversies remained, including that defined by Ribichini et al. [104]. The question of major mechanism of elastic wave creation in ferromagnetic objects was asked: is this predominantly the Lorentz force, or rather magnetostriction? Using a COMSOL FE model, they demonstrate inconsistencies in previous studies and conclude that the Lorentz force usually dominates, except for highly magnetostrictive materials, such as nickel, iron–cobalt alloys or ferrite oxides.

OTHER ENDT METHODS

Methods Involving Electromagnetic Waves

Although X-ray defectoscopy is one of the most popular NDT methods relying on electromagnetic wave propagation/absorption, it does not lend itself to standard FEA. Finite element simulation of the wave phenomena typically requires a mesh density of order of $\lambda/10\lambda/10$, so Angstrom-scale of Roentgen rays make reliable 3D calculations impossible for even most powerful computers available at present. However, waves of higher length (including microwaves) can successfully be simulated, which was demonstrated in some papers [105–107].

Electromagnetic wave propagation is the key phenomenon in electromagnetic emission (EME), sharing many features similar with the acoustic emission (AE). In case of a crack initiation and propagation, a temporary electric charge imbalance is a source of an electromagnetic wave measurable within a distance of a few millimetres [106].

Gade et al. [106] applied COMSOL commercial software to study the main characteristics of an EME sensor and obtained a very good correlation between the experiment and modelling. They claimed, that during the crack growth, an electromagnetic wave of acoustic frequency could be detected within a few mm from the surface of the studied epoxy resin object, and that EME could be a useful complementary tool for classic AE NDT.

It has to be underlined, that problems involving electromagnetic waves are more efficiently solved using other numerical approaches than FEA, such as Method of Moments (MoM) or finite difference in time domain (FDTD) [108].

ENDT Involving Static Electric Potential

NDT methods making use of an electric field as a primary factor are electrical impedance tomography (EIT) and electrical resistance tomography (ERT), being a particular case of the former. They do not involve any magnetic phenomenon and can be solved with FE (e.g. by Trefftz method) with electric potential as a single DOF. Contrary to ECT, the current sources in EIT/ERT are in electrical contact with a conducting object to be tested. The source field is produced by a direct current (DC) or by an AC source at very low frequency. Potential drops or local electric field measurements are often preferred to impedance or local magnetic field measurements, which are typical for ECT.

Although EIT/ERT was invented in the 80ties, the research results with a FEA background were first published after 2000.

Szczepanik and Rucki [109] performed a field analysis and investigated electrical models of multi-electrode impedance sensors. Two types of multi-electrode sensors, used in the impedance tomography systems, were simulated with FEA (COSMOSM software) in order to produce a simple, practical lumped-parameter model. Both the full (FEA) modelling and the simplified representation were successfully validated by the experiment.

Kahunen et al. [110] applied ERT to 3D imaging of concrete. They used an in-house software for both the forward and inverse analysis. Acceptable solutions of the inverse problem in some benchmark set-ups were presented.

Albanese et al. [64] made an initial evaluation of the usefulness of FEA in solving direct and reverse problems in an electrostatic search for buried objects. They carried out two benchmark simulations and experiments. ERT was used to retrieve the shape of a thin crack, and subsequently the geometry of a column immersed in water. Genetic algorithms were used to solve the inverse problem.

Daneshmand [111] and his co-author modelled the EIT to solve a forward problem in a biomedical context (a human thorax). The technique had been studied using FEM or BEM by other authors, but the novelty consisted in proposing a hybrid FEM-BEM strategy, combining advantages of both.

Passive MFL (MMM)

The "passive" MFL (sometimes called MMM, from "Metal Magnetic Memory") is a relatively new concept (1990ties), replacing a controlled magnetic field source with the Earth's field. Claimed cost-effectiveness of a simple portable measurement set-up has to be weighted against low reliability, leading to false-positive or false-negative findings. In spite of the official industrial standardisation (ISO 24497), there is a shortage of fundamental studies, especially those demonstrating its aptness to detect and quantify stress concentrations.

Interestingly, the magnetic FEA, useful as it proved in the standard MFL research, has rarely been used to understand, calibrate and optimize the methodology of passive stray field technique. The only published applications of numerical magnetic analyses in this context are either non-conclusive [112], or bring forth arguments against the concept [113–115].

Zurek [112] presented a preliminary study of contactless magnetostatic stress measurements on an ring made of a plain carbon steel. The results were partially successful, because a limited number of strain gauges did not allow the exact determination of stress and strain changes on the external ring surface. Both the magnetostatic (FLUX2D) and structural FEA (NASTRAN) was performed.

Usarek et al. [114] investigated the influence of the Earth's magnetic field on the distribution of stray magnetic field of a S355 steel sample, either undeformed or plastically deformed. A simple 3D model generated spatial variation of normal and tangential B component in agreement with the experimental data. The article concluded, that a reliable analysis of the stray magnetic field signal required balanced consideration of several factors, such as geometry of the element, plastic strain, and both internal and external stresses.

In a follow-up study, Usarek et al. [115] aimed at the separation of the effects of notch, magnetic permeability and macroscopic residual stress on the MFL signal characteristics. Experimental, as well as magnetostatic FEA, showed that geometrical effect, related to the notch presence, as well as degradation of magnetic permeability in the yielded zone, were mainly responsible for the specific MFL distribution above the sample.

In Augustyniak and Usarek [113] the magnetostatic 3D FEA was applied to study the influence of concurrent factors (geometrical discontinuity, magnetic permeability and remnant magnetisation) on the field strength gradient, claimed by some MMM proponents as correlated with elevated residual stress. The sensitivity study revealed, that the field gradient did not carry enough information to allow solving of an inverse quantitative problem, so determination of stress levels with a magnetic passive stray field alone is impossible. The effect of residual stresses in the studied dog-bone sample was found to be play minor role.

The published works indicate, that too many uncontrollable factors make passive MFL (or "MMM") unsuitable for a reliable quantitative assessment of stresses or material degradation. Numerical analysis served in this case to demonstrate the ambiguity of signal recorded in the passive MFL, and helped to better define the scope of applicability of the method.

SUMMARY AND PERSPECTIVES

Research Trends

The reviewed papers include four types of correlation between the simulated and measured results:

- simulation not referred to any experiment [56, 104]
- qualitative correlation with an experiment [74, 75]
- relative quantitative correlation [106, 117]
- absolute quantitative correlation [97, 113]

Relative quantitative correlation is defined here as demonstrating the same per cent variation of modelled and measured signals, with differing baseline values, typical for 2D modelling. Often it is not possible to make a sharp distinction between relative and absolute correlation, as in [106], where the measurement is presented in terms of voltage, which naturally is influenced by tunable amplification factor. If no reference to an experiment is provided, the paper should clearly demonstrate the soundness of modelling assumptions, and ideally be succeeded by another, experimental work.

In general, finite element analysis serves in every domain of ENDT as a complementary tool for better visualisation and analysis of phenomena at play, set-up optimisation/calibration, and the means of populating a matrix of results for subsequent inverse problem solution. However, application of numerical analysis in each ENDT method exhibits some unique tendencies.

In static MFL, the range of simulation roles is broad: studying the stress or defect influence on stray field [27], optimising a set-up [1–26], and finally providing a reference solution for some special alternative numerical scheme [34]. One of the most advanced works in the field was that by Babbar [27], with a non-trivial geometry, non-uniform stress distribution and non-linear B(H) curve, and an excellent agreement between FEA and experiment.

In MFL with velocity effects the defectoscopy of pipelines and its numerical representation is the grossly dominating topic in scientific papers. One of most remarkable achievements was published by Park and Park [44], who put forward a compensation scheme for the signal from a moving PIG, of direct practical use for the operators.

Research on MFL with AC excitation usually aims at a realistic description of the dynamics of magnetic induction within a magnetic circuit composed of a C-core magnet and a studied ferromagnetic object.

In ECT, the leading topic is the design and optimisation of new sensor configurations. Most of the studies feature a well-defined industrial context (aviation, power plants, steel plants).

Simulations of EMATs usually aim at enhancing the signal-to-noise ratio by reshaping a excitation/detection probe. Among several papers seeking to optimise a transducer, that by Ribichini [104] is remarkable for addressing and resolving the fundamental physical controversy concerning the source of the ultrasonic waves.

As far as the "passive" MFL is concerned, several repetitions of Villari's experiment have been published, demonstrating that some change of material state (stress, plastic strain, fatigue) causes a variation of a stray field. The break-through presented in [116] (a function allowing to quantitatively deduce a residual stress from a stray field level) was later shown to be an overinterpretation [113].

Solutions of Inverse Problems

In NDT, the direct problem consists of the evaluation of a field perturbation at measurement probe locations, for a given exciting field and sample characterization (position, shape, material properties, stress state and defects). In the inverse problem, one has to find parameters of a sample assuming measurements and a forcing field as known quantities. Fast, accurate, and reliable simulation tools are necessary for a success of any inversion procedure.

There are three fundamental obstacles making the inverse problem solution in industrial ENDT challenging. The first one is the possible non-monotonic character of some measured parameters. The second consists in coexistence of several influence factors of a comparable impact on the measured signal. Relevant examples are the confounding conductivity and permeability factors in EC, or superposition of magnetic anisotropy and stress state in MBN [50]. The third problem is the complexity of any in-situ measurements, involving uncertainties higher than those associated with laboratory setting. These issues are discussed by [118], in the context of eddy current defectoscopy, and by Karhun et al. [110], who applied ERT to detection of internal defects in concrete structures. The numerical solution of the inverse problem is usually not unique, and it is sensitive to modelling errors and a measurement noise. Some attempts were unsuccessful, for example that by Roskosz [116], because of excessive number of influence variables as compared to the amount of measured data, which made the function inversion impossible [113].

The solution of inverse problems in ENDT requires some special strategies. One of these consists in obtaining 'richer' information from the studied object by applying several methods in parallel. For example, Sabet-Shargi et al. [119] present a multi-technique work focusing on accurate determination of stress state in hole-drilling method. They employed MBN, MFL and neutron diffraction measurements, and the structural FEA (ANSYS) was used to determine stress distributions in different situations. Szielasko [120] used the same strategy, in a probe designed for characterisation of high strength steels. Finally, Uchimoto et al. [121] employed an original EMAT–EC dual probe to assess the extent of pipeline wall thinning.

Another approach to an ill-conditioned inverse problem relies on a sophisticated analysis of a signal [80], or switching from static to dynamic excitation [62]. In order to minimise the computation cost without compromising the accuracy, some authors propose a refinement of algorithms of problem inversion with design-of-experiment sampling, response-surface fitting [118], and neuron networks [122], The possible presence of several local extrema of the response function requires global optimization procedures, like genetic algorithms or artificial neural networks. The networks are first 'trained' using signals from a wide and possibly exhaustive range of defect shapes. To have a sufficient number of cases, training datasets are often generated using numerical simulations [122]. Genetic algorithms can efficiently be used in conjunction with FEM of the forward problem. In defectoscopy, a set of facets or elements forming a crack could be assuming as unknowns. In any case, to reduce the number of unknowns, some a priori assumptions have to be made on a class of defects under investigation [64].

Some successful solutions of inverse problems in ENDT were described already in the 80ties. In majority, they concerned ECT [123, 124] and static MFL. In Atherton and Czura [22], there is an attempt to assess the order of magnitude of stress change from the magnetic stray field characteristics. The direct 2D problem is solved first, with an assumption of stress-permeability relationship, and then a computation of a stray field for a given non-uniform permeability. In Atherton [24], the corrosion pits are characterised on the basis of the MFL signal. The direct problem is solved first in 2D, with determination of MFL patterns of various defect geometries. Kim and Park [125] came up with an efficient algorithm of solving the MFL inverse problem, e.g. determining the 3D sizing of an axially oriented small crack from axial and circumferential MFL distributions within a pipe. Finally, one should note a mathematically rigorous paper by Tamburrion and Rubinacci [126], where a multi-frequency, EC scan over a conducting sample was successfully exploited to form a spatial map of the resistivity, thus indicating the location and allowing reconstruction of the shape of a flaw. The authors based on their previous experience with ERT, demonstrating that similar inverse algorithms can be applied to ECT.

Perspectives of Further Research

According to the presented literature review, FEA proved useful in all three constitutive domains of ENDT: defectoscopy, stress assessment and evaluation of intrinsic material properties. It has provided the possibility to simulate well-defined multi-factor scenarios and allowed to quantify physical properties at any position where actual measurements could not be performed. FEA is particularly useful when an array of solutions of a forward problem is required, and analytical formulas are insufficiently accurate. Modelling can thus be a basis for solving essential inverse problems, i.e. determining unknown structural parameters (at both micro- and macro-scale) from the measurements. Numerical analysis has been as well successful at studying and enhancing actuators and sensors, esp. in ECT and EMAT techniques.

What are the perspectives of further research?

The MFL technique with AC-excitation (mainly Barkhausen effect and MAE) is one of the most promising fields of research. Experimentally validated models, taking into account both micro- and macroscopic behaviour of the magnetic field inside the ferromagnetic object, are necessary. Assessment of bidirectional stress state, already possible with MBN, requires both more experimental and numerical studies to gain widespread acceptation and be reliably conducted.

In MFL with velocity effect, one awaits a validation and improvement of signal reconstruction scheme such as presented in [44] on a large series of industrial pipelines. Analysts and software developers should focus on overcoming numerical difficulties associated with increasing magnetic permeability and/or velocity of a probe.

In ECT there is a need for studies on non-uniform materials, especially those with a surface altered by corrosion or some intentional treatment (e.g. shot-peening). Some attempts can be found in [80] and [127]. One should note as well, that in most of papers the authors do not perform convergence considerations, i.e. the influence of meshing density is not studied. Allowable as it may be in case of MFL-related works, in case of eddy-current modelling the adequate number of elements through skin depth should be carefully determined lest magnetic field is artificially reduced within the material.

In static MFL, fundamental problems seem to have been solved, however it would be practical to extend the base of inverse problem solutions for arbitrary C-core parameters and arbitrary defects.

EMATs are the only field where simulation seems to have been sufficiently applied, and further developments (e.g. incorporating some minor nonlinear effects into the models) are not supposed to bring significant benefits.

Apart from the specific tasks listed above, some universal challenges may be identified, including:

- extension of ENDT methods to anisotropic materials (e.g. duplex steels, which exhibit strong difference in magnetic properties in orthogonal directions)
- implementation of robust element formulations with a stress-dependent hysteresis loop (preliminary results can be found in [128])
- development of some in-situ, 'intra-operational' FEA computation tools providing real-time support for measurements of complicated geometries
- creation of a NDT-dedicated FEA software, like specialist modules in ANSYS or COMSOL devoted to PCBs, MEMS or packaging

It should be clearly stated that computational simulations, regardless whether they exploit the finite element method or other schemes, can be used only when the underlying physics is well understood and the relevant material properties can be either measured or reliably estimated. For example, the passive remnant magnetic flux leakage (MFL) technique, which was misnamed by its original promoters as metal magnetic Method, cannot be relevantly studied by simulations because the underlying magneto-mechanical effect is far too complex and depends on far too many highly variable parameters.

Finally, the authors of this paper claim, that a top-quality study on ENDT with FE modelling should simultaneously be realistic, parametric, and industrial. 'Realistic' means involving a 3D, nonlinear approach taking into account necessary couplings (e.g. magneto-acoustic in MAE). A 'parametric' study enables construction of a matrix of direct solutions, being a pre-requisite for handling an inverse problem. Finally, 'industrial' work on a well-defined object of non-trivial geometry is preferable over an analysis of a generic sample in an idealised environment. Since the promising, well-structured ENDT case studies published in the 80ties, the prompt availability of the FEA software and dramatic increase of computing power have not always resulted in a proportional quality development of new studies. In particular, quantitative correlations between a simulation and industrial measurements are still rare, and attempts to solve practical inverse problems are in minority. More awareness in the ENDT community is necessary so that the potential of the electromagnetic FEA can be fully exploited.

REFERENCES

1. Christen, R., Bergamini, A., Motavalli, M.: Influence of steel wrapping on magneto-inductive testing of the main cables of suspension bridges. NDT E Int. 42(1), 22–27 (2009)
2. Mirkhani, K., Chaggares, C., Masterson, C., Jastrzebski, M., Dusatko, T., et al.: Optimal design of EMAT transmitters. NDT E Int. 37(3), 181–193 (2004)
3. Mackerle, J.: Finite element and boundary element analyses of microwave, optical and acoustic waveguides. A bibliography (1994–1996). Finite Elem. Anal. Des. 28(2), 165–175 (1997)MATH
4. Mackerle, J.: Creep and creep fracture/damage finite element modelling of engineering materials and structures: a bibliography (1980–1999). Int. J. Press. Vessels Pip. 77, 53–77 (2000)
5. Mackerle, J.: Finite-element modelling of non-destructive material evaluation: a bibliography (1976–1997). Model. Simul. Mater. Sci. Eng. 7, 107–145 (1999)
6. Tang, K.C., Ellis, D.: Why Do Electromagnetic Finite Element Analysis?. NAFEMS Publishing, London (2009)
7. Bastos, J.P.A., Sadowski, N.: Electromagnetic Modelling by Finite Element Methods. Marcel Dekker, New York (2003)
8. Bastos, J.P.A., Sadowski, N.: Magnetic Materials and 3D Finite Element Modelling. CRC Press, Boca Raton (2014)
9. Jin, J.: The Finite Element Method in Electromagnetics. Wiley, Hoboken (2014)MATH
10. Ida, N.: Numerical Modelling for Electromagnetic Non-destructive Evaluation. Chapman & Hall, London (1995)
11. Sabbagh, H.A., Murphy, R.K., Sabbagh, E.H., Aldrin, J.C., Knopp, J.S.: Comutational Electromagnetics and Model Based Inversion. Springer, Berlin (2013)
12. Zhang, Q., Wu, X., Li, J., Sun, P.: Analysis of pulsed eddy current testing for ferromagnetic metallic materials with flat-bottom hole defect. In: Chen, Z., et al. (eds.) Electromagnetic Nondestructive Evaluation (XVIII), pp. 95–102. IOS Press, Amsterdam (2015)
13. Xin, W., Lei, Y.Z.: Analysis of a pulsed eddy current testing method for ferromagnetic pipeline wall thickness determination considering minor hysteresis. In: Chen, Z.,

et al. (eds.) Electromagnetic Nondestructive Evaluation (XVIII), pp. 117–124. IOS Press, Amsterdam (2015)

14. Miorelli, R., Dubois, A., Bilicz, S., Theodoulidis, T., Reboud, C.: Simulation of pulsed eddy current testing via surrogate models. In: Capova, K., et al. (eds.) Electromagnetic Nondestructive Evaluation (XVII), pp. 120–127. IOS Press, Amsterdam (2014)
15. Gao, Y., Wang, P., Tian, G.: Numerical simulation of magnetic flux leakage for crack geometrical characterization based on spatial magnetic field vectors. In: Chen, Z., et al. (eds.) Electromagnetic Nondestructive Evaluation (XVIII), pp. 125–132. IOS Press, Amsterdam (2015)
16. Zhou, H., Yang, Z., Li, Y.: Numerical simulation of EMAT for structures of ferromagnetic material with consideration of electromagneto–mechanical coupling effect. In: Chen, Z., et al. (eds.) Electromagnetic Nondestructive Evaluation (XVIII), pp. 110–116. IOS Press, Amsterdam (2015)
17. Mihalache, O., Yamaguchi, T., Ueda, M.: Validations of multifrequency ECT algorithms for helical SG tubes of FBR. In: Capova, K., et al. (eds.) Electromagnetic Nondestructive Evaluation (XVII), pp. 109–119. IOS Press, Amsterdam (2014)
18. Al-Naemi, F.I., Hall, J.P., Moses, A.J.: FEM modelling techniques of magnetic flux leakage-type NDT for ferromagnetic plate inspections. J. Magn. Magn. Mater. 304(2), 790–793 (2006)
19. Augustyniak, M., Usarek, Z.: Control of magnetic circuit parameters in quasi-static magnetic non-destructive testing. Part I: analytical and 2-D model. Zesz. Naukowe Wydz. Elektrotech. Autom. Politec. Gdan. 40, 17–21 (2014). (Article in Polish)
20. Li, Y., Tian, G.Y., Ward, S.: Numerical simulation on magnetic flux leakage evaluation at high speed. NDT E Int. 39(5), 367–373 (2006)
21. Hwang, J.H., Lord, W.: Finite element modeling of magnetic field/defect interactions. J. Test. Eval. 3(1), 21–25 (1975)
22. Atherton, D.L., Czura, W.: Finite element calculations on the effects of permeability variation on magnetic flux leakage signals. NDT E Int. 20(4), 239–241 (1987)
23. Atherton, D.L., Daly, M.G.: Finite element calculation of magnetic flux leakage detector signals. NDT E Int. 20(4), 235–238 (1987)
24. Atherton, D.L.: Finite element calculations and computer measurements of magnetic flux leakage patterns for pits. Brit. J. Nondestr. Test. 30(3), 159–162 (1988)
25. Katoh, M., Masumoto, N., Nishio, K., Yamaguchi, T.: Modeling of the yoke-magnetization in MFL-testing by finite elements. NDT E Int. 36(7), 479–486 (2003)
26. Gaunkar, N.P., Kypris, O., Nlebedim, I.C., Jiles, D.C.: Optimization of sensor design for Barkhausen noise measurement using finite element analysis. J. Appl. Phys. 115, 1–4 (2014)
27. Babbar, V., Bryne, J., Clapham, L.: Mechanical damage detection using magnetic flux leakage tools: modeling the effect of dent geometry and stresses. NDT E Int. 38(6), 471–477 (2005)
28. Gloria, N.B.S., Areiza, M.C.L., Miranda, I.V.J., Rebello, J.M.A.: Development of a magnetic sensor for detection and sizing of internal pipeline corrosion defects. NDT E Int. 42(8), 669–677 (2009)
29. Singh, W.S., Rao, B.P., Thirunavukkarasu, S., Mahadevan, S., Mukhopadhyay, C.K., et al.: Development of magnetic flux leakage technique for examination of steam generator tubes of prototype fast breeder reactor. Ann. Nucl. Energy 83, 57–64 (2015)
30. Mao, W., Clapham, L., Atherton, D.L.: Effects of alignment of nearby corrosion pits on MFL. NDT E Int. 36(2), 111–116 (2003)
31. Kikuchi, H., Murakami, H., Ara, K.: Feasibility study for nondestructive evaluation of magnetic properties and hardness of two-layered specimens by magnetic single-yoke probe. NDT E Int. 46, 1–6 (2012)
32. Coughlin, C.R., Clapham, L., Atherton, D.L.: Effects of stress on MFL responses from elongated corrosion pits in pipeline steel. NDT E Int. 33(3), 181–188 (2000)
33. Priewald, R.H., et al.: Fast magnetic flux leakage signal inversion for the reconstruction of arbitrary defect profiles in steel using finite elements. IEEE Trans. Magn. 49(1), 506–516 (2013)
34. Snarskii, A.A., Zhenirovskyy, M., Meinert, D., Schulte, M.: An integral equation model for the magnetic flux leakage method. NDT E Int. 43(4), 343–347 (2010)
35. Mahendran, V., Philip, J.: Naked eye visualization of defects in ferromagnetic materials and components. NDT E Int. 60, 100–109 (2013)
36. Mukhopadhyay, S., Srivastava, G.P.: Characterisation of metal loss defects from magnetic flux leakage signals with discrete wavelet transform. NDT E Int. 33(1), 57–65 (2000)
37. De Gersem, H., Hameyer, K.: Finite element simulation of a magnetic brake with a soft magnetic solid iron rotor. COMPEL Int. J. Comput. Math. Electr. Electron. Eng. 21(2), 296–306 (2002)MATH
38. Gay, S.E., Ehsani, M.: Parametric analysis of eddy-current brake performance by 3-D finite-element analysis. IEEE Trans. Magn. 42(2), 319–328 (2006)
39. Shannon, R.W.E., Jackson, L.: Flux leakage testing applied to operational pipelines. Mater. Eval. 46(2), 1516–1524 (1988)
40. Rodger, D., Karaguler, T., Leonard, P.J.: A formulation for 3D moving conductor eddy current problems. IEEE Trans. Magn. 25(5), 4147–4149 (1989)
41. Davis, R.J., Nestleroth, J.B.: The effects of flux leakage magnetizer velocity on volumetric defect signals. In: Thompson, D., Chimenti, D. (eds.) Review of Progress in Quantitative Nondestructive Evaluation, vol. 14, pp. 491–498. Springer, Berlin (1995)
42. Katragadda, G., Si, J.T., Lord, W., Sun, Y.S., Udpa, S., Udpa, L.: 3D finite element methods for modeling MFL inspection of pipelines. In: Thompson, D., Chimenti, D. (eds.) Review of Progress in Quantitative Nondestructive

Evaluation, vol. 15A, pp. 561–567. Springer, Berlin (1996)

43. Katragadda, G., Sun, Y.S., Lord, W., Udpa, S., Udpa, L.: Velocity effects and their minimization in MFL inspection of pipelines—a numerical study. In: Thompson, D., Chimenti, D. (eds.) Review of Progress in Quantitative Nondestructive Evaluation, vol. 14, pp. 499–505. Springer, Berlin (1995)

44. Park, G.S., Park, S.H.: Analysis of the velocity-induced eddy current in MFL type NDT. IEEE Trans. Magn. 40(2), 663–666 (2004)

45. Ireland, R.C., Torres, C.R.: Finite element modelling of a circumferential magnetiser. Sensors Actuators A Phys. 129(1–2), 197–202 (2006)

46. Ireland, R.C., Torres, C.R.: Challenges in circumferential magnetisation: a FEA point of view. In: Proc. IPC, pp. 945–955 (2004)

47. Nestleroth, J.B., Davis, R.J.: Application of eddy currents induced by permanent magnets for pipeline inspection. NDT E Int. 40(1), 77–84 (2007)

48. Kim, H.M., et al.: Defect estimation of a crack in underground pipelines by CMFL type NDT system. In: Proceedings of International Conference on Electrical Machines and Systems, Busan (2013)

49. Wang, P., Gao, Y., Tian, G., Wang, H.: Velocity effect analysis of dynamic magnetization in high speed magnetic flux leakage inspection. NDT E Int. 64, 7–12 (2014)

50. Lindgren, M., Lepistö, T.: On the stress vs. Barkhausen noise relation in a duplex stainless steel. NDT E Int. 37(5), 403–410 (2004)

51. Kleber, X., Barroso, S.P.: Investigation of shot-peened austenitic stainless steel 304L by means of magnetic Barkhausen noise. Mater. Sci. Eng. A. 527, 6046–6052 (2010)

52. Augustyniak, B.: Correlation between acoustic emission and magnetic and mechanical Barkhausen effects. J. Magn. Magn. Mater. 196–197, 799–801 (1999)

53. Zapperi, S., Durin, G.: New perspectives for Barkhausen noise. Comput. Mater. Sci. 20(3–4), 436–442 (2001)

54. Yamaguchi, K., Tanaka, S., Watanabe, H., Nittono, O., Yamada, K., Takagi, T.: Analysis of Barkhausen noise using Monte Carlo simulation for nondestructive evaluation. J. Mater. Process. Technol. 161(1–2), 338–342 (2005)

55. Augustyniak, M., Augustyniak, B., Piotrowski, L., Chmielewski, M.: Evaluation by means of magneto-acoustic emission and Barkhausen effect of time and space distribution of magnetic flux density in ferromagnetic plate magnetised by a C-core. J. Magn. Magn. Mater. 304, 552–554 (2006)

56. Augustyniak, M., Augustyniak, B., Chmielewski, M., Sadowski, W.: Numerical evaluation of spatial time-varying magnetisation of ferritic tubes excited with a C-core magnet. J. Magn. Magn. Mater. 320, 1053–1056 (2008)

57. Augustyniak, M., Augustyniak, B., Piotrowski, L., Chmielewski, M.: Determination of magnetisation conditions in a double-core Barkhausen noise measurement set-up. J. Nondestruct. Eval. 34, 16 (2015)

58. Augustyniak, B., Piotrowski, L., Augustyniak, M., Chmielewski, M., Sablik, M.J.: Impact of eddy currents on Barkhausen and magnetoacoustic emission intensity in a steel plate magnetized by a C-core electromagnet. J. Magn. Magn. Mater. 272–276, E543–E545 (2004)

59. Kasai, N., Takada, A., Fukuoka, K., Aiyama, H., Hashimoto, M.: Quantitative investigation of a standard test shim for magnetic particle testing. NDT E Int. 44(5), 421–426 (2011)

60. Nagata, S., Enokizono, M.: Numerical simulations and experiments of eddy current tests under various excitation methods. J. Mater. Process. Technol. 161(1–2), 353–358 (2005)

61. Martínez-Ortiz, P., Pérez-Benitez, J.A., Espina-Hernández, J.H., Caleyo, F., Hallen, J.M.: On the estimation of the magnetic easy axis in pipeline steels using magnetic Barkhausen noise. J. Magn. Magn. Mater. 374, 67–74 (2015)

62. Sophian, A., Tian, G.Y., Zairi, S.: Pulsed magnetic flux leakage techniques for crack detection and characterisation. Sensors Actuators A Phys. 125(2), 186–191 (2006)

63. Novotny, P., Sajdl, P., Machac, P.: A magneto-optic imager for NDT applications. NDT E Int. 37, 645–649 (2004)

64. Albanes, R., Calabró, G., Lombardo, G., Reitano, G., Fresa, R., Morabito, P.: Error bounds for inverse electromagnetic problems in soil mechanics. Math. Comput. Model. 37(5–6), 603–613 (2003) MathSciNetMATH

65. Testing Electromagnetic Analysis Methods. International Compumag Society.http://www.compumag.org/jsite/team.html. Accessed 30 Nov 2015

66. Chari, M.V.K., Kincaid, T.: The application of finite element method analysis to eddy current nondestructive evaluation. IEEE Trans. Magn. 15(6), 1956–1960 (1979)

67. Ida, N., Lord, W.: A finite element model for three-dimensional eddy current NDT phenomena. IEEE Trans. Magn. 21(6), 2635–2643 (1985)

68. Allen, B., Ida, N., Lord, W.: Finite element modeling of pulse eddy current NDT phenomena. IEEE Trans. Magn. 21(6), 2250–2253 (1985)

69. Udpa, L., Lord, W.: Impedance and mesh structure considerations in the finite element analysis of eddy current NDT probe phenomena. IEEE Trans. Magn. 21(6), 2269–2272 (1985)

70. Palanisamy, R., Lord, W.: Prediction of eddy current probe signal trajectories. IEEE Trans. Magn. 16(5), 1083–1085 (1980)

71. Palanisamy, R., Lord, W.: Prediction of eddy current signals for nondestructive testing of condenser tubing. IEEE Trans. Magn. 19(5), 2213–2215 (1983)

72. Nath, S., Wincheski, B., Fulton, J.P., Namkung, M.: Study of the new eddy current non-destructive testing sensor on ferromagnetic materials. IEEE Trans. Magn. 30(6), 4644–4646 (1994)

73. Thollon, F., Burais, N.: Geometrical optimization of sensors for eddy currents. Non destructive testing and evaluation. IEEE Trans. Magn. 31(3), 2026–2031 (1995)

74. Huang, H., Sakurai, N., Takagi, T., Uchimoto, T.: Design of an eddy-current array probe for crack sizing in steam generator tubes. NDT E Int. 36, 515–522 (2003)

75. Robaina, R.R., Alvarado, H.T., Plaza, J.A.: Planar coil-based differential electromagnetic sensor with null-offset. Sensors Actuators A Phys. 164(1–2), 15–21 (2010)

76. Dyck, D.N., Gilbert, G., Forghani, B., Webb, J.P.: An NDT pulse shape study with TEAM problem 27. IEEE Trans. Magn. 40(2), 1406–1409 (2004)

77. Rosell, A., Persson, G.: Finite element modelling of closed cracks in eddy current testing. Int. J. Fatigue 41, 30–38 (2012)

78. Ma, X., Peyton, A.J., Zhao, Y.Y.: Measurement of the electrical conductivity of open-celled aluminium foam using non-contact eddy current techniques. NDT E Int. 38(5), 359–367 (2005)

79. Augustyniak, B., Chmielewski, M., Sablik, M.J., Augustyniak, M.: A new eddy current method for nondestructive testing of creep damage in austenitic boiler tubing. Nondestruct. Test. Eval. 24(1–2), 121–141 (2009)

80. Sablik, M., Augustyniak, M.: Nonlinear harmonic amplitudes in air coils above and below a steel plate as a function of tensile strength via finite-element simulation. IEEE Trans. Magn. 40(4), 2182–2184 (2004)

81. Lord, W., Sun, Y.S., Udpa, S.S., Nath, S.: A finite element study of the remote field eddy current phenomenon. IEEE Trans. Magn. 24, 435–438 (1988)

82. Hoshikawa, H., Koyama, K., Koido, J., Ishibashi, Y.: Characteristics of remote-field eddy current technique. Mater. Eval. 47(1), 93–97 (1988)

83. Atherton, D.L., Szpunar, B., Sullivan, S.: The application of finite element calculations to the remote field inspection technique. Mater. Eval. 44(9), 1083–1086 (1987)

84. Nath, S., Lord, W., Sun, Y.S.: Theoretical and experimental studies of the remote field eddy current effect. In: Thompson, D., Chimenti, D. (eds.) Review of Progress in Quantitative Nondestructive Evaluation, vol. 8, pp. 267–274. Springer, Berlin (1989)

85. Kim, D., Udpa, L., Udpa, S.: Remote field eddy current testing for detection of stress corrosion cracks in gas transmission pipelines. Mater. Lett. 58(15), 2102–2104 (2004)

86. Kasai, N., Matsuzaki, S., Sakamoto, T.: Experimental and analytical study for detectability of the back-side flaws of flat ferromagnetic plates by RFECT. NDT E Int. 44(8), 703–707 (2011)

87. Clauzon, T., Thollon, F., Nicolas, A.: Flaws characterization with pulsed eddy currents NDT. IEEE Trans. Magn. 35(3), 1873–1876 (1999)

88. Homberg, C., Henneron, T., Clenet, S.: Stochastic model in eddy current non destructive testing. In: Proc. 2011 14th Eur. Conf. Power Electron. Appl., pp. 1–7 (2011)

89. Kinoshita, K.: Estimation of fatigue evolution of aluminum alloy plated with electroless Ni–CO–P by using electromagnetic impedance method. J. Magn. Magn. Mater. 375, 80–86 (2015)

90. Fetzer, J., Kurz, S., Lehner, G.: The coupling of boundary elements and finite elements for nondestructive testing applications. IEEE Trans. Magn. 33(1), 677–681 (1997)

91. Ida, N.: Eddy current nondestructive evaluation—the challenge of accurate modeling. In: DAS—2014 Int. Conf. Dev. Appl. Syst., pp. 97–102 (2014)

92. Sabbagh, H.: A model of eddy-current probes with ferrite cores. IEEE Trans. Magn. 23(3), 1888–1904 (1987)

93. Sabbagh, L.D., Sabbagh, H.A.: A computer modelof eddy-current probe-crack interaction. In: Review of Progress in Quantitative Nondestructive Evaluation, Brunswick (July 1989)

94. Murphy, K., Sabbagh, H.A.: Boundary-integral equations in eddy-current calculations. In: Review of Progress in Quantitative Nondestructive Evaluation 14, pp. 267–274

95. Garcia-Martín, J., Gómez-Gil, J., Vázquez-Sánchez, E.: Non-destructive techniques based on eddy current testing. Sensors 11, 2525–2565 (2011)

96. Thompson, R.B.: A model for the electromagnetic generation and detection of Rayleigh and Lamb waves. IEEE Trans. Sonics Ultrason. SU–20(4), 340–346 (1973)

97. Kawashima, K.: Theory and numerical calculation of the acoustic field produced in metal by an electromagnetic ultrasonic transducer. J. Acoust. Soc. Am. 60, 1089–1099 (1976)

98. Hao, K.-S., Huang, S.-L., Zhao, W., Duan, R.-J., Wang, S.: Modeling and finite element analysis of transduction process of electromagnetic acoustic transducers for nonferromagnetic metal material testing. J. Cent. South Univ. Technol. 18(3), 749–754 (2011)

99. Wang, S., Kang, L., Zhichao, L., Zhai, G., Zhang, L.: 3-D modeling and analysis of meander-line-coil surface wave EMATs. Mechatronics 22(6), 653–660 (2012)

100. Dhayalan, R., Balasubramaniam, K.: A hybrid finite element model for simulation of electromagnetic acoustic transducer (EMAT) based plate waves. NDT E Int. 43(6), 519–526 (2010)

101. Kim, I.K., Kim, W., Kim, Y.Y.: Magnetostrictive grating with an optimal yoke for generating high-output frequency-tuned SH waves in a plate. Sensors Actuators A Phys. 137(1), 141–146 (2007)

102. Dutton, B., Boonsang, S., Dewhurst, R.J.: A new magnetic configuration for a small in-plane electromagnetic acoustic transducer applied to laser-ultrasound measurements: Modelling and validation. Sensors Actuators A Phys. 125(2), 249–259 (2006)

103. Huang, S., Zhao, W., Zhang, Y., Wang, S.: Study on the lift-off effect of EMAT. Sensors Actuators A Phys. 153(2), 218–221 (2009)

104. Ribichini, R., Nagy, P.B., Ogi, H.: The impact of magnetostriction on the transduction of normal bias field EMATs. NDT E Int. 51, 8–15 (2012)

105. Grimberg, R., Savin, A., Steigmann, R.: Electromagnetic imaging using evanescent waves. NDT E Int. 46, 70–76 (2012)

106. Gade, S.O., Weiss, U., Peter, M.A., Sause, M.G.R.: Relation of electromagnetic emission and crack dynamics in epoxy resin materials. J. Nondestruct. Eval. 33(4), 711–723 (2014)

107. Shibata, T., Hashizume, H., Kitajima, S., Ogura, K.: Experimental study on NDT method using electromagnetic waves. J. Mater. Process. Technol. 161, 348–352 (2005)

108. Klysz, G., Balayssac, J.P., Ferričres, X.: Evaluation of dielectric properties of concrete by a numerical FDTD model of a GPR coupled antenna—parametric study. NDT E Int. 41(8), 621–631 (2008)

109. Szczepanik, Z., Rucki, Z.: Field analysis and electrical models of multi-electrode impedance sensors. Sensors Actuators A Phys. 133(1), 13–22 (2007)

110. Karhunen, K., Seppänen, A., Lehikoinen, A., Monteiro, P.J.M., Kaipio, J.P.: Electrical resistance tomography imaging of concrete. Cem. Concr. Res. 40, 137–145 (2010)

111. Daneshmand, P.G., Jafari, R.: A 3D hybrid BE–FE solution to the forward problem of electrical impedance tomography. Eng. Anal. Bound. Elem. 37(4), 757–764 (2013)MathSciNetMATH

112. Zurek, Z.H.: Magnetic contactless detection of stress distribution and assembly defects in constructional steel element. NDT E 38(7), 589–595 (2005)

113. Augustyniak, M., Usarek, Z.: Discussion of derivability of local residual stress level from magnetic stray field measurement. J. Nondestruct. Eval. 34, 21 (2014)

114. Usarek, Z., Augustyniak, B., Augustyniak, M., Chmielewski, M.: Influence of plastic deformation on stray magnetic field distribution of soft magnetic steel sample. IEEE Trans. Magn. 50(4), 1–4 (2014)

115. Usarek, Z., Augustyniak, B., Augustyniak, M.: Separation of the effects of notch and macro residual stress on the MFL signal characteristics. IEEE Trans. Magn. 50(11), 1–4 (2014)

116. Roskosz, M., Bieniek, M.: Evaluation of residual stress in ferromagnetic steels based on residual magnetic field measurements. NDT E Int. 45, 55–62 (2012)MATH

117. Pérez-Benítez, J.A., et al.: Analysis of the influence of some magnetizing parameters on magnetic Barkhausen noise using a microscopic model. J. Magn. Magn. Mater. 347, 51–60 (2013)

118. Pávó, J., Gyimóthy, S.: Adaptive inversion database for electromagnetic nondestructive evaluation. NDT E Int. 40(3), 192–202 (2007)

119. Sabet-Sharghi, R., Clapham, L., Atherton, D.L., Holden, T.M.: The effect of defect introduction vs. load application sequencing on defect-induced stress distributions in steel samples. NDT E Int. 33(4), 201–212 (2000)

120. Szielasko, K., Youssef, S., Niese, F., Weikert, M., Sourkov, A., Altpeter, I., Herrmann, H.-G., Dobmann, G., Boller, C.: Multi-method probe design for the electromagnetic characterization of advanced high strength steel. In: Capova, K., et al. (eds.) Studies in Applied Electromagnetics and Mechanics 39: Electromagnetic Nondestructive Evaluation (XVII), pp. 52–59. IOS Press, Amsterdam (2014)

121. Uchimoto, T., Guy, P., Takagi, T., Courbon, J.: Evaluation of an EMAT–EC dual probe in sizing extent of wall thinning. NDT E Int. 62, 160–166 (2014)

122. Bouchala, T., Abdelhadi, B., Benoudjit, A.: Novel coupled electric field method for defect characterization in eddy current non-destructive testing systems. J. Nondestruct. Eval.33(1), 1–11 (2014)

123. Lee, H.-B., Jung, H.-K., Hahn, S.-Y., Chung, J.K.: An inverse analysis for crack identification in eddy current NDT of tubes. IEEE Trans. Magn. 30(5), 3403–3406 (1994)

124. Putek, P., Crevecoeur, G., Slodicka, M., Gawrylczyk, K., van Keer, R., Dupre, L.: Two-level approach for solving the inverse problem of defect identification in eddy current testing-type NDT. In: IET 8th International Conference on Computation in Electromagnetics (CEM 2011), pp. 1–2 (11–14 Apr 2011)

125. Kim, H.M., Park, G.S.: A study on the estimation of the shapes of axially oriented cracks in CMFL type NDT system. IEEE Trans. Magn. 50(2), 109–112 (2014)

126. Tamburrino, A., Rubinacci, G.: Fast methods for quantitative eddy-current tomography of conductive materials. IEEE Trans. Magn. 42(8), 2017–2028 (2006)

127. Yusa, N., Janousek, L., Rebican, M., Chen, Z., Miya, K., Chigusa, N.: Detection of embedded fatigue cracks in Inconel weld overlay and the evaluation of the minimum thickness of the weld overlay using eddy current testing. Nucl. Eng. Des. 236(18), 1852–1859 (2006)

128. Ossart, F., Hirsinger, L., Billardon, R.: Computation of electromagnetic losses including stress dependence of magnetic hysteresis. J. Magn. Magn. Mater. 196–197, 924–926 (1999)

CHAPTER 4 A MODELING METHOD FOR SMALL PACKING PARTICLES IN ELECTROMAGNETIC SIMULATION

Yi Cheng Zhong, Yu Jian Cheng and Yong Fan

INTRODUCTION

Most natural substances, such as building materials and minerals, present in the mixed state, especially in the state of packing particles. The propagation and reflection properties of the electromagnetic wave in inhomogeneous materials have been widely studied [1, 2]. The model of packing particles plays an important role in the research of the ground-penetrating radar, the microwave heating, the microwave desulfurization, and so forth. Therefore, there is an urgent need for building an accurate model of packing particles to study the propagation characteristics of the electromagnetic wave in nature [3–5].

The particle in this paper is defined as the smallest unit of substances with identical electromagnetic properties. Electromagnetic properties of the individual particle are usually easy to recognize by experiment [6–8]. However, when the material is formed by a number of solid packing particles, it is difficult to investigate electromagnetic properties of this mixed-particle material because of the large number and random distribution of these particles [9].

Previous researchers have done rigorous and impressive job on this issue and proposed different methods. Effective complex permittivity of dielectric mixtures can be evaluated by different formulations and rules from theoretical models and experiments [10–14]. However, the theoretical model can only be used in the particles with simple and regular shapes. Besides, the experimental method is affected by the random distribution of particles, and most of the time, the accurate formulations of mixtures cannot be gotten from the experimental results.

Many numerical studies have also been performed to analyze the effective permittivity of dielectric mixtures [15–17]. The dynamical solution method (FDTD) is firstly applied to the two-dimensional dielectric mixture [17]. This method is also used in the three-dimensional analysis in this paper. Besides, there are some other kinds of modeling software, such as PFC 3D, used in fields outside of the electromagnetism [18–23]. They are employed in the model of packing particles in architecture and civil engineering. Their principle is to calculate movements of the particles through boundary conditions and physical properties until a stable state is achieved. The available shapes of particles are only the sphere and the combination of several spheres. Thus, particles with complex shapes are difficult to be modeled. The modeling software also needs a long time as well as a lot of computing resources for the modeling of a large number of particles.

This paper aims to establish an accurate model through a simple and quick process to simulate various packing particles, such as sand, soil, and coal powder. Furthermore, the model can be imported into some kinds of commercial electromagnetic simulation software easily. Currently, there are a variety of methods for modeling stacked particles in the region of architecture, mechanics, and geophysics. But very few of them are proper to the electromagnetic research and none of them can be imported into simulation software directly.

This paper will focus on how to model contacted balls and regular polyhedrons to simulate the packing particles. It can also be employed in the modeling of packing particles with complex shapes. The accuracy of the model is validated by experiments of packing density and effective permittivity.

Citation: Yi Cheng Zhong, Yu Jian Cheng, and Yong Fan, "A Modeling Method for Small Packing Particles in Electromagnetic Simulation," International Journal of Antennas and Propagation, vol. 2016, Article ID 1785751, 7 pages, 2016. doi:10.1155/2016/1785751.

MODELING

Model of Packing Spheres

The completed modeling process is shown in Figure 1. The first step is to distribute balls uniformly within the desired region. Supposing the desired region with a length of L, a width of S, and a height of H, the number of spheres, W, can be calculated by floor $[(L \times S \times H \times 2)/(a+b)]$. Floor is the function of the rounding down. The diameter range of spheres is (a, b). Secondly, centers of spheres are moved and radii of spheres are gradually increased to make them contact each other. Finally, the modeling will be ended if all the spheres contact at least two other spheres or the edge of the analyzed region. When the process is over, the packing density of the spheres becomes relatively stable and within the possible region validated by experiments [24]. Thus, this model is considered to be consistent with the actual distribution of the packing spheres.

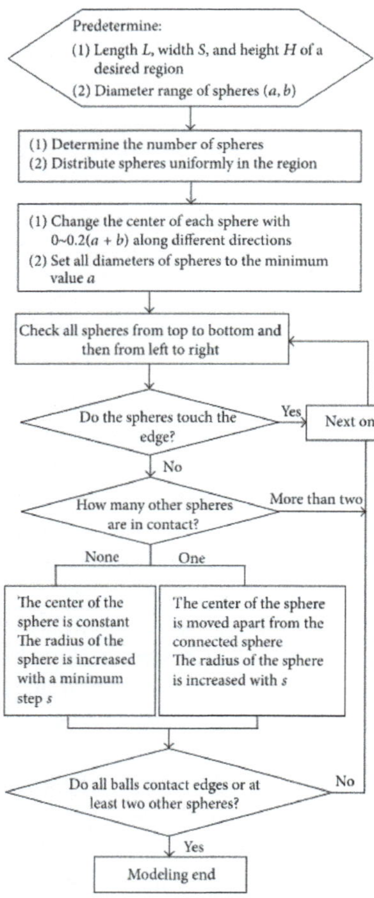

Figure 1: Modeling process.

The modeling method is different from these rigorous modeling and analytical methods in architecture and civil engineering. That is because it does not consider the gravity and friction of the balls. However, the modeling method is particularly used for the electromagnetic analysis of relatively small packing particles; thus it can be accepted without the gravity and the friction in the modeling process. Besides, in the electromagnetic simulation of accumulated particles, the more important influence factors are the shape, size, and packing density of the investigated material [16]. The effects of the gravity and the friction are small, which are usually excluded in the analysis of electromagnetic properties of packing particles.

Figure 2 demonstrates some key steps in the modeling process with different cycle time, P. The present modeling was carried out with balls, which have diameters ranging in 8~10 mm, within a cubic region of 75 × 75 × 105 mm^3. The total modeling time is limited in 10 seconds.

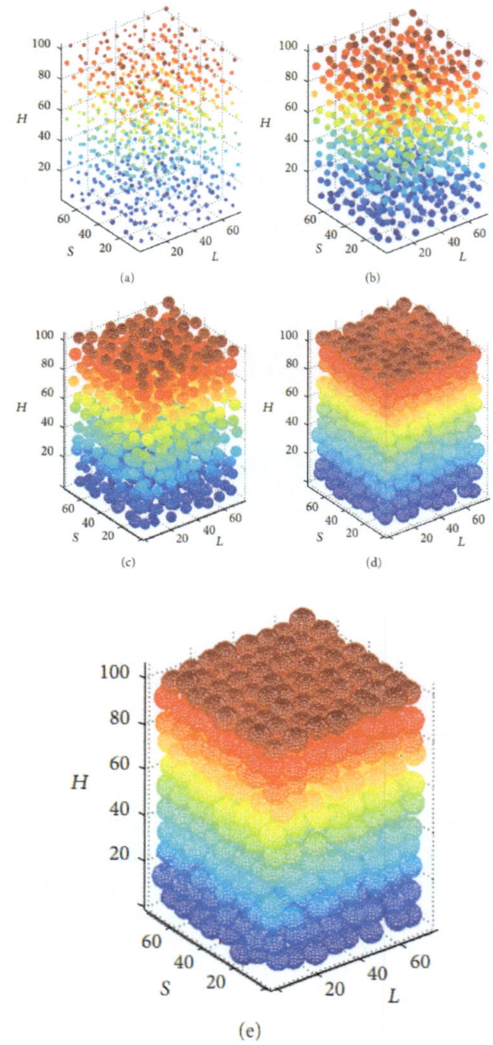

Figure 2: Modeling process with different cycle times (mm). (a) P=1; (b) P = 20; (c) P = 40; (d) P = 50; and (e) P = 55

In the electromagnetic analysis of packing particles, the shape, size, and packing density are the most important parameters. The shape and size of particles in the model should be consistent with the actual situation. The criterion for the reasonable modeling is that the packing density of the model is in good agreement with the experimental result.

The experimental result on the packing density of steel balls in cylindrical containers is published by Scott [24]. The relationship between the packing density and the diameter of the cylindrical container is investigated by experiment. The packing density of steel balls is from 0.56 to 0.6 in the cylindrical container. The diameter of the cylindrical container is 10.5 times compared with the diameter of the steel ball. When the container becomes bigger, the packing density will rise to 0.58~0.62.

The packing density of the modeling method proposed in this paper should comply with the experimental result. Several models are built in the same condition as described in [24] in order to validate the correctness of our method. The results are shown in Figure 3. Modeling is defined within a cylinder with a diameter of 105 mm and a height of 120 mm. The diameter of balls is limited in the range of 9~11 mm and 4.5~5.5 mm, respectively. This is because the packing density in a limited container is related to the size of balls.

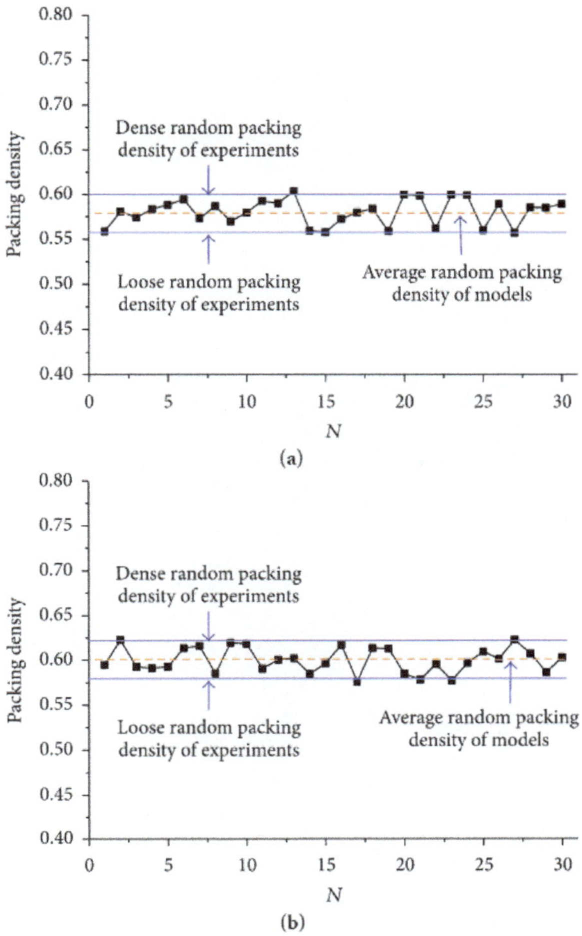

Figure 3: Comparison of the packing density between the model and the experiment. is the modeling number. (a) Diameter of balls ranging from 9 mm to 11 mm. (b) Diameter of balls ranging from 4.5 mm to 5.5 mm.

Figure 3 shows a good agreement of the packing density between the modeling result and the experimental result in [24]. Modeling has been implemented 30 times in these two cylindrical containers. Most packing density of models agrees with the result described in [24]. The average packing density of 30 models is 0.585 and 0.601, respectively. They are nearly middle values in the packing density range. These indicate the stability and reliability of the modeling method.

There are some other modeling methods applied in other areas [25, 26], such as PFC 3D. Figure 4 shows the time to model packing balls by use of our method and PFC 3D. Both of the modeling methods are carried out with balls having 8~10 mm diameter in the cube having the bottom area of 75 × 75 mm. The density of balls is defined as 2000 kg/m^3 and the stiffness of all balls and boundary is 10^8 N/m. The friction between balls is set as 1. The modeling time is defined to be from the beginning of the modeling to reach a steady state. The comparison shows that the proposed method has much less modeling time, which is more and more obvious with the increase of the number of balls.

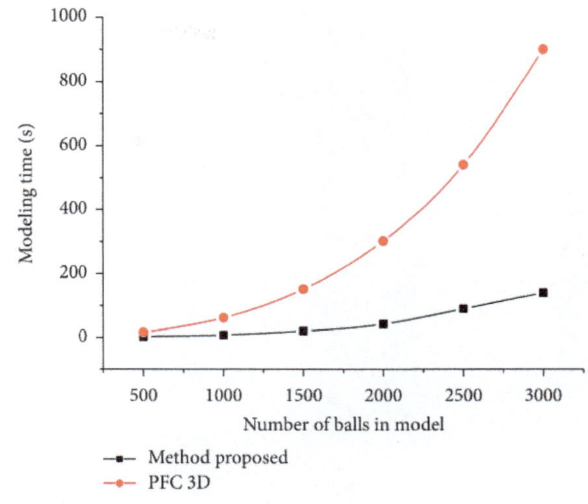

Figure 4: Comparison of the modeling time between PFC 3D and the method proposed in this paper.

Model of Packing Polyhedrons

Generally speaking, the particles are not of the spherical type in most modeling cases. In this modeling method, spheres can be conveniently converted to polyhedrons.

As an example, these spheres are deformed into polyhedrons for the modeling of packing fragmented FR4 substrate. The shape of particles close to reality can increase the accuracy of the following electromagnetic analysis.

As shown in Figure 5, the first step is to choose the edge number, N, of polygons according to the shape of the modeling particle. Then, the balls are divided uniformly by N polygons with N edges. Finally, all the polygons are connected to constitute a polyhedron. This method can transform the spheres into polyhedrons in terms of N, which is good for the modeling of some ball-like particles.

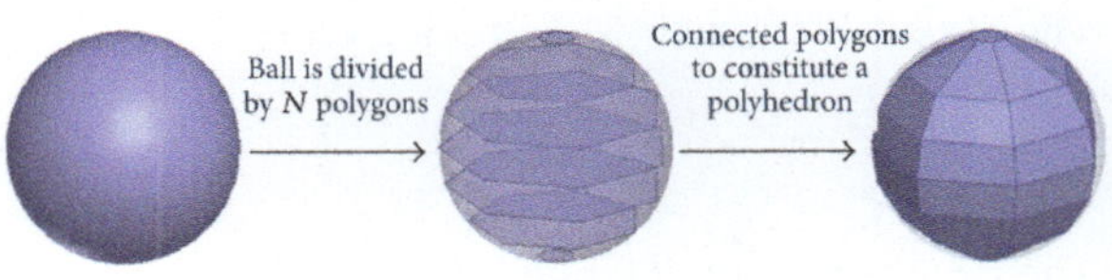

Figure 5: Change spheres into polyhedrons.

This model is built in Matlab software according to the modeling process. The model can be imported into ANSYS HFSS by use of HFSS-MATLAB-Scripting-API directly. Now, the simulation can be started in HFSS.

CORRECTNESS VALIDATION OF MODELING METHOD

To verify the stability and reliability of our modeling method, the model of packing fragmented FR4 as shown in Figure 6 is imported into HFSS and then simulated. The simulated and experimental results will be compared. The investigated frequency band is from 10 GHz to 15 GHz. For a FR4 dielectric substrate, the relative permittivity, ε_r, approximately linearly decreases from 4.29 to 4.21 within the interested frequency band, while the loss tangent, $\tan\delta$, approximately linearly increases from 0.020 to 0.024.

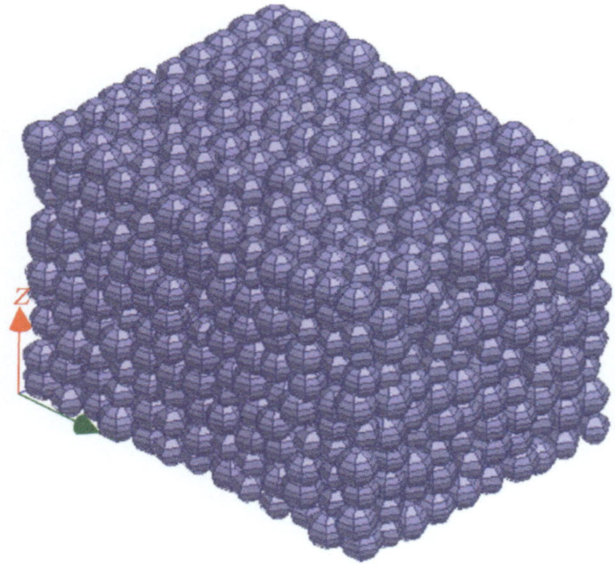

Figure 6: Accumulation of polyhedrons.

The particles of FR4 are filled into a standard rectangular waveguide (19.050 mm × 9.525 mm) to test the S parameters [11]. The equivalent ε_r and $\tan\delta$ of the packing particles can be calculated from the S parameters and compared with the results obtained by HFSS. The fragmented FR4 is cut off from a 1 mm thick FR4 substrate with the size of 1 × 1 mm^2. Then, these cube particles are mutually extruded and rubbed to convert them to the ball-like polyhedrons with the diameter of mainly 1.2~1.5 mm.

First, three famous mixing rules about the effective permittivity will be introduced and their results will be compared with simulations and experiments. The Looyenga model [27, 28], which is widely used to calculate ε_r of random distributed balls, is

$$\sqrt[3]{\varepsilon_m} = v_1\sqrt[3]{\varepsilon_1} + v_2\sqrt[3]{\varepsilon_2}. \tag{1}$$

The Maxwell-Garnett (MG) mixing rule [29] is

$$\frac{\varepsilon_m - \varepsilon_2}{\varepsilon_m - 2\varepsilon_2} = v_1\frac{\varepsilon_1 - \varepsilon_2}{\varepsilon_1 + 2\varepsilon_2}. \tag{2}$$

The Bruggeman mixing rule [30] is

$$v_1\frac{\varepsilon_1 - \varepsilon_m}{\varepsilon_1 + 2\varepsilon_m} = -v_2\frac{\varepsilon_2 - \varepsilon_m}{\varepsilon_2 + 2\varepsilon_m}, \tag{3}$$

where ε_m is the equivalent dielectric constant of the packing particles and v_1 and v_2 are the volume ratios of the particles and the homogeneous environment, respectively. The volume ratios are calculated by the total volume and the weight of the packing particles. The theoretical ε_m of packing FR4 particles packing in the air ($v_1 = 0.58, \varepsilon_1 = 4.21\sim4.29; v_2 = 0.42, \varepsilon_2 = 1$) of three famous mixing rules is shown in Figure 9.

The measured packing particles in the rectangular waveguide are shown in Figure 7. The vector network analyzer Agilent N5244A is employed to test the S parameters of the waveguide filled with the FR4 particles. Each end of the waveguide is connected to a coaxial converter as shown in Figure 8. A transparent plastic paper with the thickness of 0.08 mm is clipped between the waveguide and the converter in order to confine the particles in the waveguide.

Figure 7: FR4 particles in a rectangular waveguide.

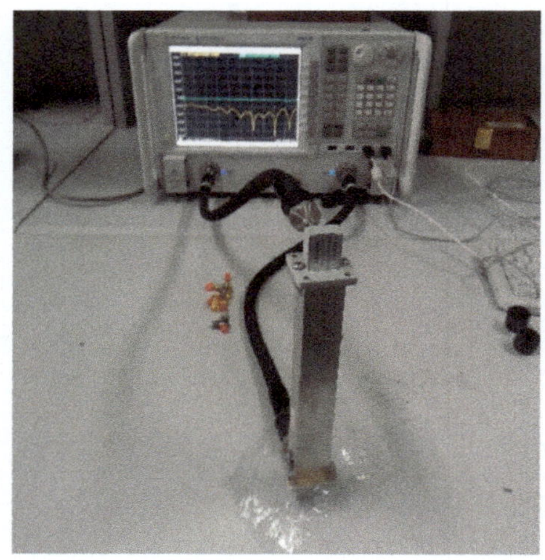

Figure 8: Measurement of electromagnetic properties of packing particles in the rectangular waveguide.

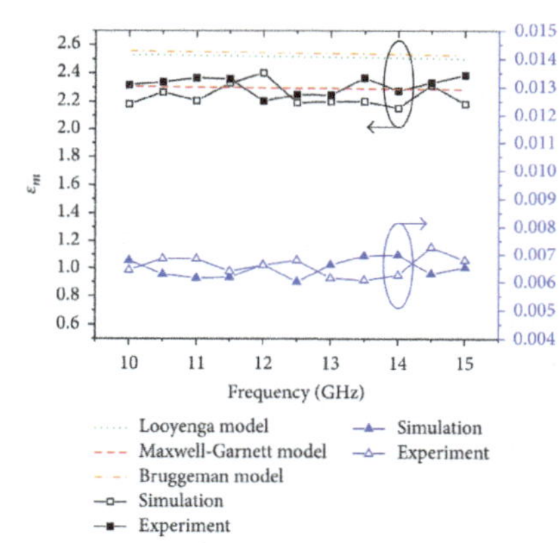

Figure 9: Theoretical, experimental, and simulated results of ε_m and $\tan\delta_m$ in the frequency band with H=3 mm.

The ambient temperature is 25°C and the relative humidity is 65% in the laboratory. The packing depth, H, should be longer than $0.1\lambda_g$ (λ_g (λ_g is the guide wavelength). Otherwise, the measured result is inaccurate. Besides, H should not be the integer times of /2 because the resonant effect in the waveguide will seriously affect the measured result [6, 10]. In the testing frequency band, the range of λ_g is 14.72 mm~30.48 mm. Thus, the effective range of H is selected as 3~10 mm in the measurement.

The theoretical, experimental, and simulated results of ε_m and $\tan\delta_m$ calculated from S parameters [6, 31, 32] are shown in Figures 9 and 10. These results are the average value of 30 repeated simulations and experiments. In Figure 9, the simulated results of the effective permittivity show good agreement with experimental data in the whole band at H=3 mm. The MG mixing rule agrees well with the experimental results of the packing polyhedrons with smaller packing depth H. In Figure 10, the simulated and experimental effective permittivities still agree well especially in the range of large packing depth, and both increase with the packing depth. This is because the random distribution of particles has a significant effect on the result in the case of fewer particles. With the increase of H, the difference between the simulated and experimental results becomes smaller. Both of them get closer to the results of the Looyenga rule and the Bruggeman rule. They still fall between the predictions of the MG rule and the Bruggeman rule, which is consistent with the result in [17] for two-dimensional dielectric mixture by using the FDTD method similar to this paper. It is worth pointing out that the simulated and experimental results satisfy the Wiener bounds and the Hashin-Shtrikman bounds as well, since they are often looser than the bound between the MG rule and the Bruggeman rule. The influence of the random distribution of particles will be weakened with the increasing of number of packing particles.

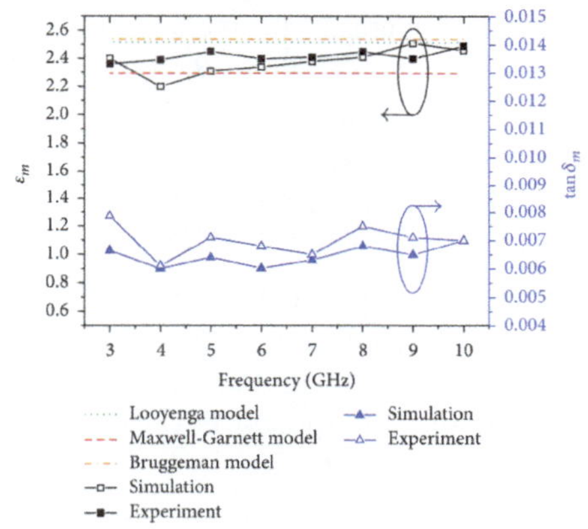

Figure 10: Theoretical, experimental, and simulated results of ε_m and $\tan\delta_m$ at the center frequency, 12.5 GHz, with different packing depth H of the particles.

The advantages of modeling and numerical simulation are embodied in the analysis above. Compared with the empirical formula, it takes more conditions into consideration such as shape, size and distribution of the particles, the thickness of packing, and the frequency. Besides, the modeling and simulation seem to be helpful to select the suitable formulation for the effective permittivity. For example, the MG mixing rule shows better agreement with the simulated and experimental results than the other two rules in the analysis of the packing polyhedrons with smaller packing depth H.

CONCLUSION

A modeling method for the packing particles is proposed in this paper. The packing particles in the shape of sphere or sphere-like polygon can be modeled easily and imported into the electromagnetic simulation software directly. It is much faster than PFC 3D to construct a model with a number of packing balls. It has been verified by experiment and agrees well with formulations and rules of the mixture in literatures. Besides, it is more accurate than the empirical formula to recognize equivalent electromagnetic properties of packing particles due to the fact that more conditions are considered. The modeling method has unique advantages in the field of electromagnetic simulation of packing particles.

ACKNOWLEDGMENTS

This work is supported in part by National Basic Research Program of China under Grant no. 2012CB214900, in part by Program for New Century Excellent Talents in University under Grant no. NCET-13-0089, and in part by National Program for Support of Top-Notch Young Professionals.

REFERENCES

1. H. Looyenga, "Dielectric constants of heterogeneous mixtures," Physica, vol. 31, no. 3, pp. 401–406, 1965.
2. J. C. Santamarina, A. Klein, and M. A. Fam, "Soils and waves: particulate materials behavior, characterization and process monitoring," Journal of Soils and Sediments, vol. 1, no. 2, p. 130, 2001.·
3. M. Meng and F. Wang, "Theoretical analyses and experimental research on a cement concrete dielectric model," Journal of Materials in Civil Engineering, vol. 25, no. 12, pp. 1959–1963, 2013.
4. S. Lambot, I. Van den Bosch, B. Stockbroeckx, P. Druyts, M. Vanclooster, and E. Slob, "Frequency dependence of the soil electromagnetic properties derived from ground-penetrating radar signal inversion," Subsurface Sensing Technologies and Applications, vol. 6, no. 1, pp. 73–87, 2005.
5. M. I. Mishchenko and L. Liu, "Electromagnetic scattering by densely packed particulate ice at radar wavelengths: exact theoretical results and remote-sensing implications," Applied Optics, vol. 48, no. 13, pp. 2421–2426, 2009.
6. A. M. Nicolson and G. F. Ross, "Measurement of the intrinsic properties of materials by time-domain techniques," IEEE Transactions on Instrumentation and Measurement, vol. 19, no. 4, pp. 377–382, 1970.· ·
7. Y.-J. Cheng, Z.-X. Xia, L. Wang, and Y. Fan, "Frequency reconfigurable microwave reaction experiment apparatus for coal desulphurization," Journal of the University of Electronic Science and Technology of China, vol. 43, no. 1, pp. 31–35, 2014.
8. Y. J. Cheng and X. L. Liu, "W-band characterizations of printed circuit board based on substrate integrated waveguide multi-resonator method," IEEE Transactions on Microwave Theory and Techniques, vol. 64, no. 2, pp. 599–606, 2016.
9. D. A. Robinson and S. P. Friedman, "The effective permittivity of dense packings of glass beads, quartz sand and their mixtures immersed in different dielectric backgrounds," Journal of Non-Crystalline Solids, vol. 305, no. 1–3, pp. 261–267, 2002.
10. A. M. Shutko and E. M. Reutov, "Mixture formulas applied in estimation of dielectric and radiative

characteristics of soils and grounds at microwave frequencies," IEEE Transactions on Geoscience and Remote Sensing, vol. 20, no. 1, pp. 29–32, 1982.

11. G. Y. Chernyak and H. Heimann, Dielectric Methods for Investigating Moist Soils, Israel Program for Scientific Translations, Jerusalem, Israel, 1967.

12. G. C. Topp, J. L. Davis, and A. P. Annan, "Electromagnetic determination of soil water content: measurements in coaxial transmission lines," Water Resources Research, vol. 16, no. 3, pp. 574–582, 1980.

13. D. A. Robinson and S. P. Friedman, "Electrical conductivity and dielectric permittivity of sphere packings: measurements and modelling of cubic lattices, randomly packed monosize spheres and multi-size mixtures," Physica A: Statistical Mechanics and its Applications, vol. 358, no. 2–4, pp. 447–465, 2005.

14. Z. X. Xia, Y. J. Cheng, and Y. Fan, "Frequency-reconfigurable TM010-mode reentrant cylindrical cavity for microwave material processing," Journal of Electromagnetic Waves and Applications, vol. 27, no. 5, pp. 605–614, 2013.

15. B. Sareni, L. Krähenbühl, A. Beroual, and C. Brosseau, "Effective dielectric constant of periodic composite materials," Journal of Applied Physics, vol. 80, no. 3, pp. 1688–1696, 1996. ··

16. B. Sareni, L. Krähenbühl, A. Beroual, and C. Brosseau, "Effective dielectric constant of random composite materials," Journal of Applied Physics, vol. 81, no. 5, pp. 2375–2383, 1997. ··

17. K. K. Karkkainen, A. H. Sihvola, and K. I. Nikoskinen, "Effective permittivity of mixtures: numerical validation by the FDTD method," IEEE Transactions on Geoscience and Remote Sensing, vol. 38, no. 3, pp. 1303–1308, 2000.

18. M. Suzuki and T. Oshima, "Verification of a model for estimating the void fraction in a three-component randomly packed bed," Powder Technology, vol. 43, no. 2, pp. 147–153, 1985.

19. L. Oger, J. P. Troadec, D. Bideau, J. A. Dodds, and M. J. Powell, "Properties of disordered sphere packings I. Geometric structure: statistical model, numerical simulations and experimental results,"Powder Technology, vol. 46, no. 2-3, pp. 121–131, 1986. · ·

20. P. R. Rios, "Comparison between a computer simulated and an analytical grain size distribution,"Scripta Materialia, vol. 40, no. 6, pp. 665–668, 1999.

21. R. Al-Raoush and M. Alsaleh, "Simulation of random packing of polydisperse particles," Powder Technology, vol. 176, no. 1, pp. 47–55, 2007.

22. M. Elimelech, J. Gregory, and X. Jia, Particle Deposition and Aggregation: Measurement, Modelling and Simulation, Butterworth-Heinemann, 2013.

23. D. He, N. N. Ekere, and L. Cai, "Computer simulation of random packing of unequal particles,"Physical Review E, vol. 60, no. 6, pp. 7098–7104, 1999. ·

24. G. D. Scott, "Packing of spheres: packing of equal spheres," Nature, vol. 188, no. 4754, pp. 908–909, 1960.

25. Y. Konakawa and K. Ishizaki, "The particle size distribution for the highest relative density in a compacted body," Powder Technology, vol. 63, no. 3, pp. 241–246, 1990.

26. J. Rodríguez, C. H. Allibert, and J. M. Chaix, "A computer method for random packing of spheres of unequal size," Powder Technology, vol. 47, no. 1, pp. 25–33, 1986.

27. L. D. Landau, J. Bell, M. Kearsley, L. Pitaevskii, E. Lifshitz, and J. Sykes, Electrodynamics of Continuous Media, Elsevier, Philadelphia, Pa, USA, 1984.

28. J. A. Reynolds and J. M. Hough, "Formulae for dielectric constant of mixtures," Proceedings of the Physical Society, Section B, vol. 70, no. 8, pp. 769–775, 1957.

29. J. C. Maxwell-Garnett, "Colours in metal glasses and in metallic films," Philosophical Transactions of the Royal Society of London, vol. 203, no. 359–371, pp. 385–420, 1904.

30. Ć. Blanchard, J. A. Portí, J. A. Morente, A. Salinas, and E. A. Navarro, "Determination of the effective permittivity of dielectric mixtures with the transmission line matrix method," Journal of Applied Physics, vol. 102, no. 6, Article ID 064101, 2007.

31. C. J. F. Böttcher, O. C. van Belle, P. Bordewijk, and A. Rip, Theory of Electric Polarization, Elsevier Science, Philadelphia, Pa, USA, 1978.

32. W. B. Weir, "Automatic measurement of complex dielectric constant and permeability at microwave frequencies," Proceedings of the IEEE, vol. 62, no. 1, pp. 33–36, 1974.

CHAPTER 5

COMPUTATIONALLY-EFFICIENT FDTD MODELING OF SUPERCONTINUUM GENERATION IN PHOTONIC CRYSTAL FIBERS

Tomasz Karpisz, Bartlomiej Salski, Ryszard Buczynski, Pawel Kopyt and Adam Pacewicz

Keywords: Supercontinuum generation, Finite difference methods, Fiber nonlinear optics, Dispersion

INTRODUCTION

Supercontinuum (SC) generation in photonic crystal fibers (PCFs) has become one of promising ways to produce cost-effective broadband infrared radiation sources applicable in broadly-understood spectroscopy (Hult et al. 2007; Mandon et al. 2008). The process of SC generation relies on appropriate balance between dispersion and various third-order nonlinear effects, such as self-phase modulation, soliton generation/fission, and four-wave mixing (Agrawal2007). However, PCFs are optically large structures, especially due to their length which extends well beyond a wavelength scale. Consequently, electromagnetic (EM) modeling of SC evolution is usually undertaken with the aid of approximate techniques, like the split-step-Fourier method (SSFM), which solves the one-dimensional nonlinear Schrodinger equation representing propagation along a fiber with a pre-defined mode characteristic (Travers et al.2010). However, SSFM is an implicit method solving for the evolution of the pulse envelope in a one-dimensional EM transmission line. As a result, the method is not immune to spurious solutions (Bosco et al. 2000; Sinkin et al. 2003). In addition, the concept of slowly varying pulse envelope exploited in the nonlinear Schrodinger equation becomes questionable in the case of ultra-short pulses, the duration of which becomes comparable with an optical period (Agrawal2007). As it will be shown later in this paper, the same issue occurs when the pulse break up into a noise-like signal, which leads to a rapid increase of spectral bandwidth.

The alternative is to apply a rigorous full-wave approach in order to solve such a nonlinear problem. One of prospective candidates is the finite-difference time-domain (FDTD) method as a versatile numerical tool naturally adapted to broadband EM analysis (Taflove and Hagness2005). Although several valuable papers focused on FDTD modelling of third-order nonlinear effects, that stay behind SC generation, have been published (Fujii et al. 2004; Lubin et al.2011), large computational effort and memory requirements involved in the full-wave EM analysis of centimeter-long PCFs make the method inapplicable (Hu et al. 2008). For that reason, a novel approach that takes the advantage of the accuracy of the FDTD method, while substantially alleviating its prohibitive computational requirements, is proposed in this paper.

The EM field in guiding structures, with geometry invariant along a propagation axis, can be represented as a product of a transverse vector function and a scalar longitudinal one:

$$\vec{E}\,(x,y,z) = \vec{E}_t\,(x,y)\,E_f\,(z), \tag{1}$$

As a result, the analysis can be split in two independent EM problems, namely, vector two-dimensional (V2D) one solving for E_t (x, y), and scalar one-dimensional (S1D) solving for E_f (z). The transverse field distribution, E_t (x, y), can be represented as a discrete set of well-defined functions (modes):

$$\vec{E}_t\,(x,y) = \sum_n \vec{E}_{t,n}\,(x,y), \tag{2}$$

where n denotes the order of the mode with n = 1 indicating the fundamental mode with the lowest cut-off frequency.

If the transverse field distribution (mode) does not have its analytical representation, unlike homogeneously filled

Citation: Tomasz Karpisz, Bartlomiej Salski, Ryszard Buczynski, Pawel Kopyt and Adam Pacewicz, Computationally-efficient FDTD modeling of supercontinuum generation in photonic crystal fibers, doi:10.1007/s11082-016-0428-y

rectangular or circular metallic waveguides (Collin 1990), it has to be computed with the aid of a rigorous (full-wave) numerical technique, such as a finite element method (FEM) or FDTD. Consequently, the analysis reduced to the cross-section of a waveguide's structure allows evaluating properties of the considered mode such as effective refractive index, group velocity dispersion, effective mode area, or confinement loss (Karpisz et al. 2014).

In the case of V2D-FDTD (Gwarek et al. 1993), the mode analysis can account for material dispersion represented as a multi-pole Lorentz model implemented with the aid of auxiliary differential equations (ADE) (Taflove and Hagness 2005). Recently, it has been shown that V2D-FDTD can also solve for third-order nonlinear Kerr-Raman effects, which allows studying the impact of pulse intensity on a dispersion characteristic of the mode propagating in a photonic crystal fiber (Salski et al. 2015).

However, electromagnetic study of SC generation in PCFs requires considering nonlinear effects occurring along the propagation. It is proposed in this paper to replace SSFM with S1D-FDTD in the analysis of longitudinal propagation of a given mode, as a direct consequence of (1). There are several advantages of such an approach. First of all, computational effort of S1D-FDTD simulation is several orders of magnitude smaller than complete 3D-FDTD computation. For instance, the model of a 1-cm-long PCF, discretized into 25-nm-long FDTD cells, consists of merely 4×10^5 cells, while a complete 3D model would have required at least dozen of billions of FDTD cells, which is unsolvable at the moment. Second, FDTD as a rigorous numerical method can give a deep insight into the dynamics of SC generation, thus, contributing to a better understanding of the phenomena.

In Section II, the concept of the hybrid V2D-S1D FDTD modeling of SC generation in PCFs is discussed in details, while in Section III the elaborated technique is validated against experimental and numerical data available in literature.

CONCEPT

Despite the ultimate choice of the method solving for nonlinear propagation of the pulse along the PCF (SSFM or FDTD), mode characteristics are usually computed with the aforementioned rigorous numerical techniques (FEM or FDTD). Subsequently, the obtained effective refractive index n_{eff} of a given mode can be treated as material dispersion. As a result, S1D-FDTD simulation of the pulse propagating in such artificial (equivalent) material can be undertaken.

Transverse Properties

Transverse (mode) properties of the PCF can be computed with the V2D-FDTD method accounting for dispersion of the applied material (Salski et al. 2015). The real part of the refractive index of glass contained in the PCFs is usually represented with a Sellmeier equation, which has to be transformed into a Lorentz model with its representation in the FDTD method (Karpisz et al. 2014) given as:

$$\varepsilon = \varepsilon_\infty + \sum_{n=1}^{N} \frac{\left(\varepsilon_{s,n} - \varepsilon_\infty\right) \omega_{p,n}^2}{\omega_{p,n}^2 - \omega^2 + j\,\omega\,\omega_{c,n}}, \quad (3)$$

where ε_∞ is optical permittivity, ε_S is static permittivity, $\omega_{p,n}$ stands for the nth pole frequency, and $\omega_{c,n}$ represents relaxation frequency of the nth pole.

Longitudinal Properties

Once spectral properties of the considered mode, $E_{t,n}$ (x, y), are determined, the next step is to represent it as material dispersion in a S1D-FDTD model. The most straightforward way to do that is to fit the aforementioned Lorentz model to the obtained effective refractive index n_{eff} of the mode. However, practical realization of that concept is not straightforward as it may look at first glance. First, SC generation is sensitive not only to n_{eff}, but even more to dispersion D, which is very vulnerable to fluctuations of n_{eff}. Consequently, as it will be shown later in this paper with specific examples, special attention must be paid to represent waveguide dispersion accurately up to its second order derivative with the wavelength.

Consider as an example a PCF fabricated from lead–bismuth–gallium-oxide glass (PBG-08) proposed in (Sobon et al. 2014), which is well-suited for SC generation (see Fig. 1). Solid lines in Figs. 2a and 3 show the characteristics of the corresponding effective refractive index n_{eff} and dispersion D, respectively, computed in (Sobon et al. 2014) in the range from 500 nm to 4000 nm. In the first attempt, a triple-pole Lorentz model will be fitted to the aforementioned curves using the Levenberg–Marquardt algorithm executed with no bounds imposed on the Lorentz parameters. According to (Sobon et al. 2014), SC evolves in the considered PCF in the 500–2400 nm range (see Fig. 5 therein), so the fitting has been limited to that range. The obtained Lorentz model (see the 2nd column in Table 1) provides the effective refractive index (dispersion) with a normalized root-mean-square deviation (NRMSD) of 0.047 % (0.777 %) in that spectrum. Due to negligibly small discrepancy, these curves are not plotted in Figs. 2 and 3.

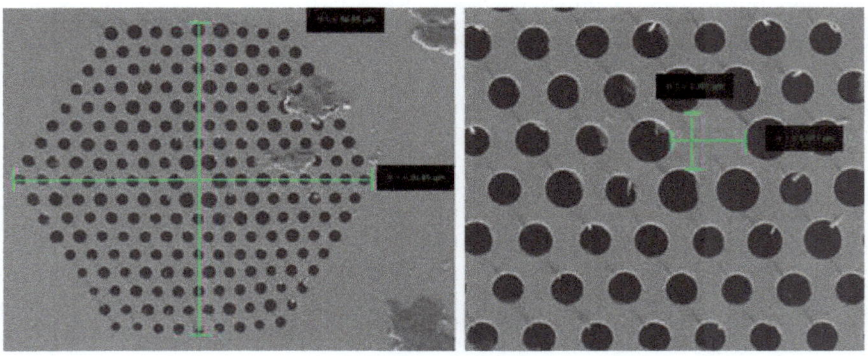

Figure. 1 SEM images of a PCF made of PBG-08 glass (Sobon et al. 2014)

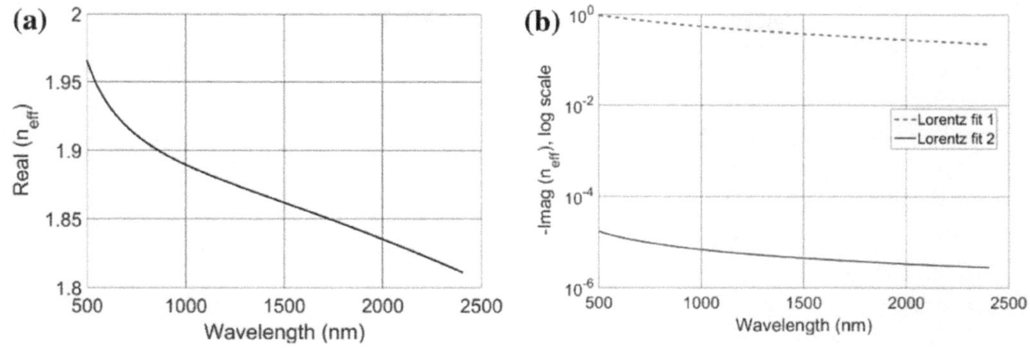

Figure. 2 a Real part of effective refractive index of the PCF computed in (Sobon et al. 2014) and **b** imaginary components of the corresponding Lorentz model obtained without imposing any restrictions (Lorentz fit 1) and assuming $f_{c,n} < 10^{-6}f_{p,n}$ (Lorentz fit 2).

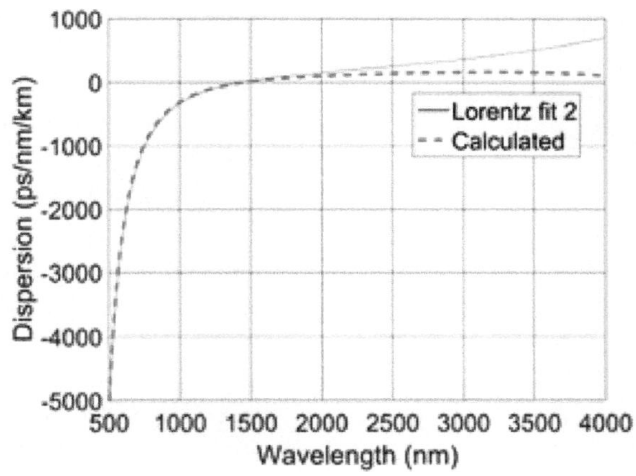

Figure. 3 Dispersion of the PCF computed in (Sobon et al. 2014) (solid line) as compared to the Lorentz model obtained imposing $f_{c,n} < 10^{-6}f_{p,n}$ (Lorentz fit 2).

Table 1 Material representation of the PCF as in Fig. 1

Parameters	Lorentz_fit1	Lorentz_fit2	Metamaterial[a]
ε_{∞}	−16.1205	1.0	1.6708 (1.6708)
ε_{S1}	−10.5325	3.5	1.8942 (1.8942)
ε_{S2}	−2.0442	1.03994	47.068 (47.068)
ε_{S3}	42.867×10^{6}	10.99085	1.9621 (1.9621)
f_{p1} (THz)	8564.65	1821.74	1238.48 (1238.48)
f_{p2} (THz)	2457.18	888.409	4.94804 (4.94804)
f_{p3} (THz)	0.0103	20.3595	52.4071 (52.4071)
f_{c1} (THz)	69441.51	10.0×10^{-3}	6.238×10^{-3} (-6.238×10^{-3})
f_{c2} (THz)	851.514	10.0×10^{-3}	1.749×10^{-3} (-1.749×10^{-3})
f_{c3} (THz)	−35.0282	0.1	202.2252 (−202.2252)

[a]Permeability coefficients are given in brackets

It can be noticed that the lack of restrictions imposed on the Lorentz parameters results in a decent agreement with the properties of the fundamental mode computed rigorously. Practically, the third Lorentz pole with a very small resonant frequency $f_{p,3} = 0.0103$ THz and large and negative relaxation frequency $f_{c,3} = -35.0282$ THz can be approximated with a Drude model. Unfortunately, large relaxation frequencies $f_{c,n}$, that have been obtained (see the 2nd column in Table 1), contribute to prohibitively large propagation loss correlated with an imaginary part of (3). It is confirmed by the plot shown in Fig. 2b (Lorentz fit 1). Moreover, the condition $\varepsilon_{s,n} < \varepsilon_{\infty}$ makes the FDTD analysis numerically unstable. For that reason, it has been assumed in a subsequent step of curve fitting that $f_{c,n} < 10^{-6}f_{p,n}$, thus, imposing substantially reduced losses (see the 3rd column in Table 1). As a result, the NRMSD of the effective refractive index (dispersion) has increased to 0.3363 % (0.7611 %) in the 500–2400 nm range, which is still within reasonable limits.

Several tests undertaken by the authors have shown that the multi-pole Lorentz model cannot accurately represent dispersion-flattened characteristics, which are decreasing with the wavelength at some spectral range. An example of such a scenario is shown in Fig. 3 (solid line) for $\lambda > 2400$ nm. Consequently, if that spectral range was essential for SC generation, other representation of mode dispersion should be applied instead. It should be noted, however, that the decrease of D with the wavelength usually coincides with the increase of confinement loss (Stepniewski et al. 2014), so that the evolution of spectrum does not extend beyond that limit.

The conclusion that can be derived from attempts to conveniently represent material properties undertaken so far is that suitable fitting of waveguide dispersion can be obtained only if no restrictions are imposed on an imaginary part of the Lorentz model. However, such an approach cannot be applied due to resulting erroneously large propagation loss leading to a rapid attenuation of the signal along the fiber. It seems that these losses may be suppressed if metamaterial representation is applied, where the effective refractive index is the square root of the product of complex permittivity as in (3), and complex permeability given in a similar Lorentz form:

$$\mu = \mu_\infty + \sum_{n=1}^{N} \frac{\left(\mu_{s,n} - \mu_\infty\right) \omega_{pp,n}^2}{\omega_{pp,n}^2 - \omega^2 + j\,\omega\,\omega_{cc,n}} \tag{4}$$

where μ_∞ is optical permeability, μ_S is static permeability, $\omega_{pp,n}$ stands for the nth magnetic pole frequency, and $\omega_{cc,n}$ represents the corresponding magnetic relaxation frequency.

If (3) and (4) are conjugates of each other ($\varepsilon' = \mu'$, $\varepsilon'' = -\mu''$), the resulting effective refractive index of such single-negative metamaterial will be purely real:

$$n = \sqrt{\left(\varepsilon' - j\varepsilon''\right)\left(\varepsilon' + j\varepsilon''\right)} = \sqrt{\varepsilon'^2 + \varepsilon''^2}, \tag{5}$$

The fourth column in Table 1 depicts the Lorentz parameters of the obtained metamaterial model, which resembles waveguide dispersion of the considered mode. The NRMSD of the effective refractive index (dispersion) is as low as 2.17 % (2.63 %) in the 500–2400 nm range, so there is no significant improvement when compared with the third column in Table 1. However, if the spectrum above the 2400 nm limit was also essential in the analysis, metamaterial representation would be a much better choice than the fitting done with the permittivity function only. Due to negligibly small discrepancy in the whole 500–4000 nm range, the plot of the corresponding metamaterial curves is omitted in Figs. 2a and 3. Unfortunately, tests undertaken by the authors have shown that the FDTD simulation is quickly becoming unstable if such metamaterial is computed. Apparently, instability is due to a large level of the positive imaginary part of (3) (loss), which is suppressed by the negative imaginary part of (4) (gain). It seems that the undertaken attempt at compensating the imaginary part of (5) improves the quality of the fit but it does not suppress numerical noise of the FDTD algorithm. Consequently, the noise is quickly amplified by the FDTD routines associated with the permeability model given by (4). Although that subject goes beyond the main scope of this paper, it can be expected that the presented technique with metamaterial representation of PCF waveguide dispersion may be useful in future studies if the aforementioned stability issues are resolved. In this paper, the fiber will be represented with the already discussed Lorentz permittivity model (see the third column in Table 1).

Nonlinear Properties

Dispersive properties of the fiber will be supplemented in the S1D-FDTD model with a third-order nonlinear polarization model derived on the basis of the Born–Oppenheimer approximation (Fujii et al. 2004; Hellwarth 1977):

$$P^{(3)}(t) = \varepsilon_0\, \chi_0^{(3)}\, E(t)\, \left[g(t) \otimes E^2(t)\right], \tag{6}$$

where ε_0 is free-space permittivity, $\chi_0^{(3)}$ is third-order susceptibility, and g(t) stands for a nonlinear response function consisting of both Kerr and Raman contributions. It is assumed that nonlinear properties of glass are not dispersive, which is quite common in EM modelling of SC generation (Agrawal 2007; Travers et al. 2010; Sobon et al. 2014).

The details of the corresponding FDTD model are discussed in (Salski et al. 2015), so only major features will be pointed out hereafter. Equation (3) can be implemented in FDTD with the aid of ADE [14]. However, an FDTD solution of (3) suffers from either low computational efficiency or substantially deteriorated accuracy (Maksymov et al. 2011). In (Salski et al. 2015), a new approach that combines the accuracy of a rigorous FDTD update scheme and the computational speed of an approximate solution has been proposed and successfully validated. In that approach, rigorous FDTD update equations are applied only in those nonlinear cells, where instantaneous amplitude of an electric field is relatively high, while all the remaining nonlinear FDTD cells are updated with approximate update equations. In the case of short femtosecond pulses propagating in a long transmission line, most of FDTD cells are populated with low electric field's amplitude, and only a small fraction of FDTD cells requires more rigorous computation. As a result, FDTD simulation of nonlinear materials can be substantially speeded up with no deterioration of accuracy. As shown in (Salski et al. 2015) in selected cases the improvement can exceed 80 %.

RESULTS

Once the material properties of the fiber are already represented with the Lorentz model (see previous Section) and the corresponding computationally efficient S1D-FDTD method is elaborated, as in (Salski et al. 2015), an FDTD model can be defined. In addition to the aforementioned time-efficient computation of the nonlinear FDTD cells, appropriate definition of the model allows further speeding up FDTD analysis of long transmission lines, like the one discussed in this paper. It can be achieved by taking advantage of the fact that a pulse spatial length is much shorter than a total length of the fiber, so that periodic boundary conditions (PBCs) with a zero phase shift ($\psi_x = 0$ rad) can be applied at both ends of the model (see Fig. 4) (Salski et al. 2007). Consequently, pulse repetitively circulates in the model enabling the reduction of its length L from a few centimeters, which are typical for SC generation, down to a few multiples of the pulse spatial length.

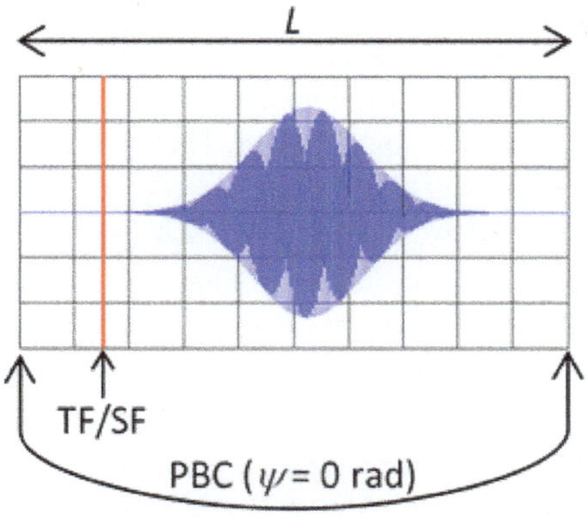

Figure. 4 The S1D-FDTD model representing femtosecond pulse propagation in a photonic crystal fiber.

The length of the model with PBCs is L_p = 500 µm, while an FDTD cell size is a = 30 nm and a time step is Δt = 0.05 fs. The pulse is excited with a total-field/scattered-field (TF/SF) soft source (Taflove and Hagness 2005; Salski et al. 2007) in an anomalous dispersion regime atλ = 1.56 µm (compare Fig. 3), and its duration is 400 fs (see Fig. 5a). An FDTD simulation has been carried out until the pulse reaches the end of a 5-cm-long fiber [compare Fig. 6 in Sobon et al. (2014)]. Figure 5 depicts the evolution of pulse shape along the fiber. At first, self-phase-modulation induced by large amplitude modifies the shape of the pulse, so that its front is stretched by long-wavelength components, while its rear part is compressed (see Fig. 5b). After some distance, a soliton is being created (see Fig. 5c) and, afterwards, modulation instability prevails (see Fig. 5d), finally leading to soliton fission.

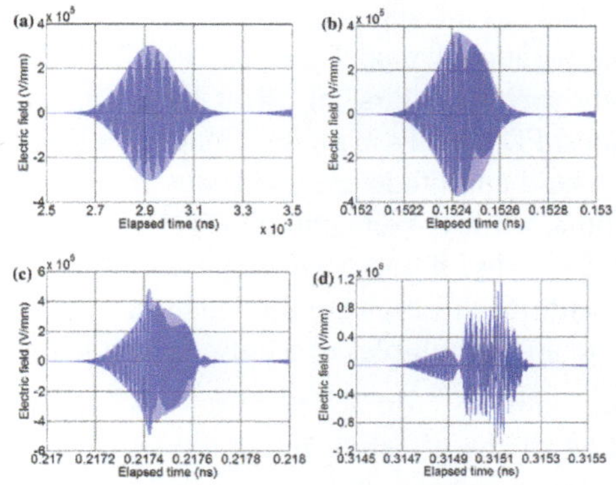

Figure. 5 Pulse shape at consecutive distances along the fiber: **a** L = 0 cm, **b** L = 2.3 cm, **c** L = 3.3 cm and **d** L = 4.9 cm.

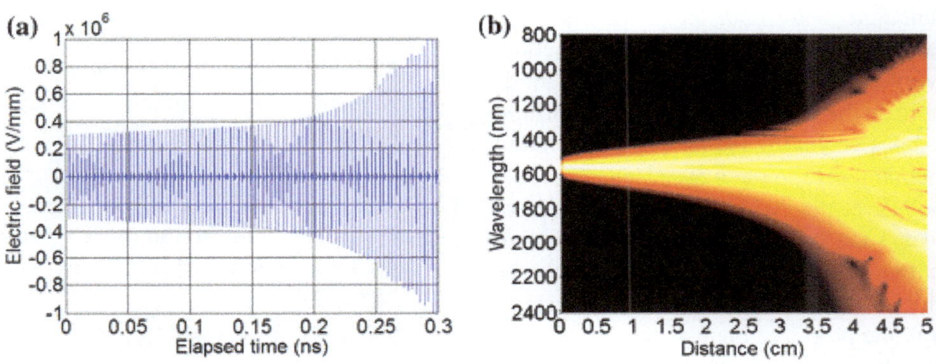

Figure. 6 a Pulse train during the whole S1D-FDTD simulation and **b** evolution of supercontinuum spectrum along the fiber for the peak power of 30 kW assuming 6 % coupling efficiency into the fiber.

For instance, if pulse duration is 400 fs, then its spatial length is about 60 µm, assumingn_{eff} = 2. Consequently, the FDTD model can be as short as L_p = 500 µm in order to avoid overlapping of the pulse circulating between the two PBC boundaries. The reduction of computational effort, which adds up to the previously mentioned speed-up factor due to the hybrid solution of the nonlinear FDTD cells, is at the level of L/L_p = 100.

Figure 6a shows the results of the whole FDTD simulation in one plot, so that the evolution of the envelope can be observed, indicating that the amplitude is increasing with the distance. Consequently, nonlinear effects are being intensified so that the soliton is finally destroyed into a spectrally broadband signal. It is confirmed in Fig. 6b depicting the evolution of pulse spectrum, which is in good agreement with the corresponding plot shown in Fig. 6a (Sobon et al. 2014). Similarity of both plots obtained with completely different methods validates the approach proposed in this paper. However, the advantage of the FDTD method is that it rigorously treats the considered model of third-order nonlinearities assisted by mode dispersion. The proposed approach can, thus, be easily extended to additional effects, like multi-pole representation of Raman response, which may further enhance the capability and accuracy of the FDTD analysis. As in the case of SSFM, S1-FDTD does not take into account coupling efficiency to a given mode but, if needed, it can be computed with full-wave 3D EM methods. Although, coupling between the modes is not accounted for in the S1-FDTD method, there are no fundamental limitations that prevent extending the method in that direction.

The whole FDTD simulation executed on Intel Core i7-3930 K CPU took 10 h with the speed of 182 iterations per second, which is about 2 times faster than rigorous computation without a hybrid approach proposed in (Salski et al. 2015). What needs to be stressed out, the model elaborated without the use of PBCs would extend the S1D-FDTD analysis to about 15 days. There is still a lot of room for additional acceleration with such techniques as GPU computing or multi-threading (Rudnicki and Sypniewski 2012), but that subject goes beyond the main scope of this paper. In comparison, similar simulation undertaken with SSFM takes about 4 min. Nevertheless, it can be clearly seen that rigorous FDTD modeling of the whole supercontinuum evolution in microstructured optical fibers performed in reasonable time limits on a personal computer is well within reach.

CONCLUSION

A novel hybrid FDTD method applicable to a rigorous analysis of supercontinuum generation in photonic crystal fibers has been proposed in this paper. It takes the advantage of the separation of variables method. The analysis is split into vector-2D and scalar-1D electromagnetic problems accounting, respectively, for transverse and longitudinal properties of the photonic crystal fibers. It allows substantially speeding up the FDTD analysis of supercontinuum generation provided that the computational model and the FDTD algorithm are appropriately specified. First of all, it has been proposed to solve nonlinear FDTD cells in a hybrid manner, where cells with low electric field amplitude are updated with an approximate FDTD formula, while the remaining FDTD cells are computed rigorously. Secondly, it has been proposed to apply periodic boundary conditions to truncate the model from a centimeter scale down to a few multiples of a pulse spatial length.

Another novelty discussed in this paper is the representation of the effective refractive index of the mode with a dispersive model. It has been shown that a multi-pole Lorentz model represents fiber mode characteristics very well, provided that dispersion D is not flattened too much in the considered spectrum. The alternative is to apply metamaterial representation with gain of a permeability function compensated by loss of a permittivity function. The latter option is very promising, although it still requires thorough studies focused on the stability of an FDTD algorithm representing such model. Eventually, the obtained simulation results of supercontinuum generation in a 5-cm-long photonic crystal fiber are in good agreement with literature data. Future studies can be focused on such issues as hardware acceleration, multi-pole representation of Raman nonlinear response or vector 1D FDTD, which accounts for polarization effects in the analysis. In general, the proposed method opens room for rigorous studies of nonlinear effects occurring in optically-long transmission lines.

ACKNOWLEDGMENTS

This work was supported by the project TEAM/2012-9/1 operated within the Foundation for Polish Science Team Programme co-financed by the European Regional Development Fund, Operational Program Innovative Economy 2007–2013. Authors would like to thank Prof. Wojciech Gwarek for fruitful discussions on the elaborated FDTD model.

REFERENCES

1. Agrawal, G.P.: Nonlinear Fiber Optics, 4th edn. Academic Press, London (2007)
2. Bosco, G., Carena, A., Curri, V., Gaudino, R., Poggiolini, P., Benedetto, S.: Suppression of spurious tones induced by the split-step method in fiber systems simulation. IEEE Photonics Technol. Lett. 12(5), 489–491 (2000)
3. Collin, R.: Field Theory of Guided Waves, 2nd edn. Wiley, New York (1990)
4. Fujii, M., Tahara, M., Sakagami, I., Freude, W., Russer, P.: High-order FDTD and auxiliary differential equation formulation of optical pulse propagation in 2-D Kerr and Raman nonlinear dispersive media. IEEE J. Quantum Electron. 40(2), 175–182 (2004)
5. Gwarek, W.K., Morawski, T., Mroczkowski, C.: Application of the FDTD method to the analysis of the circuits described by the two-dimensional vector wave equation. IEEE Trans. Microw. Theory Tech. 41(2), 311–316 (1993)
6. Hellwarth, R.W.: Third-order optical susceptibilities of liquid and solids. J. Prog. Quantum Electron. 5, 1–8 (1977)
7. Hu, J.J., Shum, P., Ren, G., Yu, X., Wang, G., Lu, C.: Three-dimensional FDTD method for the analysis of optical pulse propagation in photonic crystal fibers. Optoelectron. Adv. Mater. Rapid Commun. 2(9), 519–524 (2008)
8. Hult, J., Watt, R.S., Kaminski, C.F.: High bandwidth absorption spectroscopy with a dispersed supercontinuum source. Opt. Exp. 15(18), 11385–11395 (2007)
9. Karpisz, T., Salski, B., Szumska, A., Klimczak, M., Buczynski, R.: FDTD analysis of modal dispersive properties of nonlinear photonic crystal fibers. Opt. Quantum Electron. 47(1), 99–106 (2014)
10. Lubin, Z., Greene, J.H., Taflove, A.: fdtd computational study of ultra-narrow TM non-paraxial spatial soliton interactions. IEEE Microw. Wirel. Compon. Lett. 21(5), 228–230 (2011)
11. Maksymov, I.S., Sukhorukov, A.A., Lavrinenko, A.V., Kivshar, Y.S.: Comparative study of FDTD-adopted numerical algorithms for Kerr nonlinearities. IEEE Antennas Wirel. Propag. Lett. 10, 143–146 (2011)
12. Mandon, J., Sorokin, E., Sorokina, I.T., Guelachvili, G., Picqué, N.: Supercontinua for high-resolution absorption multiplex infrared spectroscopy. Opt. Lett. 33(3), 285–287 (2008)
13. Rudnicki, J., Sypniewski, M.: Large 3D object FDTD simulations. In: 19th International Conference on Microwaves, Radar and Wireless Communications: MIKON 2012, May 2012
14. Salski, B., Celuch, M., Gwarek, W.Ł.: Enhancements to FDTD modeling for optical metrology applications. In: SPIE Optical Metrology—18th International Congress on Photonics in Europe, Munich, June 2007
15. Salski, B., Karpisz, T., Buczynski, R.: Electromagnetic modeling of third-order nonlinearities in photonic crystal fibers using a vector two-dimensional FDTD algorithm. IEEE J. Lightwave Technol. 33(13), 2905–2912 (2015)
16. Sinkin, O.V., Holzlöhner, R., Zweck, J., Menyuk, C.R.: Optimization of the split-step Fourier method in modeling optical-fiber communications systems. J. Lightwave Technol. 21(1), 61–68 (2003)
17. Sobon, G., Klimczak, M., Sotor, J., Krzempek, K., Pysz, D., Stepien, R., Martynkien, T., Abramski, K., Buczynski, R.: Infrared supercontinuum generation in softglass photonic crystal fibers pumped at 1560 nm. Opt. Mater. Exp. 4(1), 7–15 (2014)
18. Stepniewski, G., Klimczak, M., Bookey, H., Siwicki, B., Pysz, D., Stepien, R., Kar, A.K., Waddie, A.J., Taghizadeh, M.R., Buczynski, R.: Broadband supercontinuum generation in normal dispersion all-

solid photonic crystal fiber pumped near 1300 nm. Laser Phys. Lett. 11(5), 055103 (2014)

19. Taflove, A., Hagness, S.C.: Computational electrodynamics—the finite-difference time-domain method. Artech House, Boston-London (2005)

20. Travers, J.C., Frosz, M.H., Dudley, J.M.: Nonlinear fibre optics overview. In: Dudley, J.M., Taylor, J. (eds.) R.Supercontinuum Generation in Optical Fibers, pp. 32–61. Cambridge University Press, Cambridge (2010)

CHAPTER 6 FDTD-SPICE FOR CHARACTERIZING METAMATERIALS INTEGRATED WITH ELECTRONIC CIRCUITS

Zhengwei Hao, Soheil Saadat and Hossein Mosallaei

INTRODUCTION

In recent years, the subject of metamaterials integrated with electronic circuits, nonlinear elements, and non-Foster transistor configurations have been drawing significant attention in research communities. Metamaterials are typically periodic structures created by array of resonant elements which are small and stationary compared to the wavelength of electromagnetic wave. The combination of metamaterials with circuit elements can bring novel and tunable characteristics to these materials. For instance combining loop-based metamaterials with varactors provides a means to electronically tune the performance by applying a DC voltage or a high-power signal without modifying the structure [1–4]. Alternatively, a non-Foster transistor configuration can be introduced to enhance the metamaterial bandwidth [5]. Active components can be also integrated with transmission line metamaterials to compensate the loss [6], and finally metasurface can be integrated with circuit elements to tailor their reflection phase performance [7].

The complexity of the structure requires a powerful computational technique to bridge the physics of electromagnetic metamaterial with circuit configuration. In this paper, the Finite Difference Time Domain (FDTD) technique is combined with SPICE-like software to provide a capable numerical technique for time-domain full wave analysis of the composite structure. The technique is applied to the design and simulation of metamaterials integrated with circuit elements. Broadband performance is obtained with the focus on bandwidth and tunable behavior engineering.

Section 2 provides the details of FDTD-SPICE hybridization and shows the validation through some examples. In Section 3, the following three applications are explored: (a) loop-based metamaterial integrated with lumped capacitors to tune the resonance performance, (b) metamaterial integrated with non-Foster transistor configuration for bandwidth manipulation, and (c) a metasurface combined with varactors featuring tunable reflection phase.

FDTD-SPICE COMPUTATIONAL MODEL

Hybridization of FDTD with SPICE

As a major computational electromagnetic modeling scheme, Finite Difference Time Domain technique can tackle problems by providing a full wave solution in time. It is a capable method for modeling a variety of scientific and engineering problems [8, 9]. On the other hand, the general-purpose analog electronic circuit simulator, SPICE [10], can solve complex circuit topologies in time domain. It makes great sense to combine FDTD with SPICE to feature a powerful numerical engine for solving metamaterials integrated with circuit configurations.

FDTD uses Maxwell's equations (1) and (2) in time domain to recursively calculate the electromagnetic fields at each position in three-dimensional space:

$$\varepsilon\frac{\partial E}{\partial t} + J(E) = \nabla \times H, \tag{1}$$

Citation: Zhengwei Hao, Soheil Saadat, and Hossein Mosallaei, "FDTD-SPICE for Characterizing Metamaterials Integrated with Electronic Circuits," International Journal of Antennas and Propagation, vol. 2012, Article ID 282159, 7 pages, 2012. doi:10.1155/2012/282159.

$$\mu\frac{\partial H}{\partial t}+M(H)=\nabla\times E. \qquad (2)$$

While both equations can provide the coupling between FDTD and SPICE according to reference [8], the first equation is used in this paper. After some simple manipulations, Ampere's Law of (1) can be written as [11]

$$C\frac{dV}{dt}+I(V)=I. \qquad (3)$$

Equation (3) describes a circuit model as shown in Figure 1, where $C=\varepsilon A/dl$ is a capacitance that abstracted from the model (note that here C is not the intrinsic lattice capacitance of FDTD cell), dl is the cell size, and A is the area of one FDTD unit cell. $(V)=A^*J(E)$ is the current across the SPICE element, and $V=E^*dl$ is the voltage across the element, $I=A^*\nabla\times H$ is the whole current of the current source. Thus, for the cell containing SPICE element, instead of using (1) to update the electric field, SPICE can carry out (3) and obtain the voltage across the element and then pass it to FDTD program. From the voltage, it is easy to obtain E field afterward.

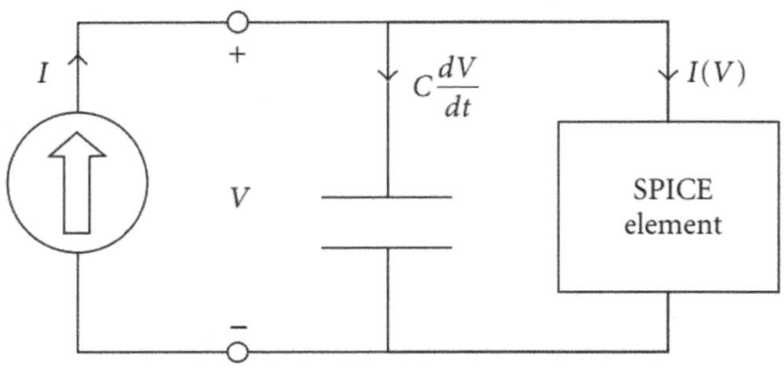

Figure 1: Circuit model for Ampere's Law equation that connects FDTD and SPICE.

To determine the voltage across the corresponding element, SPICE software needs to run a transient simulation for a period of Δt, which should be the same as FDTD time step. Then, in the next time step, regular FDTD method will update all the fields except the cell with SPICE element. For the SPICE element, SPICE program needs to run another Δt, starting from where it stopped last time. Before each run of SPICE, FDTD sends data that contains circuit information to SPICE. After each time step of SPICE simulation, it passes results back to FDTD. In this way, the FDTD and SPICE engines run alternatively until the end of the whole simulation time. With the help of SPICE, FDTD is ready to simulate any complex integrated configuration as long as it is provided in SPICE's library. Since conventional FDTD can only model very simple circuit elements, SPICE function adds a big advantage to FDTD.

The SPICE software used here is SPICE3f5. It has an interactive mode that can receive commands from and send results back to the interface. It also has "stop" and "resume" commands that can set a breakpoint after running some time steps in a transient simulation and then resume it. These features make it very suitable to be connected to FDTD.

To exchange data between FDTD and SPICE, interprocess communication should be developed. Figure 2 presents a complete picture of our FDTD-SPICE model. When a simulation begins, FDTD main program calls fork function which spawns two processes: the son process that will run SPICE, and the parent process which is the original FDTD process itself. Two sockets: Socket[0] and Socket[1], are created in the main program before it forks so that both of the two children have access to the sockets. Here these sockets act like two-way pipe, where FDTD process sends SPICE commands to Socket[0] and reads results from Socket[1], while SPICE process reads message from Socket[0] and sends results back to Socket[1]. FDTD process will be waiting for SPICE outputs while SPICE process is running and vice versa. These two processes then communicate through the two sockets at each time step until the end of the simulation.

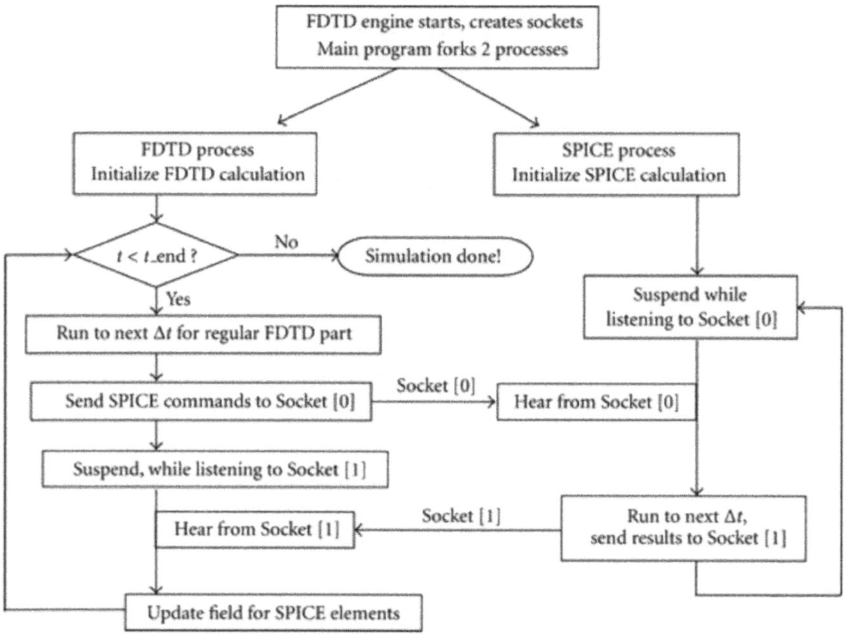

Figure 2: Flowchart for FDTD-SPICE simulation.

While DC solution is needed in SPICE for many cases but not very trivial to be obtained with FDTD, literature [12, 13] addressed DC solution problem by separating the AC and DC components, which is helpful for the metasurface simulations of diode varactors with bias that presented in Section 3.

Our FDTD code has been written in a very powerful manner and it has many features such as modeling periodic structures and linear dispersive metamaterials [14]. By applying auxiliary differential equation (ADE) method [15], it can also model nonlinear dispersive materials as well as gain materials. The absorbing wall is based on CPML [16] where one can truncate the computational domain very close to the structure. The code has been applied successfully to a variety of metamaterials composites [9, 17–19].

Test Results

Two cases are studied to validate the FDTD-SPICE model. All simulations in this paper use uniform gridding and are run on Sun Ultra45 workstation with 2 GHz CPU and 4 GB memory.

In the first case as depicted in Figure 3(a), one parallel RLC circuit is connected to a Gaussian pulse voltage source in order to obtain the impedance. There are about 8000 time steps in the calculation with each time step Δt=0.2ps, cell size Δx=0.15 mm. Including CPML boundary, the whole computational domain is modeled by 26×26×26 Yee's cells. The FDTD process takes 456 seconds, while, since the circuit is simple, SPICE process takes only 10 seconds. Figure 3(b) shows both the FDTD-SPICE results and the analytical calculation for the real and imaginary parts of the impedance Z, which agree well with each other.

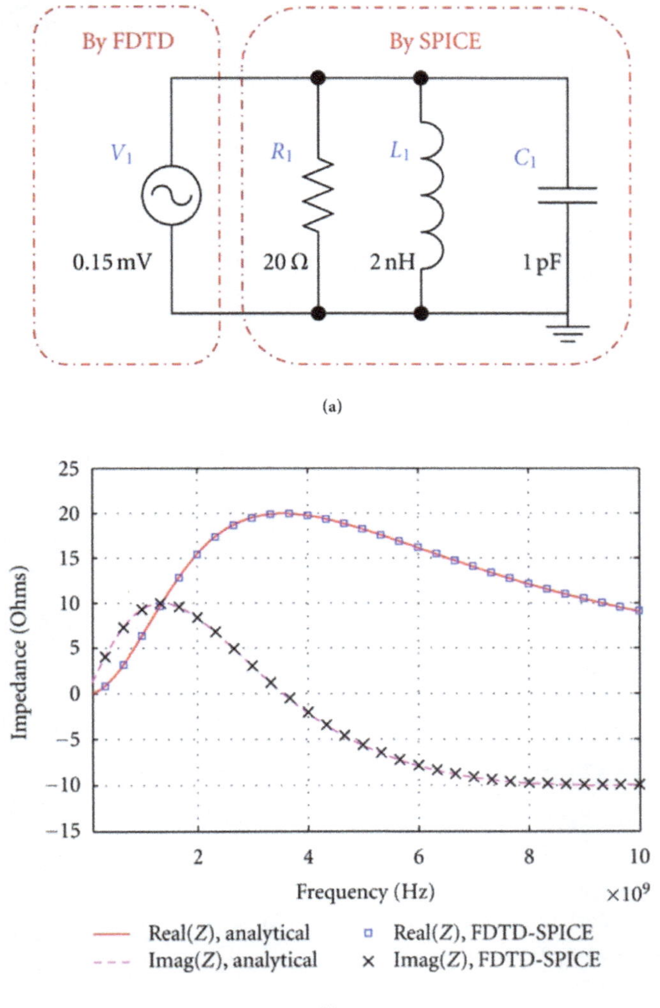

Figure 3: (a) RLC circuit configuration and (b) its impedance Z behavior. Great comparison between FDTD results and analytical model is illustrated.

In the second case as shown in Figure 4(a), one BFG403W transistor from Discrete Semiconductor is utilized in a single frequency simulation with a sinusoidal voltage source. The BJT has a saturation current IS=5.554 aA and the forward active current gain is BF=145. The calculation has 18000 time steps with each time step Δt=54.82 fs and cell size Δx=30 μm. Because of the complex property of BJT circuit, the cell size and time step are much smaller than those used in previous simulation. The whole computational domain has a size of 25×37×32 Yee's cells. FDTD and SPICE processes take 1734 seconds and 41 seconds, respectively. Figure 4(b) shows the comparison for the waveform of voltage across resistor R2 in time domain, between FDTD-SPICE and complete SPICE simulation. An excellent agreement is demonstrated.

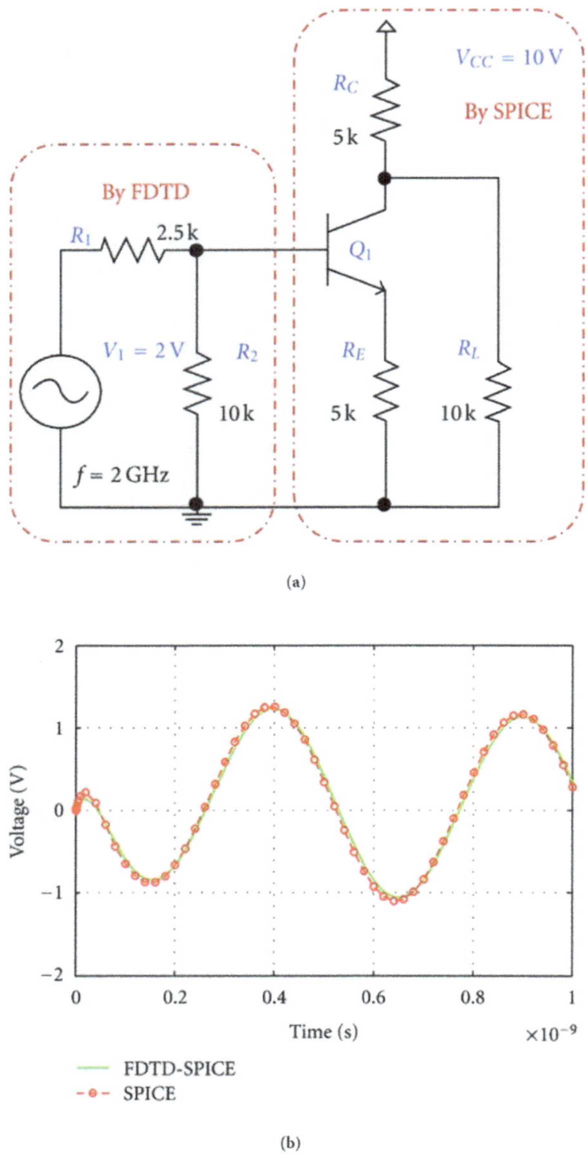

Figure 4: (a) Transistor circuit configuration. (b) Output voltage waveform in time. The FDTD-SPICE simulation matched very well with pure SPICE simulation.

APPLICATIONS ON METAMATERIALS

This section presents applications of FDTD-SPICE method to some recent metamaterial designs, namely, loop-based composite metamaterial with capacitors, metamaterial integrated with non-Foster circuits, and nonlinear elements loaded metasurface. In all of the cases, uniform gridding, perfectly matched layers, and periodic boundary conditions are used. Since they all deal with plan wave incidence, total field/scattered field method is also utilized [17].

Metamaterial Loops Integrated with Passive Elements

A periodic structure made from metallic loops can artificially create a medium with permeability characteristics (positive and negative) [19]. The metamaterial structure to be simulated here is shown in Figure 5(a). It is a four-layer structure in $x$direction where each layer has loops periodic in y-z directions and the loops are terminated to capacitor Cp. The unit loop dimension is 4.8 mm and 1.8 mm in x and$z$ directions, respectively. The excitation is a plane wave that has Ez/Hy components and propagates in $-x$ direction with a Gaussian signature. A transmission line analogy can be obtained for plane wave propagation through the medium [19], which is embedded with a loop that has a self-inductance Lp and is terminated to a lumped capacitor Cp. By changing the circuit elements (Lp and/or Cp) in the equivalent transmission line model, the effective parameter of the medium can be tuned, as well as the property of the metamaterial.

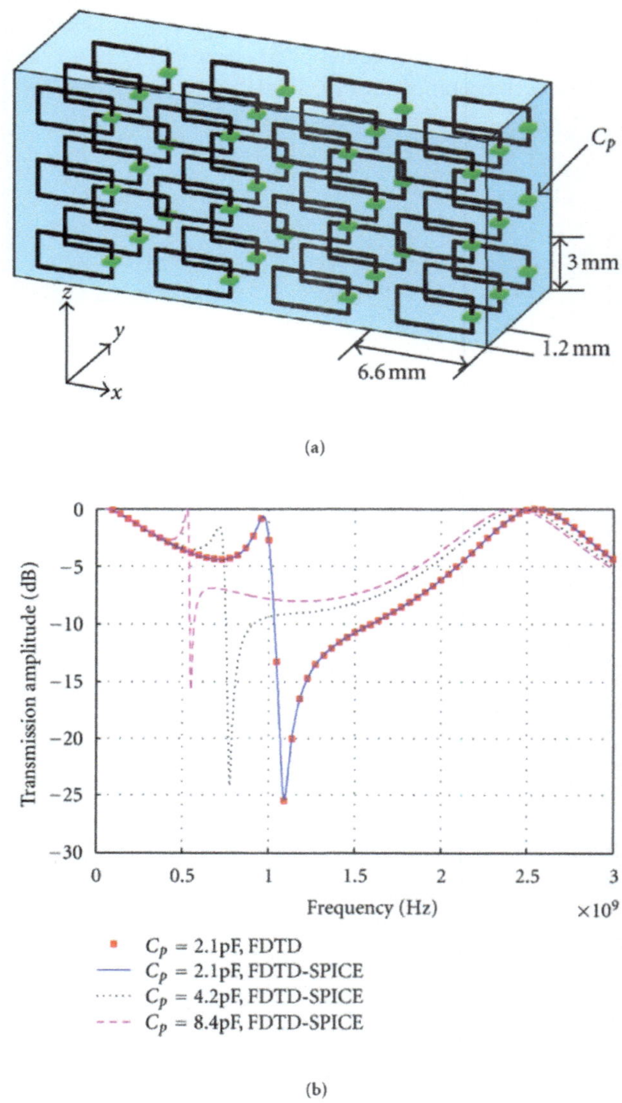

Figure 5: (a) Metamaterial constructed from array of loops terminated to capacitors Cp and (b) its transmission coefficient.

FDTD-SPICE method is applied to model the structure and obtain the transmission coefficient as shown in Figure 5(b). The loops are modeled in FDTD and the capacitors in SPICE. There are 15700 time steps in the calculation with each time step Δt=0.2 ps, cell size Δx=0.15 mm. The whole computational domain has a size of 216×8×20 Yee's cells. FDTD and SPICE processes take 1669 seconds and 1319 seconds, respectively. Because of the multiple-port interprocess communication, SPICE process needs considerable simulation time, which may be improved by introducing batch data communication between processes. Figure 5(b) shows 4 simulation results. The red dots represent results from pure FDTD simulation with Cp=2.1 pF (Cp is also modeled in FDTD). Then the blue line comes from FDTD-SPICE simulation with the same parameter, and it matches well with red dots, which show the accuracy of our model. Changing the capacitor Cp allows tuning the resonance and as well as the equivalent permeability of the metamaterial. Figure5(b) also shows the transmission coefficients for the metamaterial loops terminated to Cp=4.2 pF and Cp=8.4 pF lumped capacitors, depicted in black and magenta, respectively. Doubling capacitance would half the resonance frequency. As can be seen from Figure 5(b), the resonance frequency is shifted from 1.1 GHz down to 0.78 GHz and 0.55 GHz, respectively, according the value of capacitors.

Integration with Non-Foster Circuit Elements

Non-Foster elements have recently attracted significant interests due to their novel properties in offering negative impedance performance and successfully engineering the frequency dispersion characteristics of antennas, transmission lines, and metamaterials towards the goals of interests [20–23]. A non-Foster element can be realized using a Negative Impedance Converter (NIC) circuit by translating a given positive load into a negative one. NIC circuit is made from transistor elements and one would need a comprehensive modeling scheme to characterize them while they are integrated with metamaterials. It is also worth mentioning that the design of NIC circuit involves several challenging issues such as noise, biasing, and stability and special care must be taken to obtain a wideband performance for non-Foster circuit. The NIC stability (which is not the subject of this paper) has also been checked with the help of pole-zero analysis to ensure that the system has absolutely no pole in the Right-Half-Plane (RHP) of the complex S-plane.

Here we explore the integration of NIC circuit with metamaterial loops [5] and its FDTD-SPICE simulation to investigate the bandwidth enhancement of the structure. A two-layer loop-based composite metamaterial with capacitors as shown in Figure 5(a) is utilized. The loops are terminated to capacitors Cp and inductors (4 nH). By integrating NIC designs offering −2 nH inductance, we reduce the effect of the positive inductance and enhance the bandwidth. The NIC circuit used in this paper is based on Linvill's model [24], and it is an example of Voltage Inversion NIC (VINIC). As shown in Figure 6(a), the circuit consists of two N-type BJTs and a lumped passive inductor $L1$=2 nH as the load. After adding the non-Foster circuit to the loop, the equivalent circuit model [19] for the transmission line is depicted in Figure 6(a), where the NIC configuration is included in a negative inductor Ln.

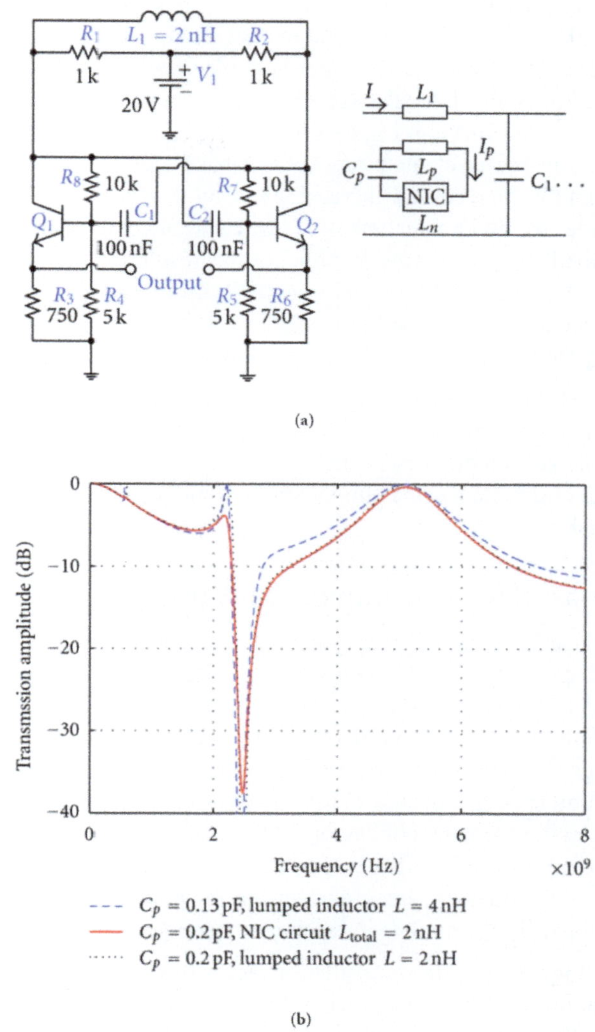

Figure 6: (a) Basic non-Foster element: NIC configuration based on Linvill's design and equivalent circuit model for the loop-based metamaterial integrated with NIC negative inductor. (b) Transmission coefficient for metamaterial of loops terminated with (i) Cp=0.13 pF, L=4 nH; (ii) Cp=0.2 pF, L=4 nH, NIC circuit Ln=−2 nH; (iii) Cp=0.2 pF, L=2 nH.

The simulation has 11600 time steps with each time step Δt=0.274 ps, cell size Δx=0.15 mm. The whole computational domain has a size of 128×8×20 Yee's cells. FDTD and SPICE processes take 686 seconds and 133 seconds, respectively. Because of the complexity of the NIC circuit model, SPICE needs more time than the test examples from previous section. Here three cases are performed. Firstly, metamaterial with loop terminated to capacitor Cp=0.13 pF and lumped inductor L=4 nH is simulated. As depicted by the blue curve in Figure 6(b), it has a resonance around 2.4 GHz. Secondly, to broaden the bandwidth, a NIC circuit Ln=−2 nH is added to inductor L=4 nH, making the whole inductance 2 nH. In order to make a fair comparison, the capacitor is made Cp=0.2 pF so that the two simulations have the same resonance. As can be observed from the red curve in Figure 6(b), the metamaterial integrated with non-Foster element has a wider bandwidth than the case without it (blue curve). Increasing the value of $-Ln$ can enable a wider bandwidth performance (as has been comprehensively detailed in [5]). Lastly, to validate the results of the second case, NIC circuit Ln=−2 nH together with inductor L=4 nH is replaced by one lumped inductor L=2 nH. As shown in Figure 6(b), the black curve follows the behavior of the second case.

At the end we want to mention that the reason we have considered a positive 4 nH inductor in series with the −2 nH non-Foster element is to make sure the time-domain SPICE modeling is stable (the equivalent total inductance is positive, +2 nH).

Metasurface Integrated with Varactor Diodes

Metasurfaces have great potential for making low profile antennas [25], and achieving an electronically tunable reflection phase is of remarkable interest in this regard. This section presents the simulation for a structure of metasurface integrated with varactors. The metasurface used in this paper is similar to the one in [7], which is designed to act as an artificial magnetic conductor (AMC) and uses active NIC circuits as negative elements to increase the bandwidth. Here varactor diodes are integrated in the metamaterial and make the metasurface tunable in resonance frequency.

Figure 7(a) shows a metasurface structure consisting of periodic square metal patches and PEC ground plane with dielectric slab in between. It also presents the unit cell of the metasurface loaded with varactor diodes. The structure is periodic in y and z directions and the incidence plane wave is $-x$ direction. The FDTD-SPICE method is applied to simulate the structure. There are 7000 time steps in the calculation with each time step Δt=0.953 ps, cell size Δx=0.5 mm. The whole computational domain has a size of 45×21×21 Yee's cells. FDTD and SPICE processes take 386 seconds and 8 seconds, respectively. Figure 7(b) shows the reflection coefficient phase for the metasurface without varactor diodes, as well as those with diodes that have different biases. The green line, as a reference frequency response, represents the one without varactor diode, while the others are for results that have diode with different bias. The varactor diode model used here is Skywork's SMV1232-079 Hyperabrupt Tuning Varactors [26]. When the diode is added to the structure, it increases the total capacitance therefore decreases the resonance frequency of the metasurface. As can been seen from Figure 8, typical capacitance values of the varactor diode decrease when the reverse bias voltage increases. The capacitance is a maximum while there is no reverse bias voltage. Thus for reverse bias with smaller voltage it should give larger shift in resonance frequency, which agrees well with the simulation results.

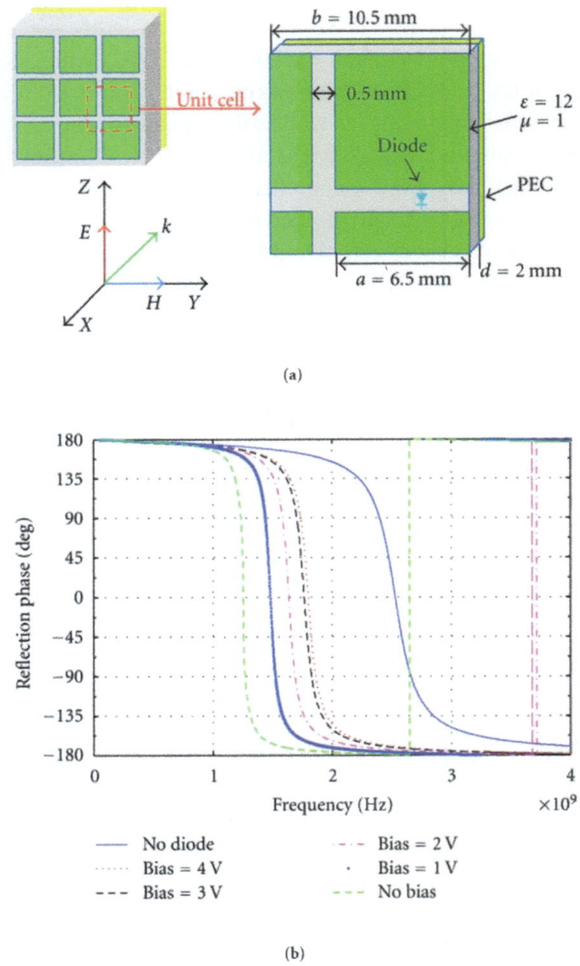

Figure 7: (a) Metasurface structure with periodic square metal patches and its unit cell that loaded with varactor diodes. (b) Reflection phase for the metasurface. Controlling the bias of the diode tunes the resonance frequency.

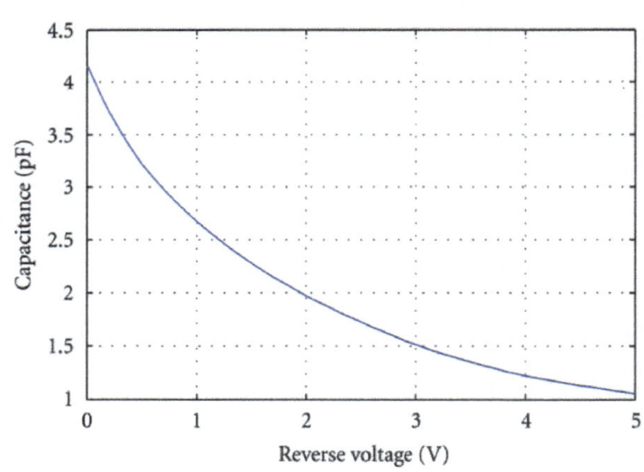

Figure 8: Performance data for SMV1232 Diode Varactor.

CONCLUSION

This paper develops a hybridized FDTD-SPICE model and applies it to the simulations of some novel metamaterials integrated with electronic circuits. The Maxwell's equations are linked with SPICE through the interchange between electromagnetic fields and voltages across the SPICE element nodes. Interprocess communication technique with socket is used as it is crucial for FDTD and SPICE to exchange data at proper running time. Diagrammatic flowchart is discussed that demonstrates the detail of the execution process as well as the data exchanging with the help of sockets. To show the validity of the model, two simple cases that have analytical solution or pure SPICE results are investigated. Then, three kinds of metamaterial are studied and modeled by the FDTD-SPICE engine. First, loop-based metamaterial with lumped capacitor is investigated where changing the capacitance can vary the resonance frequency. Then, integration of metamaterial with non-Foster circuit element, which provides negative impedance and increases the bandwidth, is studied and characterized. Lastly, metasurface simulation with integrated nonlinear varactor elements shows a tunable reflection phase property for this metamaterial. The technique is very powerful and its accuracy is validated well. We expect our scheme to be of great engine for comprehensive simulation of metamaterials integrated with active and nonlinear circuits and allow exploring new artificial materials combined with a lot of potential complex circuits.

ACKNOWLEDGMENTS

This work was supported in part by the US Office of Naval Research (ONR) Grant no. N00014-10-1-0264. The authors would like to thank Allen Taflove from Northwestern University for his kind discussion on the FDTD-SPICE model theory and acknowledge Vincent Thomas from Los Alamos National Laboratory for his helpful information on SPICE and sockets programming.

REFERENCES

1. D. A. Powell, I. V. Shadrivov, Y. S. Kivshar, and M. V. Gorkunov, "Self-tuning mechanisms of nonlinear split-ring resonators," Applied Physics Letters, vol. 91, no. 14, Article ID 144107, 3 pages, 2007.
2. I. V. Shadrivov, A. B. Kozyrev, D. W. Van Der Weide, and Y. S. Kivshar, "Tunable transmission and harmonic generation in nonlinear metamaterials,"Applied Physics Letters, vol. 93, no. 16, Article ID 161903, 3 pages, 2008.
3. D. Huang, E. Poutrina, and D. R. Smith, "Analysis of the power dependent tuning of a varactor-loaded metamaterial at microwave frequencies," Applied Physics Letters, vol. 96, no. 10, Article ID 104104, 3 pages, 2010.
4. I. V. Shadrivov, S. K. Morrison, and Y. S. Kivshar, "Tunable split-ring resonators for nonlinear negative-index metamaterials," Optics Express, vol. 14, no. 20, pp. 9344–9349, 2006. · ·
5. S. Saadat, M. Adnan, H. Mosallaei, and E. Afshari, "Composite Metamaterial and Metasurface Integrated with Non-Foster Active Circuit Elements: A Bandwidth-Enhancement Investigation," IEEE Transactions on Antennas and Propagation. In press.
6. T. Jiang, K. Chang, L.-M. Si, L. Ran, and H. Xin, "Active microwave negative-index metamaterial transmission line with gain," Physical Review Letters, vol. 107, no. 20, Article ID 205503, 5 pages, 2011.
7. D. J. Kern, D. H. Werner, and M. J. Wilhelm, "Active negative impedance loaded EBG structures for the realization of ultra-wideband artificial magnetic conductors," in Proceedings of the IEEE International Antennas and Propagation Symposium and USNC/CNC/URSI North American Radio Science Meeting, vol. 2, pp. 427–430, June 2003.
8. A. Taflove and S. C. Hagness, Computational Electrodynamics: The Finite-Difference Time-Domain Method, House, Norwood, Mass, USA, 3rd edition, 2005.
9. H. Mosallaei and Y. Rahmat-Samii, "Broadband characterization of complex periodic EBG structures: an FDTD/Prony technique based on the split-field approach," Electromagnetics, vol. 23, no. 2, pp. 135–151, 2003.
10. http://bwrc.eecs.berkeley.edu/classes/icbook/spice/.
11. V. A. Thomas, M. E. Jones, M. Piket-May, A. Taflove, and E. Harrigan, "Use of SPICE lumped circuits as sub-grid models for FDTD analysis," IEEE Microwave and Guided Wave Letters, vol. 4, no. 5, pp. 141–143, 1994.
12. G. Kobidze, A. Nishizawa, and S. Tanabe, "Ground bouncing in PCB with integrated circuits," in Proceedings of the IEEE International Symposium on Electromagneti Compatibility, vol. 1, pp. 349–352, August 2000.
13. N. Orhanovic, R. Raghuram, and N. Matsui, "Full wave analysis of planar interconnect structures using FDTD-SPICE," in Proceedings of the 51st Electronic Components and Technology Conference, pp. 489–494, June 2001.
14. H. Mosallaei, "FDTD-PLRC technique for modeling of anisotropic-dispersive media and metamaterial devices," IEEE Transactions on Electromagnetic Compatibility, vol. 49, no. 3, pp. 649–660, 2007.
15. M. Fujii, M. Tahara, I. Sakagami, W. Freude, and P. Russer, "High-order FDTD and auxiliary differential equation formulation of optical pulse propagation in 2-D Kerr and Raman nonlinear dispersive media," IEEE Journal of Quantum Electronics, vol. 40, no. 2, pp. 175–182, 2004.
16. J. A. Roden and S. D. Gedney, "Convolutional PML (CPML): an efficient FDTD implementation of the CFS-PML for arbitrary media," Microwave and Optical Technology Letters, vol. 27, pp. 334–339, 2000.
17. H. Mosallaei and Y. Rahmat-Samii, "Periodic bandgap and effective dielectric materials in electromagnetics: characterization and applications in nanocavities and waveguides," IEEE Transactions on Antennas and Propagation, vol. 51, no. 3, pp. 549–563, 2003.
18. H. Mosallaei and K. Sarabandi, "Magneto-dielectrics in electromagnetics: concept and applications," IEEE Transactions on Antennas and Propagation, vol. 52, no. 6, pp. 1558–1567, 2004.
19. H. Mosallaei and K. Sarabandi, "Design and modeling of patch antenna printed on magneto-dielectric embedded-circuit metasubstrate," IEEE Transactions on Antennas and Propagation, vol. 55, no. 1, pp. 45–52, 2007. · ·
20. W. Jendernalik, "A low-voltage CMOS negative impedance converter for analogue filtering applications," Bulletin of the Polish Academy of Sciences, vol. 55, no. 4, pp. 419–423, 2007. ·

21. R. M. Rudish and S. E. Sussman-Fort, “Progress in use of non-Foster impedances to. match electrically-small antennas and arrays,” in Proceedings of the Antenna Applications Symposium, pp. 89–108, Allerton Park, Ill, USA, September 2005.

22. P. Jin and R. W. Ziolkowski, “Broadband, efficient, electrically small metamaterial-inspired antennas facilitated by active near-field resonant parasitic elements,” IEEE Transactions on Antennas and Propagation, vol. 58, no. 2, pp. 318–327, 2010.

23. S. Hrabar, I. Krois, and A. Kiricenko, “Towards active dispersionless ENZ metamaterial for cloaking applications,” Metamaterials, vol. 4, no. 2-3, pp. 89–97, 2010.

24. J. G. Linvill, “Transistor negative-impedance converters,” Proceedings of the IRE, vol. 41, pp. 725–729, 1953.

25. H. Mosallaei and K. Sarabandi, “Antenna miniaturization and bandwidth enhancement using a reactive impedance substrate,” IEEE Transactions on Antennas and Propagation, vol. 52, no. 9, pp. 2403–2414, 2004.

CHAPTER 7

EFFICIENT FREQUENCY-TRANSIENT CO-SIMULATION OF COUPLED HEAT-ELECTROMAGNETIC PROBLEMS

Christof Kaufmann, Michael Günther, Daniel Klagges, Michael Knorrenschild, Matthias Richwin, Sebastian Schöps and E Jan W ter Maten

Keywords: Inductive charging, coupled simulation, co-simulation, electromagnetic, heat, modelling, dynamic iteration

INTRODUCTION

With the recent advent of inductive power charging systems and wireless power transmission in consumer and mobile phone technology, [1], all major automotive manufacturers develop concepts to wirelessly charge electric vehicles, both plug-in and pure electric vehicles (EV). For example the prototype from the Leopold Kostal GmbH & Co. KG of such an inductive charging station is shown in Figure 1. The necessity to charge EVs with today's battery technology after every prolonged use - at least every night - is seen as one of the major drawbacks in the usability of EVs. A system to automate the charging process would reduce the burden on the driver; it could increase the acceptance of EVs, and, in the case of plug-in hybrid EVs, it could help to further reduce the CO_2 footprint since the battery of the plug-in hybrid could always be considered fully charged. This is important for the calculation of the fleet CO_2 emission.

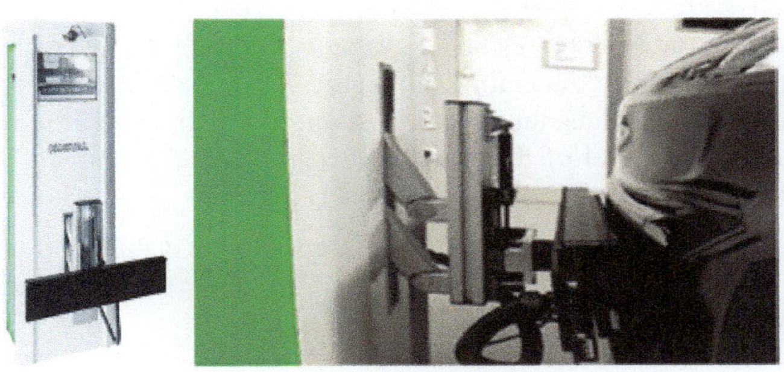

(a) Charge station. (b) Charging of a electric vehicle.

Figure 1 Prototype of an inductive charging station that charges the vehicle through its number plate (Images from Leopold Kostal GmbH & Co. KG).

A future inductive charging system does not necessarily exhibit a lower efficiency than a comparable conductive charging system, since there are only a few additional components; in a simplified view, the inductive charging system could be considered as a conductive charging system that has been cut in half in the middle of the transformer. There are, however, certain aspects that require attention and detailed design and optimization. These include positioning tolerances of the stationary ('primary') and car-mounted ('secondary') coils, magnetic stray fields, and thermal aspects.

The efficiency of both conductive and inductive charging systems is aimed well above 90%, measured from the primary AC connection to the drive train battery. But even with this high efficiency, at 3.5kW of first generation systems charging power there is a non-negligible amount of heat to be dissipated. Later generations with even higher power will further increase the heat load on the components. This heat load is the result of several different processes, namely resistance losses due

Citation: Christof Kaufmann, Michael Günther, Daniel Klagges, Michael Knorrenschild, Matthias Richwin, Sebastian Schöps, E JanW ter Maten, Efficient frequency-transient co-simulation of coupled heat-electromagnetic problems, doi:10.1186/2190-5983-4-1.

to DC resistance and proximity effects, ferrite core losses and switching losses in the active semiconductor switching components. The main effects appear at the same frequency range as the magnetic field, which is of the order of 10-100 kHz. The resulting temperature however changes on much slower timescales, in the order of minutes, determined by the heat capacity and the (relatively large) mass of the involved components. This electromagnetic-thermal problem is fully coupled, as many of the material parameters show a significant thermal dependence. Typical ferrite core losses, for instance, are minimal at temperatures around 90°C and increase below and above this temperature. This drives the equilibrium temperature of the ferrite material always close to this temperature, if the dissipated power is small enough, or makes the system thermally unstable, if the heat power is too high.

Engineering samples of such systems are expensive, heavy, possibly dangerous to operate, and take a lot of time to build and optimize. Virtual prototyping using efficient simulation methods accelerates this process. There are different methodologies and models available, [2].

The paper is structured as follows: In Section 2, we propose a particular frequency-transient model for electromagnetic-thermal problems. The electromagnetic (EM) field is considered at high frequencies, where the time scale of the heat conduction is significantly lower than the time scale of the EM field. This can be exploited in the modeling. We propose a co-simulation scheme, similar to dynamic iteration [3], and analyze its convergence properties in Section 3. The analysis exhibits interesting results, especially for high frequencies. Possibilities to generalize this model are also discussed. Section 4 validates the results by a numerical simulation of a simplified model of the inductive charging system shown in Figure 1.

In contrast to [2], where the different ways of co-simulation are discussed, we focus here on modeling and analysis of the frequency-transient model.

MODELING

In this section, we derive a model, which describes the electromagnetic field coupled to the temperature in the materials. For that, in Section 2.1 Maxwell's equations are introduced with temperature dependent material parameters. The conduction of the heat is described in Section 2.2 by the heat equation together with an electromagnetic power term as source to describe the Joule losses of the EM field. Finally, in Section 2.3 assumptions and approximations lead to the frequency-transient model [4].

Maxwell's Equations

Maxwell's equations read

$$\nabla \cdot \mathbf{D} = \rho, \tag{1}$$

$$\nabla \cdot \mathbf{B} = 0, \tag{2}$$

$$\nabla \times \mathbf{E} = -\frac{\partial \mathbf{B}}{\partial t}, \tag{3}$$

$$\nabla \times \mathbf{H} = \frac{\partial \mathbf{D}}{\partial t} + \mathbf{J}, \tag{4}$$

where **E** and **H** are the electric and magnetic field strength, **D** and **B** are the electric and magnetic flux densities, ρ and **J** are the electric charge and current densities. These laws are supported by the constitutive relations

$$\mathbf{D} = \varepsilon \mathbf{E}, \quad \mathbf{B} = \mu \mathbf{H} \quad \text{and} \quad \mathbf{J} = \sigma(T)\mathbf{E} + \mathbf{J}_{\text{src}},$$

where the permittivity ε and permeability μ parameters depend only on space while the conductivity σ may also depend on the temperature T. In this way the EM field solution is a function of the temperature distribution (parameter coupling). However, the source current density J_{src} describes an external excitation [4, 5]; it is assumed to be independent of the temperature. Now to reduce the unknowns in Maxwell's equations, we introduce the magnetic vector potential **A** and the electric scalar potential φ as

$$\mathbf{E} = -\nabla\varphi - \frac{\partial \mathbf{A}}{\partial t} \quad \text{with } \mathbf{B} = \nabla \times \mathbf{A}.$$

By using these potentials, (2) and (3) are fulfilled automatically. From (4) we get

$$\nabla \times \left(\mu^{-1}\nabla \times \mathbf{A}\right) + \varepsilon\frac{\partial^2 \mathbf{A}}{\partial t^2} + \sigma(T)\frac{\partial \mathbf{A}}{\partial t} + \varepsilon\nabla\frac{\partial \varphi}{\partial t} + \sigma(T)\nabla\varphi = \mathbf{J}_{\text{src}}.$$

With Buchholz gauge transformation (∇φ=0) this reduces to

$$\nabla \times \left(\mu^{-1}\nabla \times \mathbf{A}\right) + \varepsilon\frac{\partial^2 \mathbf{A}}{\partial t^2} + \sigma(T)\frac{\partial \mathbf{A}}{\partial t} = \mathbf{J}_{\text{src}}. \tag{5}$$

The so-called curl-curl equation will be treated in the following on a finite domain with adequate boundary conditions (BC) and initial values (IV).

For the low frequency range, where inductive effects dominate, usually the displacement current $\partial \mathbf{D}/\partial t \mathrel{\hat{=}} \omega^2 \varepsilon \hat{\mathbf{A}}$ can be disregarded. This is called magnetoquasistatic formulation. For details on these formulations we refer to [6]. Here we are interested in the high frequency range. Therefore our model is based on the full Maxwell formulation (1)-(4).

Heat Equation

The classical heat equation describes conduction of heat in materials:

$$\rho c\frac{\partial T}{\partial t} = \nabla \cdot (k\nabla T) + Q \tag{6}$$

with heat conductivity k, density ρ and specific heat capacity c, all constant in time and again on a finite domain with BC and IV. The term Q represents the Joule losses of the EM field. Hence we use a source coupling to connect the heat equation with the EM curl-curl equation (5). For simplicity we consider only eddy-current and Joule losses and thus obtain

$$Q(\mathbf{A}, T) = \mathbf{J} \cdot \mathbf{E} = \sigma(T)\mathbf{E} \cdot \mathbf{E} + \mathbf{J}_{\text{src}} \cdot \mathbf{E} = \sigma(T)\frac{\partial \mathbf{A}}{\partial t} \cdot \frac{\partial \mathbf{A}}{\partial t} - \mathbf{J}_{\text{src}} \cdot \frac{\partial \mathbf{A}}{\partial t}. \tag{7}$$

The power term is further simplified in the next section.

Frequency-transient Model

Now we aim at a model which allows an efficient simulation. The model consisting of (5) and (6) with (7) is defined in the time domain. A multirate co-simulation scheme could simulate both equations with different time steps. However, for a fast varying source current density the main part of the computational costs is caused by the simulation of (5). A discussion of single-rate and multirate schemes can be found in [2]. We will reduce these costs further by refining the model.

Since the temperature is only slowly varying in comparison to the EM field in (5) the temperature T, where σ is evaluated at, can be averaged by

$$\tilde{T}_i := \frac{1}{\tau_{i+1} - \tau_i} \int_{\tau_i}^{\tau_{i+1}} T(t)\,dt.$$

Thus we use $\sigma(T(t)) \approx \sigma(\tilde{T}_i)$ for $t \in [\tau_i, \tau_{i+1}]$ and then (5) can be approximated by

$$\nabla \times (\mu^{-1}\nabla \times \mathbf{A}) + \varepsilon \frac{\partial^2 \mathbf{A}}{\partial t^2} + \sigma(\tilde{T}_i)\frac{\partial \mathbf{A}}{\partial t} = \mathbf{J}_{\mathrm{src}}. \tag{8}$$

However, this is still in time domain. To allow a solution in frequency domain, we assume a time harmonic source current density

$$\mathbf{J}_{\mathrm{src}} = \frac{1}{2}\hat{\mathbf{J}}_{\mathrm{src}}e^{j\omega t} + \frac{1}{2}\overline{\hat{\mathbf{J}}_{\mathrm{src}}}e^{-j\omega t}, \tag{9}$$

where $\hat{J}_{\mathrm{src}}$ is a complex phasor. It follows for μ and ε independent of **A** that

$$\mathbf{A}(t) = \frac{1}{2}\left(\hat{\mathbf{A}}_c(\tilde{T}_i)e^{j\omega t} + \overline{\hat{\mathbf{A}}_c(\tilde{T}_i)}e^{-j\omega t}\right), \tag{10}$$

where the complex phasor $\hat{\mathbf{A}}_c$ is the solution for $\hat{\mathbf{J}}_{\mathrm{src}}$. This means the amplitude is constant within the time interval $[\tau_i, \tau_{i+1}]$.

Let us look at (7) again, but now with the approximation $\sigma(T(t)) \approx \sigma(\tilde{T}_i)$. The dot product is the usual real inner product.

$$\begin{aligned} Q(\mathbf{A}, \tilde{T}_i) &= \sigma(\tilde{T}_i)\frac{\partial \mathbf{A}}{\partial t}\cdot\frac{\partial \mathbf{A}}{\partial t} - \mathbf{J}_{\mathrm{src}}\cdot\frac{\partial \mathbf{A}}{\partial t} \\ &= -\sigma(\tilde{T}_i)\frac{\omega^2}{4}\left(\hat{\mathbf{A}}_c\cdot\hat{\mathbf{A}}_c e^{j2\omega t} - 2\hat{\mathbf{A}}_c\cdot\overline{\hat{\mathbf{A}}_c} + \overline{\hat{\mathbf{A}}_c}\cdot\overline{\hat{\mathbf{A}}_c}e^{-j2\omega t}\right) \\ &\quad - j\frac{\omega}{4}\left(\hat{\mathbf{J}}_{\mathrm{src}}\cdot\hat{\mathbf{A}}_c e^{j2\omega t} - \hat{\mathbf{J}}_{\mathrm{src}}\cdot\overline{\hat{\mathbf{A}}_c} + \overline{\hat{\mathbf{J}}_{\mathrm{src}}}\cdot\hat{\mathbf{A}}_c - \overline{\hat{\mathbf{J}}_{\mathrm{src}}}\cdot\overline{\hat{\mathbf{A}}_c}e^{-j2\omega t}\right), \end{aligned}$$

where $\hat{\mathbf{A}}_c = \hat{\mathbf{A}}_c(\tilde{T}_i)$. Now we are interested in the mean power loss in the time interval $[\tau_i, \tau_{i+1}]$:

$$\tilde{Q}_i := \frac{1}{\tau_{i+1} - \tau_i}\int_{\tau_i}^{\tau_{i+1}} Q(\mathbf{A}(t), \tilde{T}_i)\,dt \tag{11}$$

and thus all parts with $e^{\pm j2\omega t}$ vanish (for interval length being a multiple of the half period length, i.e. $\tau_{i+1} - \tau_i = c\frac{\pi}{\omega}$, where c ∈ N). Then we are left with

$$\tilde{Q}_i = \sigma(\tilde{T}_i)\frac{\omega^2}{2}\|\hat{\mathbf{A}}_c(\tilde{T}_i)\|_c^2 + \frac{\omega}{2}\mathrm{Im}\left(\hat{\mathbf{J}}_{\mathrm{src}}\cdot\overline{\hat{\mathbf{A}}_c(\tilde{T}_i)}\right), \tag{12}$$

where $\|\mathbf{A}\|_c := \mathbf{A}\cdot\overline{\mathbf{A}}$ is the Euclidean norm for complex vectors **A**. Now the simplified curl-curl equation (8) can be considered in frequency domain (with vector potential $\hat{\mathbf{A}}_c = \hat{\mathbf{A}}_c(\tilde{T}_i)$) along with the simplified heat equation (14) left in time domain:

$$\left(j\omega\sigma(\tilde{T}_i) - \omega^2\varepsilon\right)\hat{\mathbf{A}}_c + \nabla \times \left(\mu^{-1}\nabla \times \hat{\mathbf{A}}_c\right) = \hat{\mathbf{J}}_{\mathrm{src}}, \tag{13}$$

$$\rho c\frac{\partial T}{\partial t} = \nabla\cdot(k\nabla T) + \tilde{Q}_i. \tag{14}$$

The curl-curl equation is formulated with constant material parameters in frequency domain. Thus, only a linear, complex system has to be solved once for each time window, instead for each time step of the curl-curl equation in time domain.

ALGORITHM

We will now discuss the algorithm to simulate heat-EM problems with the frequency-transient model. After discretization, the model is solved in a Gauss-Seidel scheme, which can be interpreted as co-simulation. It is comparable with a dynamic iteration for time integration. Section 3.1 will briefly discuss the co-simulation scheme. In Section 3.2 the convergence analysis for the iteration is proved.

Method

The co-simulation scheme uses (13) and (14) in a discretized form. For simplicity we assume Finite Integration Technique (FIT) [7–9] for spatial discretization and then for time discretization the implicit Euler method. As an alternative to FIT discretization, one could use the (lowest order) Finite Element or Boundary Element Methods [10, 11] with the drawback of non-diagonal material matrices. This would complicate the following derivations. However, in FIT-notation the curl-curl equation (13) becomes

$$\left(j\omega\mathbf{M}_\sigma^l - \omega^2\mathbf{M}_\varepsilon + \mathbf{C}^\top\mathbf{M}_\nu\mathbf{C}\right)\widehat{\mathbf{a}}^{l+1} = \widehat{\widehat{\mathbf{j}}}_s, \tag{15}$$

with (diagonal) positive semi-definite matrix for electric conductivity M_σ, (diagonal) positive definite matrices for permittivity and reluctivity M_ε, M_ν, discrete curl operators **C** and $C^\top$. Note, that $\mathbf{M}_\sigma^l := \mathbf{M}_\sigma(\mathbf{T}^l)$, where T^l denotes the discretized temperature at the l th iteration step. Let n denote the number of nodes in the computational domain, then the matrices are from $\mathbb{R}^{3n\times3n}$; the discretized (facet-integrated) source current density $\widehat{\widehat{\mathbf{j}}}_s \in \mathbb{R}^{3n}$ and the (edge-integrated) magnetic vector potential $\widehat{\mathbf{a}} \in \mathbb{R}^{3n}$ are ordered in xyz-direction, e.g., $\widehat{\mathbf{a}}^\top := [\widehat{\mathbf{a}}_x^\top, \widehat{\mathbf{a}}_y^\top, \widehat{\mathbf{a}}_z^\top]$, [9]. For the heat equation, (14) gives

$$(\mathbf{M}_{\rho c} + h_i\tilde{\mathbf{S}}\mathbf{M}_k\tilde{\mathbf{S}}^\top)\mathbf{T}^{l+1} = \mathbf{M}_{\rho c}\mathbf{T}_i + h_i\frac{\omega}{2}\mathbf{P}\left[\omega\mathbf{M}_\sigma^{l+1}\widehat{\mathbf{a}}^{l+1}\circ\overline{\widehat{\mathbf{a}}^{l+1}} + \mathrm{Im}\left(\widehat{\widehat{\mathbf{j}}}_s\circ\overline{\widehat{\mathbf{a}}^{l+1}}\right)\right], \tag{16}$$

where ∘ denotes the Hadamard (element wise) product, diagonal positive definite matrices for thermal conductivity and volumetric heat capacity M_k, $M_{\rho c}$, discrete divergence and gradient operators $\tilde{\mathbf{S}}$, $-\tilde{\mathbf{S}}^\top$ on the dual grid, respectively. The matrix $\mathbf{P} \in \mathbb{R}^{n\times3n}$ averages and sums up the discrete losses, cf. (12). We use P:=[I,I,I] with $\mathbf{I} \in \mathbb{R}^{n\times n}$ being the identity matrix. However other choices are possible and in [12] a more sophisticated averaging is proposed for **P**.

To simulate this model, (15) and (16) are solved successively. This can be repeated for one time step until convergence (see Algorithm 1), similarly as done in Gauss-Seidel schemes. Here, we call it co-simulation. The scheme is also depicted in Figure 2. However, since convergence cannot be guaranteed for such schemes in general, a proof is given in Section 3.2.

Algorithm 1 Co-simulation with the frequency transient model, with discretized temperature $\mathbf{T}_i$, discretized magnetic vector potential $\widehat{\mathbf{a}}_i$ and time t_i.

1: Initialize model
2: $\mathbf{T}(t_0) \leftarrow \mathbf{T}_0$
3: $i \leftarrow 0$
4: **while** $t_i < t_{\text{end}}$ **do**
5: $\quad \mathbf{T}_{i+1}^0 \leftarrow \mathbf{T}_i$
6: $\quad l \leftarrow 0$
7: $\quad$ **while** $l < 2$ or not converged **do**
8: $\qquad$ Solve (15) for $\widehat{\mathbf{a}}_{i+1}^{l+1}$
9: $\qquad$ Solve (16) for $\mathbf{T}_{i+1}^{l+1}$
10: $\qquad l \leftarrow l + 1$
11: $\quad$ **end while**
12: $\quad \widehat{\mathbf{a}}_{i+1} \leftarrow \widehat{\mathbf{a}}_{i+1}^l$
13: $\quad \mathbf{T}_{i+1} \leftarrow \mathbf{T}_{i+1}^l$
14: $\quad i \leftarrow i + 1$
15: **end while**

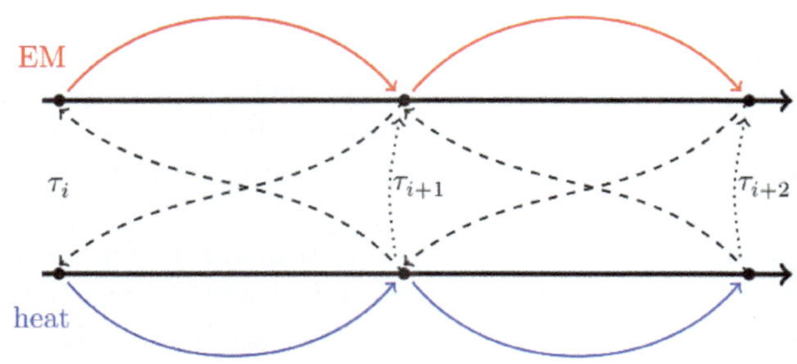

Figure 2 Frequency-transient co-simulation approach.

From Algorithm 1 it can be seen that time steps are chosen according to the time constant of the heat equation and only one (complex) linear system is solved per iteration for the curl-curl equation in frequency domain. Thus the time step size is independent of the excitation frequency ω. In contrast, a time domain solution would require many time steps per period $2\pi/\omega$ and consequently the solution of a large number of real-valued linear systems. This is the reason, why the model is very efficient for high frequencies since there is multirate behavior.

Convergence Analysis

We first verify that the Maxwell operator is bounded:

Lemma Let the EM problem (15) be given with adequate BC, IV for a frequency $\omega>\omega_0$. Then the inverse of the discrete Maxwell operator exists and is bounded

$$\left\| \left(j\omega \mathbf{M}_\sigma(\mathbf{T}) - \omega^2 \mathbf{M}_\varepsilon + \mathbf{C}^\top \mathbf{M}_\nu \mathbf{C}\right)^{-1} \right\| \leq C_0(\omega) \tag{17}$$

for some frequency dependent upper bound $C_0(\omega)$ that is independent of the temperature.

Proof Beforehand we introduce some abbreviations:

$$\mathbf{X} := \omega^2 \mathbf{M}_\varepsilon - \mathbf{C}^\top \mathbf{M}_\nu \mathbf{C}, \qquad \mathbf{Z} := -\mathbf{X} + j\omega \mathbf{M}_\sigma. \tag{18}$$

For $\omega>\omega_0$ for some ω_0 the matrix $X=X(\omega)$ is real, symmetric positive definite and $M_\sigma=M_\sigma(T)$ is real, diagonal, positive semi-definite. Then

$$\mathbf{Z} = (-\mathbf{X} + j\omega \mathbf{M}_\sigma) = \mathbf{X}^{1/2}\left(-\mathbf{I} + j\omega \mathbf{X}^{-1/2}\mathbf{M}_\sigma \mathbf{X}^{-1/2}\right)\mathbf{X}^{1/2}, \tag{19}$$

where $\mathbf{X}^{1/2} := \mathbf{U}\mathbf{\Sigma}^{1/2}\mathbf{U}^{-1}$ for an eigen decomposition $X=U\Sigma U^{-1}$.

Now let $\mathbf{A} := \mathbf{X}^{-1/2}\mathbf{M}_\sigma \mathbf{X}^{-1/2}$, which is real and symmetric positive semi-definite. It follows that the eigenvalue decomposition is $\mathbf{A} = \mathbf{Q}^{-1}\mathbf{\Lambda}\mathbf{Q}$, with $\|Q\|=1$. Then

$$\begin{aligned}\mathbf{Z}^{-1} &= \mathbf{X}^{-1/2}(-\mathbf{I} + j\omega \mathbf{A})^{-1}\mathbf{X}^{-1/2} \\ &= \mathbf{X}^{-1/2}\left(-\mathbf{Q}^{-1}\mathbf{Q} + j\omega \mathbf{Q}^{-1}\mathbf{\Lambda}\mathbf{Q}\right)^{-1}\mathbf{X}^{-1/2} \\ &= \mathbf{X}^{-1/2}\mathbf{Q}^{-1}(-\mathbf{I} + j\omega \mathbf{\Lambda})^{-1}\mathbf{Q}\mathbf{X}^{-1/2},\end{aligned}$$

$$\begin{aligned}\left\|\mathbf{Z}^{-1}\right\| &\leq \left\|\mathbf{X}^{-1/2}\right\|\left\|\mathbf{Q}^{-1}\right\|\left\|(-\mathbf{I} + j\omega \mathbf{\Lambda})^{-1}\right\|\|\mathbf{Q}\|\left\|\mathbf{X}^{-1/2}\right\| \\ &\leq \left\|\mathbf{X}^{-1/2}\right\| \cdot 1 \cdot \frac{1}{\sqrt{1+\omega^2\lambda_{\min}^2}} \cdot 1 \cdot \left\|\mathbf{X}^{-1/2}\right\| = \left\|\mathbf{X}^{-1}\right\| =: C_0(\omega),\end{aligned}$$

because **A** is semidefinite.

The Lemma guarantees solvability of Maxwell's equations independent of the temperature. However, generalizations are possible but we focus here on the high frequency case because it exhibits a distinct multirate potential. In practice the time-harmonic approach can be used over a wide range of possible excitation frequencies. In particular for low-frequencies where the displacement currents are often disregarded [13, 14]. Furthermore for other choices of gauging similar results are found. Also in pure transient simulation [15], a wide range of excitation frequencies is covered in practice.

Theorem Let the coupled problem (15)-(16) be given with adequate BC, IV for a frequency $\omega>\omega_0$. We assume for nonlinear materials, e.g., metals, positivity and differentiability for $\sigma(T)$w.r.t. temperature T and $\sigma'<0$. Let the exact (monolithic) solution be denoted by $\widehat{\mathbf{a}}^\infty$ and T^∞, then the iteration is convergent for h_i mall enough with

$$\left\|\mathbf{T}^{l+1} - \mathbf{T}^\infty\right\| \leq C(\omega) h_i \left\|\mathbf{T}^l - \mathbf{T}^\infty\right\|,$$

where $C(\omega)$ is uniformly bounded. Furthermore, we have $C(\omega) = \mathcal{O}(\frac{1}{\omega^2})$ for sufficiently large ω.

Proof Consider the inner loop of Algorithm 1 (steps 8 and 9). It consists of (15) and (16). We prove convergence for this inner loop and use the same abbreviations as introduced in (18) with the shortcut $\mathbf{Z}^l := \mathbf{Z}(\mathbf{T}^l)$. Then we deduce from (15)

$$\mathbf{Z}^l \widehat{\mathbf{a}}^{l+1} = \widehat{\widehat{\mathbf{j}}}_s \quad \Leftrightarrow \quad \widehat{\mathbf{a}}^{l+1} = \left(\mathbf{Z}^l\right)^{-1}\widehat{\widehat{\mathbf{j}}}_s \tag{20}$$

due to the Lemma above. We define $\mathbf{N} := \tilde{\mathbf{S}}\mathbf{M}_k\tilde{\mathbf{S}}^\top$, use (16) with T^∞ and $\widehat{\mathbf{a}}^\infty$ subtract it from (16)

$$\begin{aligned}(\mathbf{M}_{\rho c} + h_i \mathbf{N})\left(\mathbf{T}^{l+1} - \mathbf{T}^\infty\right) &= h_i \frac{\omega^2}{2}\mathbf{P}\left(\mathbf{M}_\sigma^{l+1}\widehat{\mathbf{a}}^{l+1} \circ \overline{\widehat{\mathbf{a}}^{l+1}} - \mathbf{M}_\sigma^\infty \widehat{\mathbf{a}}^\infty \circ \overline{\widehat{\mathbf{a}}^\infty}\right) \\ &\quad + h_i \frac{\omega}{2}\mathbf{P}\,\mathrm{Im}\left(\widehat{\widehat{\mathbf{j}}}_s \circ \overline{\left(\widehat{\mathbf{a}}^{l+1} - \widehat{\mathbf{a}}^\infty\right)}\right).\end{aligned} \tag{21}$$

Next, by adding and subtracting $h_i \frac{\omega^2}{2}\mathbf{P}\mathbf{M}_\sigma^\infty \widehat{\mathbf{a}}^{l+1} \circ \overline{\widehat{\mathbf{a}}^{l+1}}$, we have

$$\begin{aligned}&(\mathbf{M}_{\rho c} + h_i \mathbf{N})\left(\mathbf{T}^{l+1} - \mathbf{T}^\infty\right) \\ &= h_i \frac{\omega^2}{2}\mathbf{P}\left(\mathbf{M}_\sigma^{l+1} - \mathbf{M}_\sigma^\infty\right)\left(\widehat{\mathbf{a}}^{l+1} \circ \overline{\widehat{\mathbf{a}}^{l+1}}\right) \\ &\quad + h_i \frac{\omega^2}{2}\mathbf{P}\mathbf{M}_\sigma^\infty\left(\widehat{\mathbf{a}}^{l+1} \circ \overline{\widehat{\mathbf{a}}^{l+1}} - \widehat{\mathbf{a}}^\infty \circ \overline{\widehat{\mathbf{a}}^\infty}\right) + h_i \frac{\omega}{2}\mathbf{P}\,\mathrm{Im}\left(\widehat{\widehat{\mathbf{j}}}_s \circ \overline{\left(\widehat{\mathbf{a}}^{l+1} - \widehat{\mathbf{a}}^\infty\right)}\right).\end{aligned} \tag{22}$$

Now, let us consider the term

$$\left(\mathbf{M}_\sigma^{l+1} - \mathbf{M}_\sigma^\infty\right)\left(\widehat{\mathbf{a}}^{l+1} \circ \overline{\widehat{\mathbf{a}}^{l+1}}\right).$$

The conductivity matrix M_σ is a diagonal matrix for any iteration step. We assume, that the kth component σ_k $(1 \leq k \leq 3n)$ depends only on the neighboring temperature $T_k := (P^\top T)k$, see Figure 3. Thus the k th component of the term can be written as

$\widehat{a}_k^{l+1}(\sigma_k(T_k^{l+1}) - \sigma_k(T_k^{\infty}))\overline{\widehat{a}_k^{l+1}}$.

Now the mean value theorem can be applied componentwise and yields for the k th component

$$\widehat{a}_k^{l+1}(\sigma_k(T_k^{l+1}) - \sigma_k(T_k^{\infty}))\overline{\widehat{a}_k^{l+1}} = \widehat{a}_k^{l+1}\sigma_k'(\zeta)\overline{\widehat{a}_k^{l+1}}(T_k^{l+1} - T_k^{\infty}),$$

where ζ_k between T_k^{l+1} and T_k^{∞}, and σ_k' is by assumption non-positive. Hence, in matrix vector notation with $(\zeta)_k=\zeta_k$

$$(\mathbf{M}_\sigma^{l+1} - \mathbf{M}_\sigma^{\infty})(\widehat{\mathbf{a}}^{l+1} \circ \overline{\widehat{\mathbf{a}}^{l+1}}) = \mathbf{A}_{l+1}\mathbf{M}_\sigma'(\zeta)\mathbf{A}_{l+1}^{\mathsf{H}}\mathbf{P}^{\top}(\mathbf{T}^{l+1} - \mathbf{T}^{\infty}),$$

where $\mathbf{A}_l := \operatorname{diag}(\widehat{\mathbf{a}}^l)$ and the diagonal matrix $\mathbf{M}_\sigma'(\zeta)$ has only non-positive elements, i.e., it is negative semi-definite. It follows

$$\mathbf{L}(\mathbf{T}^{l+1} - \mathbf{T}^{\infty}) = h_i\frac{\omega^2}{2}\mathbf{P}\mathbf{M}_\sigma^{\infty}(\widehat{\mathbf{a}}^{l+1} \circ \overline{\widehat{\mathbf{a}}^{l+1}} - \widehat{\mathbf{a}}^{\infty} \circ \overline{\widehat{\mathbf{a}}^{\infty}}) + h_i\frac{\omega}{2}\mathbf{P}\operatorname{Im}(\widehat{\widehat{\mathbf{j}}}_s \circ \overline{(\widehat{\mathbf{a}}^{l+1} - \widehat{\mathbf{a}}^{\infty})}) \qquad (23)$$

with $\mathbf{L} := \mathbf{M}_{\rho c} + h_i\mathbf{N} - h_i\frac{\omega^2}{2}\mathbf{P}\mathbf{A}_{l+1}\mathbf{M}_\sigma'(\zeta)\mathbf{A}_{l+1}^{\mathsf{H}}\mathbf{P}^{\top}$ whose inverse exists because it holds $\|\mathbf{L}^{-1}\| \leq \|\mathbf{M}_{\rho c}^{-1}\|$, since $\mathbf{M}_{\rho c}$, $\mathbf{N}$, since $\mathbf{M}_{\rho c}$, $\mathbf{N}$ are positive definite and $\mathbf{M}_\sigma'(\zeta)$ is negative semi-definite. Multiplying (23) by the inverse of **L**, taking the norm and exploiting the upper bound $\|\mathbf{M}_{\rho c}^{-1}\|$ yields

$$\|\mathbf{T}^{l+1} - \mathbf{T}^{\infty}\| \leq \|\mathbf{M}_{\rho c}^{-1}\| \cdot \Big[h_i\frac{\omega^2}{2}\|\mathbf{P}\mathbf{M}_\sigma^{\infty}(\widehat{\mathbf{a}}^{l+1} \circ \overline{\widehat{\mathbf{a}}^{l+1}} - \widehat{\mathbf{a}}^{\infty} \circ \overline{\widehat{\mathbf{a}}^{\infty}})\|$$
$$+ h_i\frac{\omega}{2}\|\mathbf{P}\operatorname{Im}(\widehat{\widehat{\mathbf{j}}}_s \circ \overline{(\widehat{\mathbf{a}}^{l+1} - \widehat{\mathbf{a}}^{\infty})})\|\Big]. \qquad (24)$$

Now, we need estimates for (a) the linear term $\|\widehat{\mathbf{a}}^{l+1} - \widehat{\mathbf{a}}^{\infty}\|$ and (b) the quadratic term $\|\widehat{\mathbf{a}}^{l+1} \circ \overline{\widehat{\mathbf{a}}^{l+1}} - \widehat{\mathbf{a}}^{\infty} \circ \overline{\widehat{\mathbf{a}}^{\infty}}\|$.

(a) We start with the linear term: Consider (20) for the difference of exact solution and (l+1) th iterate:

$$j\omega\mathbf{M}_\sigma^l\widehat{\mathbf{a}}^{l+1} - j\omega\mathbf{M}_\sigma^{\infty}\widehat{\mathbf{a}}^{\infty} - \mathbf{X}(\widehat{\mathbf{a}}^{l+1} - \widehat{\mathbf{a}}^{\infty}) = 0$$
$$\Leftrightarrow \quad \mathbf{Z}^{\infty}(\widehat{\mathbf{a}}^{l+1} - \widehat{\mathbf{a}}^{\infty}) + j\omega(\mathbf{M}_\sigma^l - \mathbf{M}_\sigma^{\infty})\widehat{\mathbf{a}}^{l+1} = 0$$
$$\Rightarrow \quad \|\widehat{\mathbf{a}}^{l+1} - \widehat{\mathbf{a}}^{\infty}\| \leq \omega\|(\mathbf{Z}^{\infty})^{-1}\|\|\mathbf{M}_\sigma^l - \mathbf{M}_\sigma^{\infty}\|\|\widehat{\mathbf{a}}^{l+1}\| \qquad (25)$$
$$\leq \omega\|(\mathbf{Z}^{\infty})^{-1}\||\sigma_{\max}'|\|\mathbf{T}^l - \mathbf{T}^{\infty}\|\|(\mathbf{Z}^l)^{-1}\|\|\widehat{\widehat{\mathbf{j}}}_s\| \qquad (26)$$
$$\leq C_0^2(\omega)\omega|\sigma_{\max}'|\|\mathbf{T}^l - \mathbf{T}^{\infty}\|\|\widehat{\widehat{\mathbf{j}}}_s\| \qquad (27)$$
$$\leq c(\omega)\|\mathbf{T}^l - \mathbf{T}^{\infty}\|, \qquad (28)$$

where we have exploited the Lemma to obtain $c(\omega) := C_0^2(\omega)\omega|\sigma_{\max}'|\|\widehat{\widehat{\mathbf{j}}}_s\|$.

(b) We now consider the quadratic term in (24)

$$\|\widehat{\mathbf{a}}^{l+1} \circ \overline{\widehat{\mathbf{a}}^{l+1}} - \widehat{\mathbf{a}}^{\infty} \circ \overline{\widehat{\mathbf{a}}^{\infty}}\| = \|(\widehat{\mathbf{a}}^{l+1} - \widehat{\mathbf{a}}^{\infty} + \widehat{\mathbf{a}}^{\infty}) \circ \overline{(\widehat{\mathbf{a}}^{l+1} - \widehat{\mathbf{a}}^{\infty} + \widehat{\mathbf{a}}^{\infty})} - \widehat{\mathbf{a}}^{\infty} \circ \overline{\widehat{\mathbf{a}}^{\infty}}\|$$
$$\leq \|(\widehat{\mathbf{a}}^{l+1} - \widehat{\mathbf{a}}^{\infty}) \circ \overline{(\widehat{\mathbf{a}}^{l+1} - \widehat{\mathbf{a}}^{\infty})}\| + 2\|(\widehat{\mathbf{a}}^{l+1} - \widehat{\mathbf{a}}^{\infty}) \circ \widehat{\mathbf{a}}^{\infty}\|.$$

Because $\|a \circ b\| \leq \|a\|\|b\|$, this can be written as

$$\|\widehat{\mathbf{a}}^{l+1} \circ \overline{\widehat{\mathbf{a}}^{l+1}} - \widehat{\mathbf{a}}^{\infty} \circ \overline{\widehat{\mathbf{a}}^{\infty}}\| \leq \|\widehat{\mathbf{a}}^{l+1} - \widehat{\mathbf{a}}^{\infty}\|^2 + 2\|\widehat{\mathbf{a}}^{l+1} - \widehat{\mathbf{a}}^{\infty}\|\|\widehat{\mathbf{a}}^{\infty}\|$$
$$\leq c^2(\omega)\|\mathbf{T}^l - \mathbf{T}^{\infty}\|^2 + 2c(\omega)\|\widehat{\mathbf{a}}^{\infty}\|\|\mathbf{T}^l - \mathbf{T}^{\infty}\|, \qquad (29)$$

where the Lemma gives us

$$\|\widehat{\mathbf{a}}^{\infty}\| = \|(\mathbf{Z}^{\infty})^{-1}\widehat{\widehat{\mathbf{j}}}_s\| \leq C_0(\omega)\|\widehat{\widehat{\mathbf{j}}}_s\|.$$

Using the estimates (28) and (29) in (24) gives

$$\|\mathbf{T}^{l+1} - \mathbf{T}^{\infty}\| \leq h_i\frac{\omega}{2}c(\omega)\|\mathbf{P}\|\|\mathbf{M}_{\rho c}^{-1}\|\|\mathbf{T}^l - \mathbf{T}^{\infty}\|(\omega c(\omega)\|\mathbf{M}_\sigma^{\infty}\|\|\mathbf{T}^l - \mathbf{T}^{\infty}\|$$
$$+ 2\omega C_0(\omega)\|\widehat{\widehat{\mathbf{j}}}_s\|\|\mathbf{M}_\sigma^{\infty}\| + \|\widehat{\widehat{\mathbf{j}}}_s\|).$$

Now consider the asymptotic behavior of this for large ω and small h_i. It holds

$$C_0(\omega) = \|\mathbf{X}^{-1}\| = \|(\omega^2\mathbf{M}_\varepsilon - \mathbf{C}^{\top}\mathbf{M}_\nu\mathbf{C})^{-1}\| \sim \frac{1}{\omega^2}$$

and

$$c(\omega) = C_0^2(\omega)\omega|\sigma_{\max}'|\|\widehat{\widehat{\mathbf{j}}}_s\| \sim \frac{1}{\omega^3}.$$

Then, for fixed h_i and ω large enough, it follows that

$$\|\mathbf{T}^{l+1} - \mathbf{T}^{\infty}\| \leq h_i C(\omega)\|\mathbf{T}^l - \mathbf{T}^{\infty}\|,$$

with some $C(\omega) \sim \frac{1}{\omega^2}$.

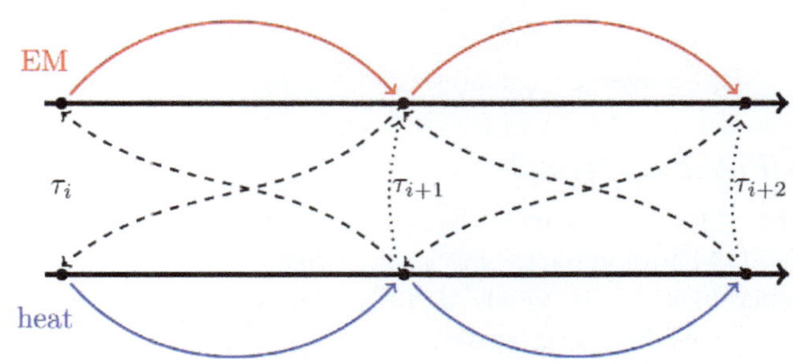

Figure 3 Mimetic discretizations of Maxwell's equations (e.g. by FIT) use a primal and dual grid. The discrete temperatures T_k are located at primary nodes, the line-integrated electric field and the magnetic vector potentials $\widehat{e}_k = \widehat{a}_k'$ are defined at primary edges and finally the facet-integrated source current density $\widehat{\widehat{j}}_k$ at dual surfaces. The material laws relate the quantities on both grids, e.g., the conductivity connects by Ohm's Law the currents and voltages, i.e. . $\widehat{\widehat{j}}_k = \sigma_k(T_k)\widehat{e}_k$ The figure shows the case for a current in x-direction.

Generalization

The frequency-transient model can be generalized in different ways. To enhance versatility, one can use a multi-frequency excitation of the EM subsystem. An application would be steel hardening of gears [16, 17], where two frequencies are necessary to obtain a homogeneous heating of the surface. Also to approximate periodic signals other than sinusoidal, multi-frequency excitation can be used. This also allows for a Harmonic Balance approach, which enables usage of a nonlinear permeability μ.

Time Dependent Phasor

The model can be improved by weakening the assumptions made. In [4] it is suggested to consider the complex phasor not as constant within one time window. This means (10) becomes

$$\mathbf{A}(t) = \frac{1}{2}(\widehat{\mathbf{A}}_c(T)e^{j\omega t} + \overline{\widehat{\mathbf{A}}_c(T)}e^{-j\omega t}) = \operatorname{Re}(\widehat{\mathbf{A}}_c(T)e^{j\omega t}). \qquad (30)$$

Thus the derivatives are

$$\frac{\partial \mathbf{A}}{\partial t} = \mathrm{Re}\left(\frac{\partial \hat{\mathbf{A}}_c}{\partial t} \cdot e^{j\omega t}\right) + \mathrm{Re}\left(j\omega \hat{\mathbf{A}}_c \cdot e^{j\omega t}\right), \tag{31}$$

$$\frac{\partial^2 \mathbf{A}}{\partial t^2} = \mathrm{Re}\left(\frac{\partial^2 \hat{\mathbf{A}}_c}{\partial t^2} \cdot e^{j\omega t}\right) + 2\,\mathrm{Re}\left(j\omega \frac{\partial \hat{\mathbf{A}}_c}{\partial t} \cdot e^{j\omega t}\right) - \mathrm{Re}\left(\omega^2 \hat{\mathbf{A}}_c \cdot e^{j\omega t}\right), \tag{32}$$

where $\hat{\mathbf{A}}_c := \hat{\mathbf{A}}_c(T)$. Substituting (30)-(32) into (5) yields

$$\nabla \times \left(\mu^{-1} \nabla \times \hat{\mathbf{A}}_c\right) + \varepsilon\left(\frac{\partial^2 \hat{\mathbf{A}}_c}{\partial t^2} + 2j\omega \frac{\partial \hat{\mathbf{A}}_c}{\partial t} - \omega^2 \hat{\mathbf{A}}_c\right) + \sigma(T)\left(\frac{\partial \hat{\mathbf{A}}_c}{\partial t} + j\omega \hat{\mathbf{A}}_c\right) = \hat{\mathbf{J}}_{\mathrm{src}},$$

which is now a second order PDE with partial derivatives with respect to time as well. Note that in [4] magnetoquasistatic formulation is used, so there the modification yields a first order system. However, since now both parts (EM and heat) have to be time integrated, the co-simulation of these can be called dynamic iteration [3].

Due to the modification (12) changes to

$$\tilde{Q}(T) = \frac{\omega^2}{2}\sigma(T)\|\hat{\mathbf{A}}_c\|_c^2 + \frac{\omega}{2}\left[2\sigma\,\mathrm{Im}\left(\frac{\partial \hat{\mathbf{A}}_c}{\partial t} \cdot \overline{\hat{\mathbf{A}}_c}\right) + \mathrm{Im}\left(\hat{\mathbf{J}}_{\mathrm{src}} \cdot \overline{\hat{\mathbf{A}}_c}\right)\right]$$
$$+ \frac{1}{2}\left[\left\|\frac{\partial \hat{\mathbf{A}}_c}{\partial t}\right\|_c^2 + \mathrm{Re}\left(\overline{\hat{\mathbf{J}}_{\mathrm{src}}} \cdot \frac{\partial \hat{\mathbf{A}}_c}{\partial t}\right)\right].$$

The convergence analysis can be extended accordingly.

MPDAE Approach

The Multirate Partial Differential Algebraic Equations (MPDAE) approach in [18] offers another type of generalization. It yields a fully time-domain model that introduces two time scales according to a fast and a slow component in the solution. For circuit simulation this can be exploited to efficiently determine envelope simulation in case of amplitude modulation. The method however also works for frequency modulation problems. For the coupled electromagnetics-heat problem the slow time scale comes from the heat equation, the fast time scale comes from the periodic source in the electromagnetic problem.

FREQUENCY-TRANSIENT CO-SIMULATION EXAMPLE

In this section the frequency-transient model is applied to an inductive charging system for electric vehicles. The charging system is modeled in a 2d-axisymmetric way by a primary coil for the station and a secondary coil as receiver in the car. Both coils include ferrite. Additionally there is a steel slice to model the part of the car body behind the secondary coil. The electric conductivity of this steel slice depends on the temperature, i.e., the system is mutually coupled. The geometry and simulation results are shown in Figure 4. For more details of the set up we refer to [2].

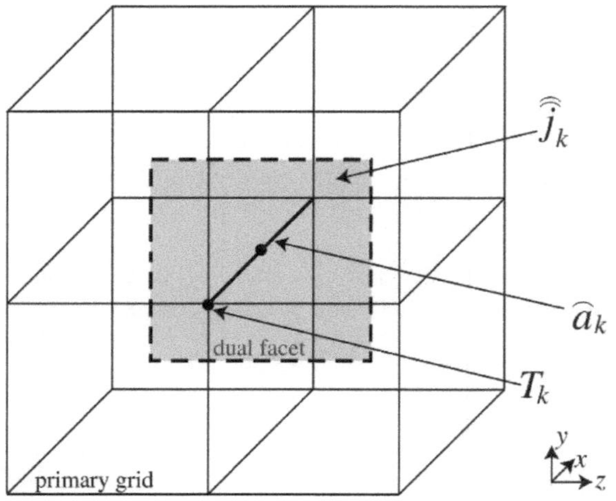

Figure 4 Model for an inductive charging system for electric vehicles (Comsol).

A constant conductivity would have lead to a single way coupled system and thus the magnetic vector potential would by a constant vector phasor. Models with a constant conductivity (according to the initial temperature) will systematically underestimate the heating. In this case the obtained maximum temperature would be 7K below the corresponding solution of the mutually coupled problem, [2].

For mutually coupled problems the frequency-transient model has proved to be very efficient. This is expected, since the main part of the computational effort - the time integration of the EM subsystem - is avoided. In this numerical example ω was chosen to be $2 \cdot \pi \cdot 10$kHz. In fact the proposed co-simulation algorithm reached the end time of t_{end}=20min by using only 17 time steps with $n \leq 14$ iterations (on average $\bar{n} \approx 5.4$ iterations). This coarse time grid sufficiently resolves the dynamics of the heat equation. For comparison, in a naive monolithic simulation with 10 time steps per period 120 million time steps would be necessary to resolve the dynamics of the EM subsystems. This underlines the computational gain.

CONCLUSIONS

A frequency-transient model tailored for heat-electromagnetic problems was derived. The time step size of the coupled system is determined by the heat subsystem only. The convergence analysis is presented in detail. Convergence for high frequencies is guaranteed. The error bound for the iteration decreases quadratically with higher frequencies. This result also applies to approaches by Driesen and Hameyer [4] and similar implementations in Comsol [19]. Thus the approach fits perfectly for applications where inductive heating either appears as losses or is intended.

ACKNOWLEDGEMENTS

This work is supported by the German BMBF in the context of the SOFA project (grant number 03MS648E). The sixth author is supported by the 'Excellence Initiative' of the German Federal and State Governments and the Graduate School of Computational Engineering at Technische Universität Darmstadt.

AUTHORS' CONTRIBUTIONS

All authors contributed to this paper as a whole. However, special merits go to DK and MR for sharing their experiences from industry, which led to the numerical example and introduction; to SS and MG for the work on the model; to CK, JtM and MK for their contribution to the analysis. All authors read and approved the final manuscript.

REFERENCES

1. Schneider D: Wireless power at a distance is still far away [Electrons unplugged]. IEEE Spectr 2010, 47(5):34-39.
2. Kaufmann C, Günther M, Klagges D, Richwin M, Schöps S, ter Maten EJW: Coupled heat-electromagnetic simulation of inductive charging stations for electric vehicles. In Progress in Industrial Mathematics at ECMI 2012. Berlin: Springer; 2014. [Mathematics in Industry.]
3. Bartel A, Brunk M, Günther M, Schöps S: Dynamic iteration for coupled problems of electric circuits and distributed devices. SIAM J Sci Comput 2013, 35(2):B315-B335.
4. Driesen J, Hameyer K: The simulation of magnetic problems with combined fast and slow dynamics using a transient time-harmonic method. Eur Phys J Appl Phys 2001, 14:165-169.
5. Schöps S, De Gersem H, Weiland T: Winding functions in transient magnetoquasistatic field-circuit coupled simulations. Compel 2013, 32(6):2063-2083.
6. Haus HA, Melcher JR: Electromagnetic Fields and Energy. New York: Prentice Hall; 1989 [http://web.mit.edu/6.013_book/www/]
7. Clemens M, Gjonaj E, Pinder P, Weiland T: Numerical simulation of coupled transient thermal and electromagnetic fields with the finite integration method. IEEE Trans Magn 2001, 36(4):1448-1452.
8. Clemens M, Gjonaj E, Pinder P, Weiland T: Self-consistent simulations of transient heating effects in electrical devices using the finite integration technique. IEEE Trans Magn 2001, 37(5):3375-3379.
9. Weiland T: Time domain electromagnetic field computation with finite difference methods. Int J Numer Model 1996, 9(4):295-319.
10. Bossavit A: Computational Electromagnetism: Variational Formulations, Complementarity, Edge Elements. San Diego: Academic Press; 1998 [http://tut.fi/~bossavit/#Books]
11. Hiptmair R, Ostrowski J, Quast R: Modelling and simulation of induction heating. Technical Report 149, SFB 382, University of Tübingen; 2000.
12. Alotto P, Freschi F, Repetto M: Multiphysics problems via the cell method: the role of Tonti diagrams. IEEE Trans Magn 2010, 46(8):2959-2962.
13. Hahne P, Weiland T: 3D eddy current computation in the frequency domain regarding the displacement current. IEEE Trans Magn 1992, 28(2):1801-1804.
14. Clemens M, Weiland T: Numerical algorithms for the FDiTD and FDFD simulation of slowly varying electromagnetic fields. Int J Numer Model Special Issue 'Finite Difference Time Domain and Frequency Domain Methods', Invited Paper 1999, 12(1/2):3-22.
15. Chen Q, Schoenmaker W, Chen G, Jiang L, Wong N: A numerically efficient formulation for time-domain electromagnetic-semiconductor co-simulation for fast-transient systems. IEEE TCAD 2013, 32(5):802-806.
16. Rudnev V: Induction hardening of gears and critical components. Part I. Gear Technol 2008, 2008:58-63.
17. Rudnev V: Induction hardening of gears and critical components. Part II. Gear Technol 2008, 2008:47-53.
18. Brachtendorf HG, Welsch G, Laur R, Bunse-Gerstner A: Numerical steady state analysis of electronic circuits driven by multi-tone signals. Electr Eng 1996, 79:103-112 [doi:10.1007/BF01232919].
19. COMSOL Multiphysics: Command reference; 2007 [www.comsol.com]

CHAPTER 8

HYBRID MULTI-PHYSICS MODELING OF AN ULTRA-FAST ELECTRO-MECHANICAL ACTUATOR

Ara Bissal, Anders Eriksson, Jesper Magnusson and Göran Engdahl

Keywords: Actuators, circuit breaker, electromagnetic launching, HVDC transmission, magneto-mechanical effects, mechanical power transmission, shock waves

INTRODUCTION

The increasing need for integrating renewable energy sources, such as off-shore wind power and power harnessed from photovoltaic cells installed in deserts, has rekindled interest in high voltage direct current (HVDC) multi-terminal grids [1]. China alone has plans to start around 30 HVDC transmission projects within the next 20 years [2]. If realized, an HVDC system will be able to transport power from remote areas to consumers with very low losses. However, its feasibility relies entirely on the existence of an HVDC breaker that is able to interrupt fault currents within very short time intervals [3]. To be able to interrupt fault currents promptly, an ultra-fast electromagnetic drive is needed to actuate the current carrying contacts by generating impulsive forces within hundreds of microseconds.

The mechanical part of a direct current (DC) circuit breaker consists of a contact system, an insulating medium, a pull rod, an ultra-fast drive and a control unit, as shown in Figure 1. The contact system consists of an inlet (1), a pair of stationary and movable contacts (2a and 2b) and an outlet (3). A pull rod (4) connects the contact system to the drive. The drive consists of an armature (5) and opening and closing coils (6a and 6b), respectively. Spring-based bistables (7) are used to keep the armature in close contact with the opening drive coil or the closing coil. When a fault current emerges, the control unit is triggered so that the DC breaker's drive can actuate the contacts promptly.

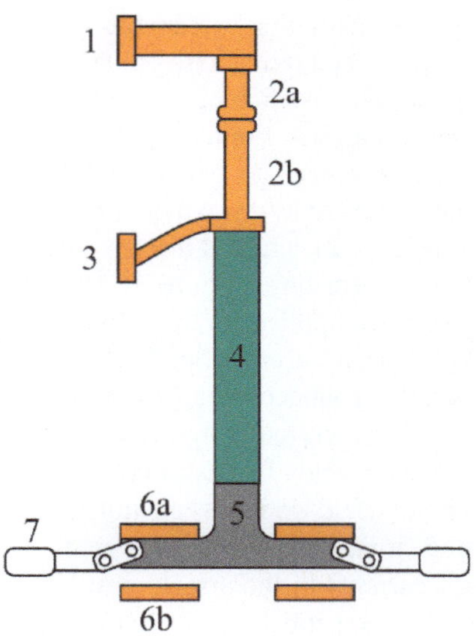

Figure 1. A sketch of a DC breaker showing the current carrying contacts (2), the pull rod (4), the armature (5), the coils (6) and the bistables (7).

After separating the contacts of a DC breaker, an arc will be initiated. To extinguish it, one can use parallel branches with semiconductors or an oscillatory circuit to inject a high frequency current forcing a current zero crossing as the contacts are separating [4]. The arc will then be extinguished at a current zero crossing. However, after contact separation, the breaker has to be able to withstand the voltage appearing across it. Otherwise, the arc might reignite, especially if the air gap separating both contacts is not sufficiently large.

The modeling of such a breaker is challenging, since it should be simulated with all of the necessary involved physics.

Citation: Ara Bissal, Anders Eriksson, Jesper Magnusson and Göran Engdahl, Hybrid Multi-Physics Modeling of an Ultra-Fast Electro-Mechanical Actuator, doi:10.3390/act4040314

As emphasized in [5], the interaction of different physics can now be understood thanks to the available computational power. Some examples of recent nonlinear coupling between mechanical stress and electric fields can be found in [6, 7]. In this paper, novel multi-physics hybrid and finite element method-based models are presented and compared.

MECHANICAL CHALLENGES OF AN HVDC BREAKER

The interest in DC breakers has increased over the past years [2, 8, 9]. One of the main challenges of an HVDC breaker is to separate the current carrying contacts within a few hundreds of microseconds. Otherwise, due to the low impedance of a DC grid [10], fault currents will rise quickly in magnitude and become harder to interrupt.

Similar linear actuators for high applications involving high accelerations are being investigated [11]. For a circuit breaker, an example of a typical requirement could be to interrupt a short circuit current by accelerating the contacts from rest up to 25 m/s within 500 μs; corresponding to a peak acceleration in the order of 10,000 g. Assuming the mass of the breaker's moving part to be around 5 kg, this would mean that the breaker will be subjected to an impulsive force of 500 kN after 250 μs. The combination of large forces and very short time scales poses a mechanical challenge, in particular when compared to a length scale in the order of meters.

In order to actuate the contact system this rapidly, an electromagnetic actuator that is capable of generating large impulsive forces is required. These actuators can be used in breakers and for other switching equipment, such as super conducting fault current limiters [12]. Several actuation schemes can be developed to generate such forces [13, 14, 15]. The actuator studied in this paper consists of a multi-turn flat spiral coil (Figure 2), situated directly under a mushroom-shaped conductive armature (Figure 3). Examples of similar actuation topologies applied to breakers can be found in [16,17,18]. Discharging a capacitor bank comprising several series and parallel connected capacitors through the drive coil results in a large current surge that, in turn, generates a substantial magnetic field. The axial component of this time-varying magnetic field induces azimuthally-directed eddy currents in the armature. Then, the vector product of the phi-directional currents in the armature along with the radial component of the magnetic flux density results in an impulsive repulsive body force, later denoted by f_{em}.

Figure 2. A picture showing a spirally-shaped coil and its contacts.

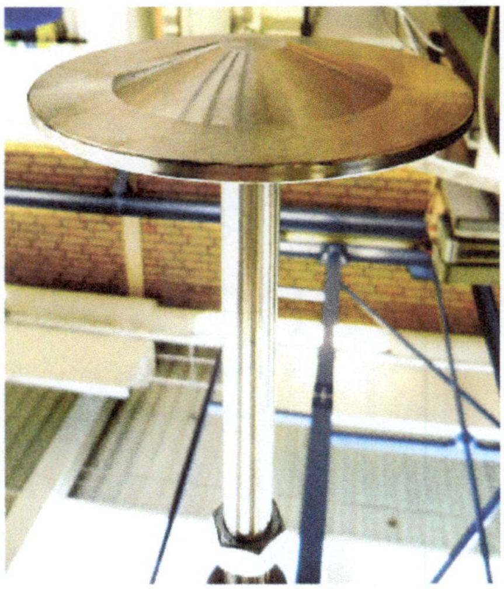

Figure 3. Mushroom-shaped conductive armature with threads on the stem's extremity to screw in a pull rod.

This actuator cannot be directly connected to the metallic contacts. One strict requirement of an HVDC breaker is to actuate its contacts using an electrically-insulating material to prevent the fault currents to flow elsewhere. An insulating material that electrically isolates the contacts from the armature and yet allows the transmission of forces between them is required. This component is the above-mentioned pull rod ((4) in Figure 1) and is usually made of composite materials or ceramics. The impulsive forces are generated in the armature and have to be first transmitted through the pull rod before they arrive at the desired location, i.e., the metal contacts. The main part of the body forces are generated within the first few millimeters inside the armature, which is located directly on top of the coil. Depending on the system voltage level, the pull rod has to have a certain minimum length, denoted by L_{Pr}. The usual length of such pull rods is around 600 mm. The time it takes for a wave to propagate through the armature to the contacts denoted by t_{travel} can be estimated using the velocity of waves (v_p) in a homogeneous isotropic medium with:

$$t_{\text{travel}} = \frac{L_{\text{Pr}}}{v_{\text{p}}} \tag{1}$$

$$v_p = \sqrt{\frac{K + 4G/3}{\rho}} = \sqrt{\frac{E(1-v)}{\rho(1+v)(1-2v)}} \tag{2}$$

where K, G, ρ, E and v are the bulk modulus, shear modulus, density of the medium, Young's modulus and Poisson's ratio.

In this study, a pull rod manufactured of fiberglass-reinforced epoxy (FR4) is investigated, since its properties are significantly superior to the traditional material used for such applications. It is electrically insulating and mechanically very strong with a Young's modulus of 24 GPa and a yield tensile stress exceeding 310 MPa. It has a shear modulus of 11 GPa and a density of 1850 kg/m^3. Furthermore, it can be machined in any desired shape and size, even if smoothness in shape should be striven for. If the travel time in the aluminum armature is neglected, then the time it takes for the wave to travel from one end of the pull rod to the other is around 160 μs, which is in the time scale of the HVDC breaker. This time may be even longer if a longer pull rod is used or the pull rod is manufactured of a material with a lower Young's modulus. Therefore, in the following sections, different models are examined in order to study the influence of the pull rod mechanics on the force generation mechanism and the displacement of the armature.

ELECTROMAGNETIC MODELING

The modeling of the actuator is divided into two parts, a circuit model and a finite element method (FEM) model, that are implemented in the software COMSOL Multiphysics (Version 4.3b, COMSOL AB, Stockholm, Sweden), as shown in Figure 4 [19]. An electrical source consisting of series and parallel-connected capacitors, a thyristor and cable leads connecting to a Thomson coil (TC) is modeled as a lumped circuit. On the other hand, the TC actuator that comprises a primary stationary coil and a mobile conductive armature is modeled using an FEM model. These two models are then coupled together and solved simultaneously.

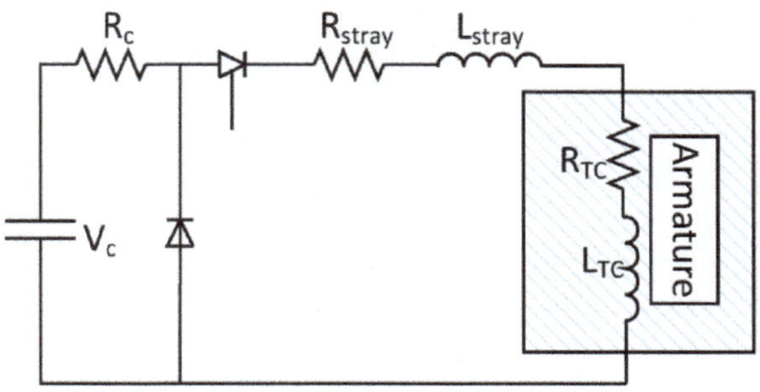

Figure 4. A SPICE circuit coupled to an FEM model for a Thomson coil. The coil and the armature are modeled using FEM, since the resistance and inductance of the coil and armature, RTCRTC and LTCLTC, are nonlinear and changing dynamically as the armature moves away. The capacitor, diode, thyristor and cables are modeled by lumped parameters.

To generate a large impulsive current, several parallel and series-connected capacitors are charged to a voltage denoted VCVC. The capacitor bank can be modeled as an effective capacitance in series with its effective resistance denoted by RCRC. As for the thyristor and the leads connecting the capacitor bank to the actuator, their impedance is lumped together and designated by R_{stray} and L_{stray}, respectively. A free wheeling diode is furthermore placed in parallel with the capacitor bank to prevent the build up of a negative voltage, since an electrolytic capacitor must be charged with only one polarity.

The finite element method is used to model the spiral coil and the armature solving the electromagnetic and mechanical equations at every time step. Concentric rectangles representing the width and depth of each coil conductor turn are drawn in a two-dimensional axi-symmetric geometry, avoiding the need to use 3D simulations and significantly reducing computation time. The voltage (V_{coil}) across the spiral coil in the FEM model serves as the connecting point to the circuit model. The total coil voltages at the terminals of the circuit are distributed across the different coil turns by:

$$\mathbf{J}_{en} = \frac{\sigma_e \mathbf{V}_n}{2\pi r_n} \tag{3}$$

$$V_{coil} = \sum_{i=1}^{n} |\mathbf{V}_i| \tag{4}$$

$$I_n = \int \mathbf{J}_n . ds \tag{5}$$

$$I_{cir} = I_1 = I_2 = ... = I_n \tag{6}$$

in such a way so as to ensure that the current (I_n) in each turn n is identical to the currents in the other turns. Here, J_{en} is the externally-applied current density due to V_n, the voltage across each coil turn, σ_e is the electrical conductivity, J_n is the current density and s is the surface area of each conductor turn. The average radius of each turn is denoted by r_n.

The induced current density in the conductive mobile armature denoted by $J_{i,\,arm}$ moving with a relative velocity v is given by:

$$\mathbf{J}_{i,arm} = \sigma_e(\mathbf{E} + \mathbf{v} \times \mathbf{B}) \tag{7}$$

where E is the electric field and B is the magnetic flux density. As for the stationary base coil, its induced current density reduces to:

$$\mathbf{J}_{i,coil} = \sigma_e \mathbf{E} \tag{8}$$

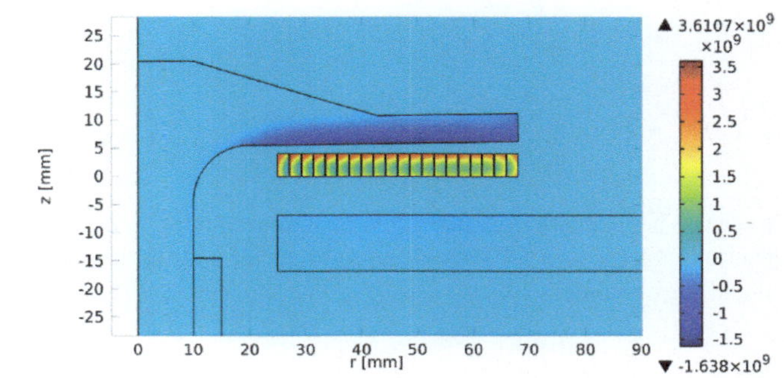

Figure 5. The current density 100 µs after the discharge of the capacitor bank is distributed in only small portions of the geometry. Positive current densities appear in the top part of the coil conductors, while negative currents are induced in a piece of the armature situated directly above the coil.

The total current density in the coil and the armature can be expressed by:

$$\mathbf{J}_{coil} = \mathbf{J}_e + \mathbf{J}_{i,coil} \tag{9}$$

$$\mathbf{J}_{arm} = \mathbf{J}_{i,arm} \tag{10}$$

respectively An example of the current density distribution can be seen in Figure 5. Based on Maxwell's equations,

$$\nabla \times \mathbf{H} = \mathbf{J} \tag{11}$$

$$\nabla \times \mathbf{E} = -\frac{\partial \mathbf{B}}{\partial t} \tag{12}$$

$$\mathbf{B} = \nabla \times \mathbf{A} \tag{13}$$

the magnetic equation for the primary coil, TC_p, can be written as:

$$\sigma_e \frac{\partial \mathbf{A}}{\partial t} + \frac{1}{\mu} \nabla \times (\nabla \times \mathbf{A}) = \mathbf{J}_e \tag{14}$$

where A is the magnetic vector potential, H the magnetic field intensity, J the current density, ∇ the nabla operator, µ the magnetic permeability and J_e the externally-applied current density. The magnetic equation for the mobile armature (TC_s) is given by:

$$\sigma_e \frac{\partial \mathbf{A}}{\partial t} + \frac{1}{\mu} \nabla \times (\nabla \times \mathbf{A}) - \sigma_e \mathbf{v} \times (\nabla \times \mathbf{A}) = 0 \quad (15)$$

The magnetic flux density upon the discharge of the capacitor bank can be seen in Figure 6. A current in the presence of a perpendicularly-oriented magnetic field results in a body force f_{em}, also know as a Lorentz force (see Figure 7), where:

$$f_{em} = \mathbf{J} \times \mathbf{B} \quad (16)$$

The force density is not homogeneously distributed and is generated only in the head of the mushroom, implying a high risk of deformations. In order to resolve these very small skin depths, triangular elements with a size around half a millimeter are used to mesh the zone where the force is generated. Although increasing the memory demands, second order elements are used to catch the gradients.

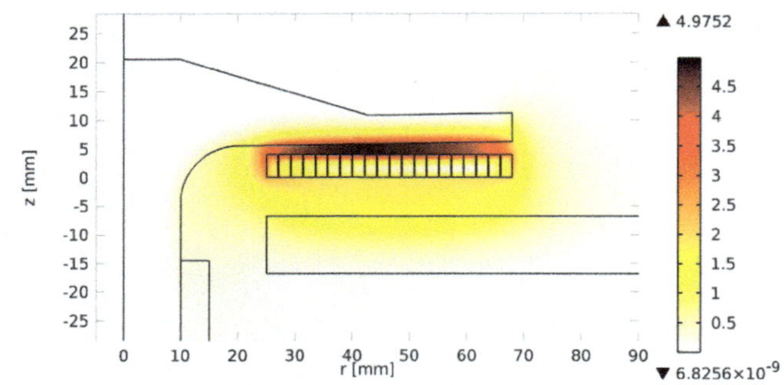

Figure 6. The magnetic flux density 100 µs after the discharge of the capacitor bank reaches 5 T and is highest in the air gap separating the coil and the armature.

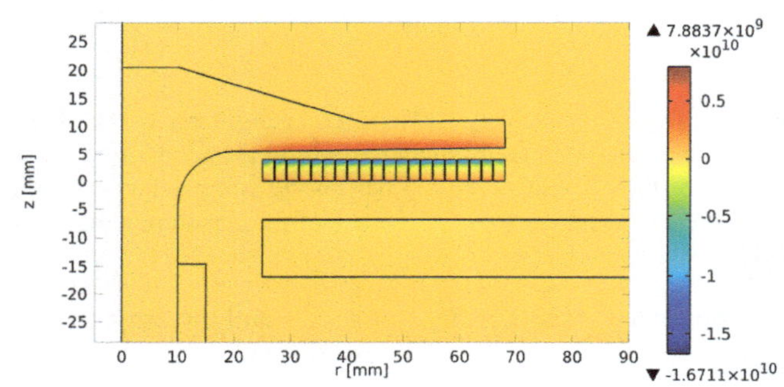

Figure 7. Force density 100 µs after the discharge of the capacitor bank. The coil is subjected to compressive forces, while the armature is subjected to an axially-directed body force that is mostly concentrated in the first few millimeters closest to the coil.

This transient may last for several milliseconds. In this model, a variable time step is implemented using very small time steps only when needed to ensure convergence. The smallest time step is around 50 ns and is only required when the voltage of the capacitor bank reaches zero, activating the highly nonlinear model of the free wheeling diode. Otherwise, a time step of around 10 µs is enough to have a good resolution.

The computed displacement is used to implement a moving mesh, which is vital to account for the movement of the armature. As the armature repels, skin depth and inductance increase. Therefore, the net resistance of the material decreases, since a larger cross-section of the conductor is utilized. The technique used is based on the arbitrary Lagrangian–Euler method, whereby the mesh is progressively stretched until a quality limit is violated. The mesh quality or a specified element stretching factor serve as criteria to stop the simulation, re-mesh and interpolate the results from the previous mesh to the newly-generated mesh.

MECHANICAL MODELING

The developed magnetic model and the computed electromagnetic forces are used as inputs to the mechanical model at every time step. Several FEM-based models are compared to determine which models can simulate the performance of the breaker with sufficient accuracy. Model 1 is used as a benchmark model, since it most accurate, but too demanding for repeated usage.

Model 1: Full Multi-Physics Model

To model the behavior of the breaker when subjected to such impulsive forces, both actuators, comprising the coil, the armature and the pull rod attached to the metal contacts, are meshed and simulated. Due to the presence of large displacements and moderate strains and introducing a notation common to mechanical formulation, even if these somewhat collide with the electro-magnetic notations above, the Green–Lagrange strain tensor EE and the second Piola–Kirchhoff stress tensor SS are used [20]. Cauchy's first equation of motion is given by:

$$\mathrm{Div}\mathbf{P} + \mathbf{F}_{em} = \rho_0 \frac{D^2\mathbf{U}}{Dt^2} \quad (17)$$

where P is the first Piola–Kirchhoff stress tensor,

$$\mathrm{DivP} = \mathrm{FS} \quad (18)$$

with F the deformation gradient and D denoting to the total derivative. The electromagnetic forces in the material reference frame,

$$J = \det\mathbf{F} \quad (19)$$

are computed by multiplying the calculated electromagnetic forces in the spatial coordinates f_{em} by:

$$J = \det F \quad (20)$$

the determinant of FF. In Equation (17), ρ_0 is the density of the reference material and UU the displacement vector defined as:

$$U = x - X \quad (21)$$

where xx and XX are position vectors in the current and reference configurations, respectively. The deformation gradient and the Green–Lagrange strain tensor are computed by:

$$\mathbf{F} = \mathrm{Grad}\mathbf{U} + \mathbf{I} \quad (22)$$

$$\mathbf{E} = \frac{1}{2}(\mathrm{Grad}^T\mathbf{U} + \mathrm{Grad}\mathbf{U} + \mathrm{Grad}^T\mathbf{U}\mathrm{Grad}\mathbf{U}) \quad (23)$$

where I is the identity matrix and Grad^T is the gradient transpose operator with respect to the material reference frame.

To be able to generate such impulsive forces without extensive deformations, the armature is manufactured out of a material that is both electrically conductive and mechanically strong. Therefore, the aluminum alloy T651 7075 was assumed for the simulations, since it is sufficiently strong, although not the most electrically conductive. It is represented as a linear elastic model:

$$S = C : E \tag{24}$$

where C is the fourth order stiffness tensor and ":" the double contraction operator. The same equations are used to model the pull rod. This stress-strain relation should be used as long as the stresses in the material do not exceed the yield stress, at which components start to deform plastically. The breaker must be designed in a way such that it does not operate in this region in order to allow several breaking events before replacement.

In COMSOL Multiphysics, the problem is simulated as an axi-symmetric continuum problem, with a model comprising of 26,000 quadratically-interpolated elements.

Model 2: Hybrid Multi-Physics First Order Model

Model 2 is referred to as a hybrid model, since the coil and the armature are modeled using FEM, i.e., by Equations (17) to (24). However, the pull rod is modeled as a first order spring-mass model with one spring of stiffness k bounded by two masses, denoted m_1 and m_2, respectively, on each of its terminals, as shown in Figure 8B. This is a simplification of the Kelvin–Voigt model [21], whereby damping is disregarded at this stage. The first mass is inserted between the spring and the stem of the armature and is given by m_1=m/2, while the second mass is inserted in the end and is represented by m_2=m/2+M, where m is the mass of the pull rod and M is the mass of the attached contact system. To get a value for the stiffness of the pull rod, the stresses in the material are considered to be purely axial, giving the linear stress-strain relation:

$$\sigma = S_{11} = E\epsilon = \frac{F}{A} \tag{25}$$

where E is the Young's modulus, ϵ is the strain in the material, F the force acting on the pull rod and A its cross-sectional area. A simplified model of the force of a spring is thereby:

$$F_{\text{spring}} = k\delta x \tag{26}$$

where k is the stiffness and δx its elongation. For a pull rod of length L, the axial stiffness can be represented by:

$$k = \frac{EA}{L} \tag{27}$$

using the reference geometry.

A generalization of this model into an n segment hybrid model is shown in Figure 8C. If the pull rod is chosen to be divided into n springs, then the stiffness of each spring will be nk, and each mass will be $\frac{m}{n+1}$. The final mass will be equal to one of the partially-distributed masses in addition to the mass of the attached copper contact.

Model 3: Pull Rod Assumed Infinitely Stiff

In this model, the pull rod is assumed to be infinitely stiff. Thus, a lumped mass representing the masses of the pull rod and the copper contacts is attached directly to the bottom of the stem of the armature. Equations (17) to (24) are used to model the response of the armature only.

Model 4: Armature and Pull Rod Assumed Infinitely Stiff

This model assumes that the entire structure is rigid and cannot be deformed. As a result, none of the developed mechanics equations are used. Instead, only the electromagnetic equations are used to compute the body forces in the armature at each time instance. Afterwards, this body force is integrated to compute the velocity and the position of the armature as:

$$\iiint \mathbf{F} r dr d\theta dz = M_{\text{tot}} \frac{d\mathbf{v}}{dt} \tag{28}$$

$$\mathbf{x} = \int \mathbf{v} dt \tag{29}$$

where M_{tot} is the summation of the mass of the armature, the pull rod and the copper contacts.

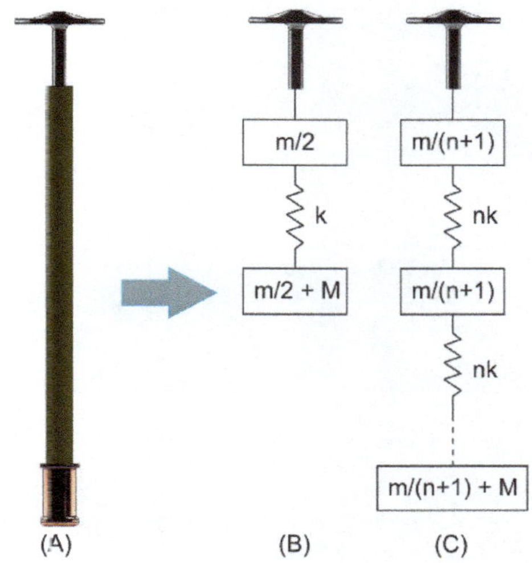

Figure 8. A sketch showing the modeling of the armature (in silver), the pull rod (in green) and the contacts (in dark red): by a full multi-physics model (A); by a first order hybrid model (B); and by a generalized n segment hybrid model (C). The pull rod's total mass is represented by m and its stiffness along elongation by k. The mass of the copper contact is represented by M.

MODEL VALIDATION BY EXPERIMENTS

One aim of the following experiments was to primarily study the influence of bending on the performance of the drive and, secondly, to see its effect on the current pulse. Another aim was to validate the full multi-physics model (Model 1) and the infinitely stiff model (Model 4) using two differently-dimensioned armatures.

To verify the simulation models, an experimental setup was built with two armature variants shown in Figure 9. The first armature is slim and is prone to bending when subjected to an impulsive force, while the other is larger and thereby more robust and rigid. Both armatures were however designed to withstand the maximum stresses.

The experimental setup can be seen in Figure 10. The mushroom-shaped armature is lying on top of a flat spiral coil, which is connected to a capacitor bank. A large steel structure was used to avoid vibrations and clamp down the setup firmly to a very heavy steel table. At this stage, the pull rod is omitted, and the masses are directly attached to the armature. Two steel masses, weighing 3.6 kg and 1.5 kg, were firmly attached to the slim and large armatures, respectively. Two capacitor banks with capacitances of 33 mF and 11 mF capable of being charged to 500 V and 900 VV were used for the slim and large armatures, respectively.

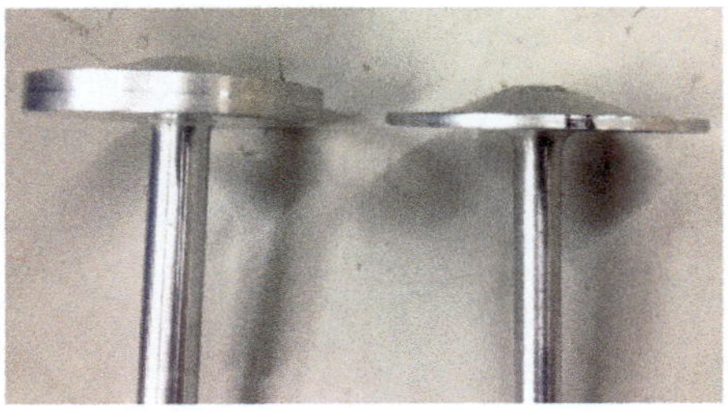

Figure 9. A picture showing the slim and large mushroom-shaped armatures. Although both are designed to withstand the mechanical stresses, one is flexible and prone to bending, while the other is more robust and stiffer.

Figure 10. A picture showing the experimental setup. The slim mushroom armature is sitting on a flat spiral coil that is connected with large cables to a capacitor bank. It is mounted with a 3.6 kg steel mass.

A Pearson probe was used to measure the current pulses. Initially, accelerometers were used to measure the acceleration of the armature. However, since high accelerations and collisions are involved in these experiments, the accelerometers break quickly. Instead, a high speed camera with frame rates of 100,000 fps was used to film the armatures. Afterwards, the images were calibrated and tracked to determine the velocity and the bending of the armature.

The simulated and measured current pulses for the slim armature are shown in Figure 11. Since bending is expected, Model 1 is used. At low voltages, the current pulse has only one peak at 750 μs and looks smooth. However, when the charging voltage, and thereby the energy, was increased, the current pulse became distorted. A clear distinct camel hump current pulse can be seen following a 500 VV discharge. Two peaks appear, one at 550 μs and one at 850 μs. The second current peak is even higher than the first one.

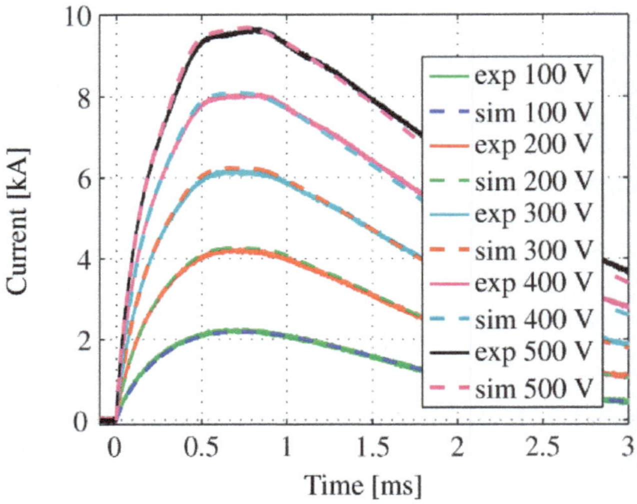

Figure 11. A picture showing the measured currents (solid lines) and the simulated currents (dashed lines) of the slim armature for increasing charging voltages in steps of 100 V.

To explain the camel hump behavior of the current pulse, the bending of the mushroom armature was measured and compared to simulations; see Figure 12. The camera was focused on the head of the mushroom only. The wing of the armature, as indicated by pt A in Figure 13, and the top flat part of the armature were tracked. The difference between these two points is defined by bending. The simulations were able to predict the experimental results with a good degree of accuracy. Although there are small differences in absolute values, the trend of the bending is captured. Two sources of error are due to filming the mushroom with an angle and the loss of focus after 800 μs. This trend can be seen in the current pulse. The second current peak, at 850 μs, is due to the wing of the armature bending back when the internal elastic forces exceeded the repulsive electromagnetic force. The mutual coupling is increased with smaller air gaps. The overall inductance seen by the system becomes smaller, causing the current to rise towards a second peak.

The measured and simulated velocities of the slim armature for different charging voltages are shown in Figure 14. It can be seen that the model predicts the behavior of the armature well.

The simulated and measured current pulses for the large armature are shown in Figure 15. In this case, since the armature was designed to minimize bending, the rigid simulation model (Model 4) was used. For low voltages, the measured and simulated current pulses are almost identical. However, for larger voltages, the simulated currents peak at an earlier time in comparison with the measured current pulses. Evidently, even such a bulky armature will bend and elongate when subjected to impulsive forces of this magnitude. In reality, any kind of elasticity will result in a larger system inductance, delaying the peak of the current pulse. Furthermore, the current pulses at all charging voltages have only one peak. Unlike the previous case, this indicates that very little bending exists.

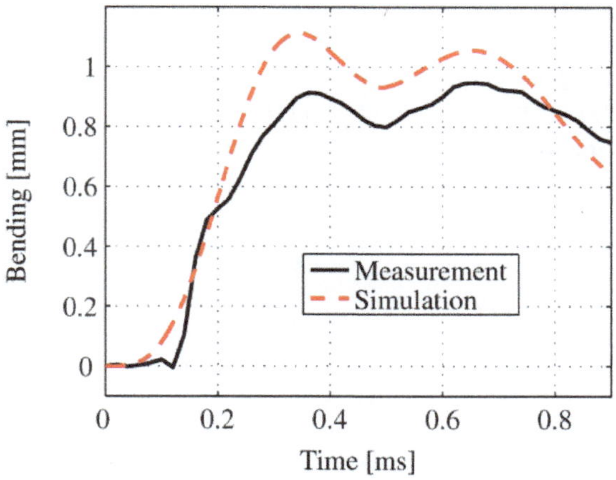

Figure 12. Comparison of the measured and simulated bending of the mushroom armature upon the discharge of a capacitor bank charged with 500 V. The bending cannot be measured for longer time scales, since the picture loses focus with large displacements, rendering the tracking unreliable.

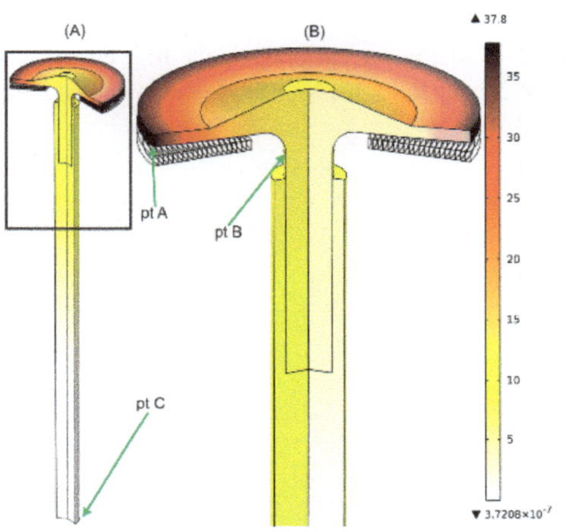

Figure 13. A 3D picture showing the velocity profile of the system in (m/s) after 150 μs. (**B**) Zoom in of the entire actuator shown in (**A**). The armature that is situated directly on top of the coil is threaded into a pull rod and attached firmly. Point A is located at the outermost extremity of the mushroom armature. Point B is located at the top of the stem of the armature, i.e., just below the rounded corner joining the head of the mushroom to its stem. Point C is located on the bottom of the pull rod, i.e., where the copper contacts are attached.

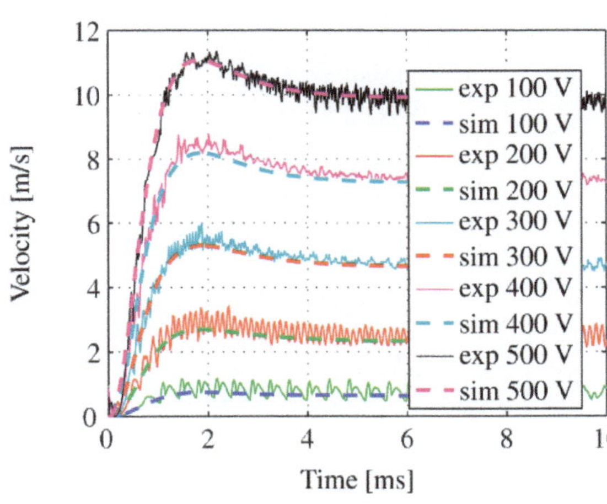

Figure 14. A picture showing the measured velocities (solid lines) and the simulated velocities (dashed lines) of the slim armature for increasing charging voltages in steps of 100 V.

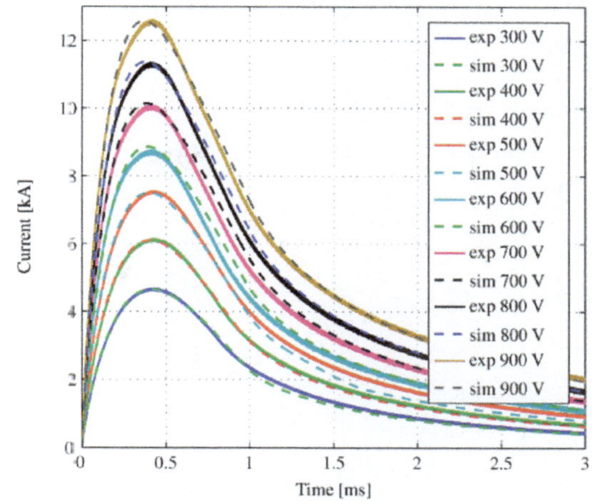

Figure 15. A picture showing the measured currents (solid lines) and the simulated currents (dashed lines) of the large armature for increasing charging voltages in steps of 100 V.

The simulated and measured armature velocities for the large armature are shown in Figure 16. For such an armature, Model 4 can definitely be used to simulate the velocity of the armature with high accuracy. A small discrepancy exists at high voltages due to the inevitable presence of small deformations.

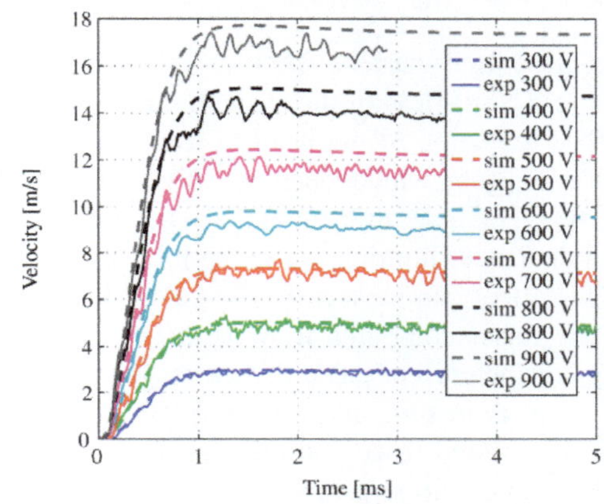

Figure 16. A picture showing the measured velocities (solid lines) and the simulated velocities (dashed lines) of the large armature for increasing charging voltages in steps of 100 V. The simulation error increases with increasing impulsive forces, since even the large armature is elastic and will therefore bend eventually.

RESULTS AND DISCUSSION

Following the model validations in the previous section, a pull rod was added, and different cases were simulated. From Model 1, the velocity distribution in the armature and the pull rod 150 μs after the discharge of a capacitor bank can be seen in Figure 13. The extremities of the mushroom attain velocities up to 38 m/s, while the velocity of the load attached at the bottom of the pull rod is still zero. This large velocity gradient was expected and was estimated initially from Equation (1).

Figure 17 shows the displacements of different points of the armature-contact system as computed by Model 1. The difference between the displacements of Points A and B show that bending occurs in the range of several millimeters, while the difference between Points B and C show that the armature and the pull rod elongate prior to any movement of the actual load. It can be seen that 300 μs after discharge, Point A of the armature is displaced by 6 mm, whereas the contacts are still stationary.

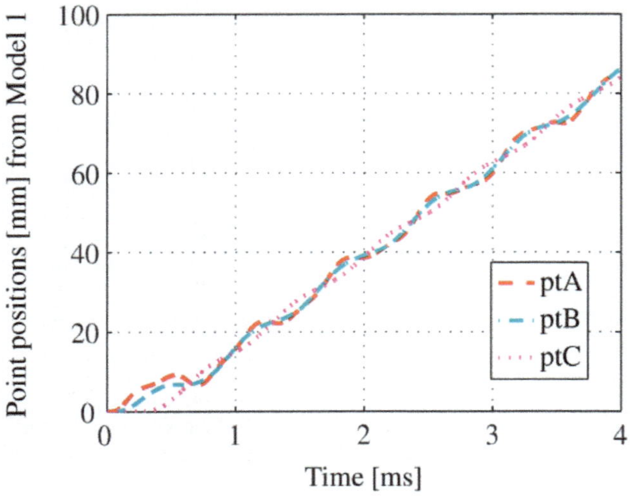

Figure 17. The displacements of three different points characterizing the dynamic motion of the breaker. The bending of the head of the mushroom can be computed by subtracting the axial displacement of ptB from that of pt A. Similarly, the elongation of the pull rod and the stem of the mushroom armature can be computed by subtracting the displacement of pt C from that of pt B.

The generated current impulse in the coil is highly sensitive to the stiffness of the system, but also to the chosen model. The repulsion of the head of the mushroom causes a premature air gap, which decreases the mutual inductance between the coil and the armature. This leads to a larger system inductance, limiting the sharp increase of the current pulse. This effect can be seen in Figure 18, when comparing the current pulses computed from Models 1 and 4. The time derivative of the current pulse decreases, leading to a shift in the current peak in time and a decrease in magnitude. If the pull rod is assumed to be infinitely stiff, then only bending and elongation of the armature are present, and the effect on the current pulse is slightly mitigated. A current peak, slightly delayed in time and lower in magnitude, occurs in Model 3. As for Models 1 and 2, they give similar current pulses. Thus, a system that is subjected to both bending and elongation yields a lower current pulse and, thereby, a considerable decrease in efficiency.

The generated force impulse is inversely proportional to the square of the length of the air gap separating the top of the coil from the bottom of the armature. As Figure 19 shows, a stiff system gives a significantly larger peak force than a system subjected to bending and elongation. Model 4 gives the highest peak force, followed by Model 3. Models 1 and 2 give similar peak force values, but the generated force impulses differ slightly in shape.

The displacements of the contacts versus time is given in Figure 20. As explained above, Models 1 and 2 give similar displacements, but oscillate out of phase. Therefore, after inspecting the current impulses, the force impulses and the displacements using Model 1 as a benchmark, it can be deduced that Model 2 gives similar results and can predict the behavior of the contacts with good accuracy. However, Models 3 and 4 are not good enough, since they overestimate the generated electromagnetic forces. Model 4 overestimated the peak force by almost 100%. Model 3 can be used if a stiff pull rod is used, while Model 4 can be used if both the armature and the pull rod are designed not to bend and elongate. They cannot, however, be considered as generally reliable for the present simulations with demands for accurate deformation and short time scales.

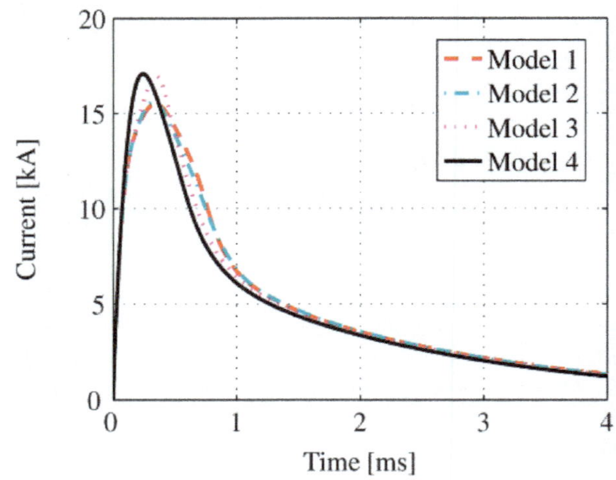

Figure 18. The arising current pulse in the four models following the discharge of the capacitor bank in the spirally-shaped coil.

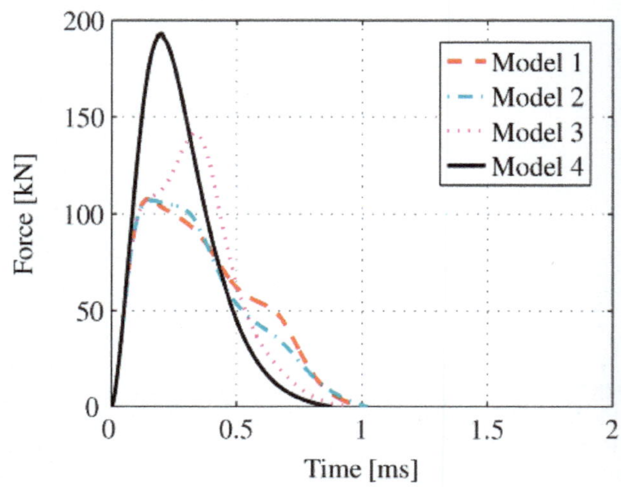

Figure 19. The arising force impulse in the four models following the discharge of the capacitor bank in the spirally-shaped coil.

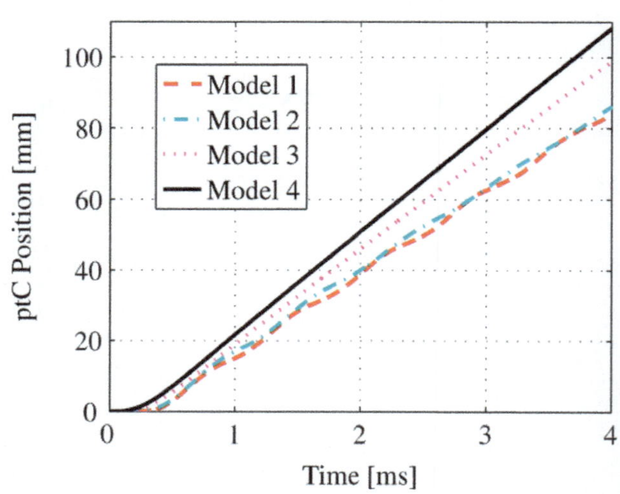

Figure 20. The displacement of the load with respect to each model.

Investigation of Using Models with Finer Segmentations

Since Model 2 gave excellent results, three more models were developed, denoted as Models 5, 6, and 7. These are hybrid models similar to Model 2, but give a more refined modeling of the pull rod by dividing it into 2, 3 and 10 spring segments; see Figure 8. Figure 21 shows a representative main displacement of the contact from all models. It can be inferred that although the complexity of the system is increased with finer segmentations of the pull rod, this does not necessarily lead to better results. The results obtained from these extra models are, however, consistently closer to those from the accurate Model 1 than the results from Models 3 and 4.

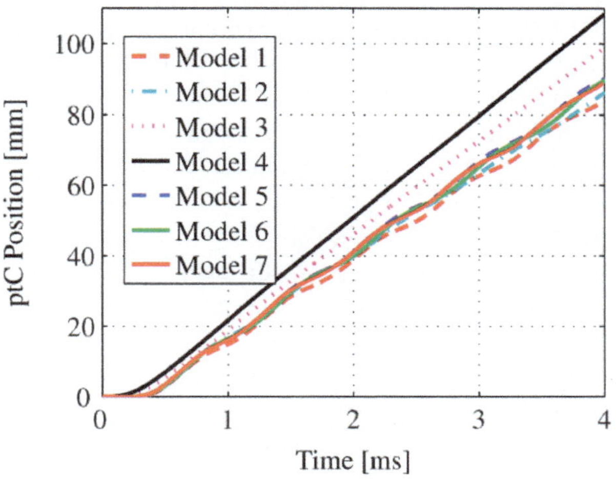

Figure 21. The displacement of the load with respect to each model. Model 2: first order hybrid model. Model 5: second order hybrid model. Model 6: third order hybrid model. Model 7: Tenth order hybrid model.

CONCLUSIONS

In an HVDC system, a breaker should be able to interrupt fault currents before they reach large values. It has been shown that the entire system, including the armature, the pull rod and the contacts, has to be taken into consideration in the design process, since this has a large influence on the magnitude of the generated force impulse and the efficiency of the actuator. From an electrical point of view, fast contact separation in the order of a couple of millimeters is crucial. With the developed multi-physics and hybrid models, the highly transient and complex magneto-mechanical interaction can be captured and modeled. The modeling can be used to build a high-end HVDC breaker with the desired specifications.

A system that is infinitely stiff will have a significantly higher efficiency, since it will fully utilize the potential of the electromagnetic drive. Deformations cause the armature to repel prematurely, causing larger air gaps and thereby limiting the build up of the force impulse before it attains its full potential. Optimizing the armature alone without consideration of the pull rod and the contacts is useless if the pull rod is allowed to elongate excessively. The breaker should be designed in such a way that deformations should be limited to fractions of a millimeter.

The developed hybrid first order multi-physics model has shown promising results and can predict the performance of the breaker with sufficient accuracy. Compared to the full multi-physics simulation model, it requires significantly less computational power and will solve the problem in a shorter period of time. Furthermore, the model does not need to be re-meshed as often, since only the armature has to be fully modeled. The increasing complexity obtained by subdividing the pull rod into several segments is shown to be of limited importance for the accuracy of the simulations. This developed model can be used to design the shape and the energizing source of the actuator to boost its efficiency. One limitation of the simplified hybrid model is that it can only be used if the push/pull rod has a linear stress-strain relationship.

Future work will be dedicated to the development of an optimization model to minimize the mass of the pull rod, while constraining its internal stresses and its elongation, to reduce the degradation of the force impulse. Furthermore, the effect of adding damping in parallel to the spring in the hybrid model will be investigated.

ACKNOWLEDGMENTS

This work was in part sponsored by EIT/InnoEnergy through the ESPE project and by SweGRIDS, the Swedish Centre for Smart Grids and Energy Storage, www.swegrids.se.

AUTHOR CONTRIBUTIONS

Ara Bissal has done all of the simulations and the experiments. Jesper Magnusson has assisted in conducting the experiments and has provided valuable feedback on the work. Göran Engdahl has supervised the electromagnetics part of this paper, being an expert in this area. Anders Eriksson, an expert in mechanics, has supervised the mechanics part of this paper. All authors have proofread the paper several times.

REFERENCES

1. Kirby, N.; Xu, L.; Luckett, M.; Siepmann, W. HVDC transmission for large offshore wind farms. Power Eng. J. 2002, 16, 135–141.
2. Wang, X.-H.; Wen, J.-L.; Peng, C.; Yao, G.-P.; Li, Y.; Zhou, J.-C.; Liu, Z.G.; Yu, K.-S. Simulation and Test Research for DC Breakers of UHVDC Transmission System. In Proceedings of the 2012 Second International Conference on Intelligent System Design and Engineering Application (ISDEA), Sanya, China, 6–7 January 2012; pp. 1480–1485.
3. Franck, C. HVDC Circuit Breakers: A Review Identifying Future Research Needs. IEEE Trans. Power Deliv. 2011, 26, 993–1007.
4. Eriksson, T.; Backman, M.; Halen, S. A Low Loss Mechanical HVDC Breaker for HVDC Grid Applications. In Proceedings of the Cigré Session, Paris, France, 24–29 August 2014.
5. Bathe, K.J.; Zhang, H.; Yan, Y. The solution of Maxwell's equations in multiphysics. Comput. Struct. 2014, 132, 99–112.
6. Bustamante, R.; Rajagopal, K.R. On a new class of electroelastic bodies. I. Proc. R. Soc. A Math. Phys. Eng. Sci. 2013, 469.

7. Bustamante, R.; Rajagopal, K.R. On a new class of electro-elastic bodies. II. Boundary value problems. Proc. R. Soc. A Math. Phys. Eng. Sci. 2013, 469.

8. Jovcic, D.; Hajian, M.; Zhang, L. DC transmission grids with fault tolerant LCL VSC converters and mechanical DC circuit breakers. In Proceedings of the 10th IET International Conference on AC and DC Power Transmission (ACDC 2012), Birmingham, UK, 4–5 December 2012; pp. 1–6.

9. Faulkner, R.; Karnes, R. Electromechanical ballistic DC breaker for use on ships. In Proceedings of the 2011 IEEE Electric Ship Technologies Symposium (ESTS), Alexandria, VA, USA, 10–13 April 2011; pp. 339–344.

10. Hafner, J.; Jacobson, B. Proactive Hybrid HVDC Breakers—A key Innovation for Reliable HVDC Grids. In Proceedings of the CIGRE, Bologna, Italy, 13–15 September 2011.

11. Meessen, K.J.; Paulides, J.J.H.; Lomonova, E.A. Analysis and design of a slotless tubular permanent magnet actuator for high acceleration applications. J. Appl. Phys. 2009, 105.

12. Lim, S.H.; Ahn, H.J.; Park, C. Study on Fault Current Limiting Characteristics of an SFCL Using Magnetic Coupling of Two Coils with Mechanical Switch Driven by Electromagnetic Repulsion Force. IEEE Trans. Appl. Superconduct. 2014, 24, 1–4.

13. Fair, H. Advances in Electromagnetic Launch Science and Technology and Its Applications.IEEE Trans. Magn. 2009, 45, 225–230.

14. Zhao, C.; He, J.; Zou, J.; Li, X.; Zhou, Z. Simulation and Optimization of the Multi-Stage Reconnection Electromagnetic Launch. IEEE Trans. Magn. 2009, 45, 506–509.

15. Cowan, M.; Cnare, E.; Duggin, B.; Kaye, R.; Tucker, T. The reconnection gun. IEEE Trans. Magn. 1986, 22, 1429–1434.

16. Steurer, M.; Frohlich, K.; Holaus, W.; Kaltenegger, K. A novel hybrid current-limiting circuit breaker for medium voltage: Principle and test results. IEEE Trans. Power Deliv. 2003, 18, 460–467.

17. Zheng, Z.; Dong, E.; Zou, J.; Zhang, Z.; Tian, P.; Chen, X. A high speed actuator for 1.14 kV rated vacuum switch. In Proceedings of the 2011 1st International Conference on Electric Power Equipment—Switching Technology (ICEPE-ST), Xi'an, China, 23–27 October 2011; pp. 661–665.

18. Tsukima, M.; Takeuchi, T.; Koyama, K.; Yoshiyasu, H. Development of a high-speed electromagnetic repulsion mechanism for high-voltage vacuum circuit breakers. Electr. Eng. Jpn. 2008, 163, 34–40.

19. Bissal, A.; Magnusson, J.; Engdahl, G. Comparison of Two Ultra-Fast Actuator Concepts.IEEE Trans. Magn. 2012, 48, 3315–3318.

20. Holzapfel, G.A. Nonlinear Solid Mechanics: A Continuum Approach for Engineering; John Wiley & Sons Ltd.: Hoboken, NJ, USA, 2000.

21. Agrawal, O.P.; Shabana, A.A. Automated visco-elastic analysis of large scale inertia-variant spatial vehicles. Comput. Struct. 1986, 22, 165–178.

CHAPTER 9 ELECTROMAGNETIC WAVE PROPAGATION IN BODY AREA NETWORKS USING THE FINITE-DIFFERENCE-TIME-DOMAIN METHOD

Jonathan N. Bringuier and Raj Mittra

Keywords: Body Area Networks, FDTD, conformal, numerical phantoms, Cole-Cole model, recursive convolution method, mutual coupling

INTRODUCTION

As wireless technologies continue to evolve, the development of personalized devices which exchange data with unprecedented ease and efficiency is certain to grow unabated. Recently, such technologies have garnered much attention from those interested in biomedical sensing and Body-Area-Networks (BANs). These technologies, which are data networks that operate through wireless transmission using the human body as a propagation medium, continue to be an active area of research in fields such as remote health monitoring and medical biosensors. However, the complexity of integrating efficient communication systems in the human body environment poses many challenges to the future of this emerging technology. The excitation of surface and space waves from radiating antennas mounted on the body can have a large impact on the performance of co-site body-centric antenna systems. Moreover, the presence of multiple antennas on the body can lead to unexpected coupling due to creeping wave interaction. Fortunately, recent developments in powerful numerical methods, such as the Finite-Difference-Time-Domain (FDTD) method running on parallel platforms, has made it feasible for us to carry out a detailed study of these body-centric antenna systems.

SIMPLE MODELS FOR BANS

For many years researchers have used simplified geometries to model the interaction of electromagnetic energy with biological tissue. Such geometries typically lend themselves to simple shapes, such as cylinders, ellipses and spheres for which past research has typically treated these as perfectly conducting objects, low loss high dielectrics, and surface impedance models. This type of treatment was convenient due to the availability of well-known analytical methods used solve these problems, such as the uniform theory of diffraction (UTD), ray tracing (RT), creeping waves, and eigenfunction analysis [1,2]. Indeed, these models have been shown in recent years to yield reasonably good results for simple cases that compare well to those derived numerically rigorously with modern CEM techniques. However, the asymptotic techniques fail to perform well when the dielectric medium they deal with is arbitrarily shaped, inhomogeneous and lossy, and such problems must be handled by using general-purpose numerical methods, such as the FDTD [3–6]. Nevertheless, these simple geometries can still provide a computationally efficient way to study the propagation around and through the human body. Therefore, these models remain useful for studying BANs.

In this section we provide the results of a simplified model for the human body torso. The proposed model is a 3-layer elliptical structure having major and minor axis of 150 cm and 120 cm, respectively. This model has been used in [7] to investigate the coupling around a 2-D ellipse using the sub-band FDTD, UTD/RT and measurement techniques at UWB frequencies. However, in that work the model was assumed to be perfectly conducting when applying the UTD/RT method while a homogeneous muscle phantom was employed for the sub-band FDTD analysis. Additionally, a conformal FDTD algorithm was not used to improve the accuracy of the curved surface whereas the author has done so in this work. Furthermore, it neither accounted for the multi-layer dielectric properties of the tissue in the human torso, nor did it simulate

Citation: Jonathan N. Bringuier and Raj Mittra, Electromagnetic Wave Propagation in Body Area Networks Using the Finite-Difference-Time-Domain Method, doi:10.3390/s120709862

the actual radiating element, i.e., viz., the monopole antenna rigorously.

The 3-layer ellipse model incorporates the skin, fat, and muscle layers. In [8] a simple 3-layer planar slab model using 3 mm skin layer, 5 mm fat layer and muscle was used to study the penetration depth of an incident plane wave for use in implantable medical devices. In this work we have used the same thickness for the skin and fat layers in the elliptical model.

Although the FDTD method enjoys a significant advantage over MoM in terms of its ability to simulate complex structures and lossy inhomogeneous materials, it has been known to require far more computational resources than are usually available to accurately model, simultaneously, the fine features of the radiating element, the layered structure of the geometry and the electrically large size of the entire structure typically encountered in the study of BANs. To circumvent this problem, most research done on BANs by utilizing the FDTD have been carried out using a point source approximation. While such an approximation may be adequate for directly estimating the path loss associated with the body, it does not rigorously account for a number of crucial antenna factors that affect the antenna performance, such as finite ground size, radiation pattern and efficiency. Furthermore, any field behavior in either the near or the intermediate region is inherently neglected in the point source approximation, whereas the physical structure of the radiating element must be included to properly model the physical system.

In this paper we have used a parallel version of FDTD that can handle such electrically large geometries as well as the fine features of the radiating element. To this end, we have used quarter-wave monopoles resonant at 2.45 GHz with a 75 mm × 75 mm square ground plane located at tangent points one to two FDTD cells away from the models analyzed in this study. Although other antennas could be used, the choice of the monopole antenna has the distinct advantage in that the radiation pattern in the plane azimuthal to the monopole is inherently stable across the bandwidth of interest. However, the monopoles used in this study do not have an impedance matching bandwidth wide enough to cover the frequencies of interest, and, therefore, the channel path loss data between observation points should be considered relative to one another for a given frequency. Nevertheless, the polarization and radiation pattern deviations of the transmitter and receivers across the frequency band can more or less be neglected when performing coupling calculations, which make these antennas useful for studying channel characteristics of the on-body propagation medium.

The setup used in the simulation is shown in Figure 1 where the receiver is located first in the source plane and then displaced vertically by 210 mm and 400 mm, respectively, from the above plane. For each observation plane the receiver is moved around the elliptical trunk model and theS_{21} is recorded. The mode of propagation is known to be a creeping wave and the direct ray paths are shown. For each observation plane the S_{21} was calculated along the elliptical path in the level plane and plotted for frequencies between 0.8–6 GHz as shown in Figures 2–4.

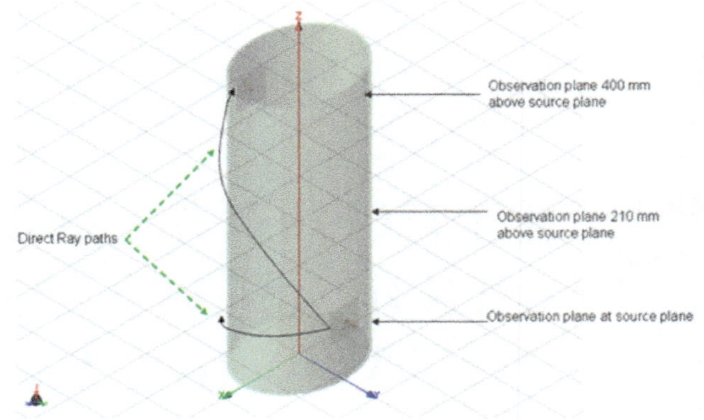

Figure 1. 3-layer ellipse model of the human torso with transmitting antenna at the front and receiving antenna at the back.

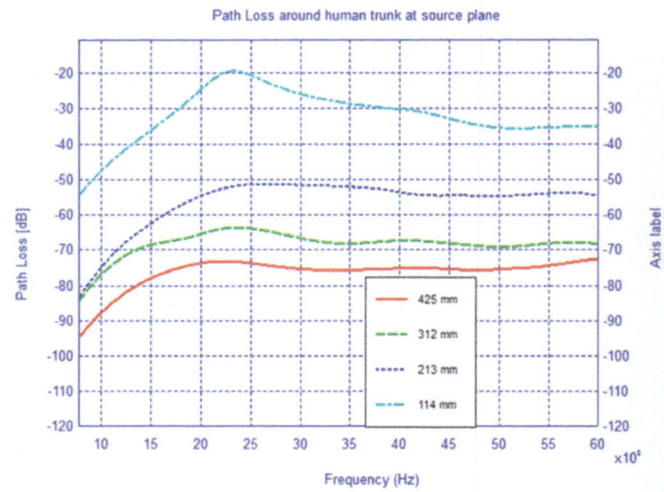

Figure 2. Path loss around the cylindrical human trunk model at the source plane.

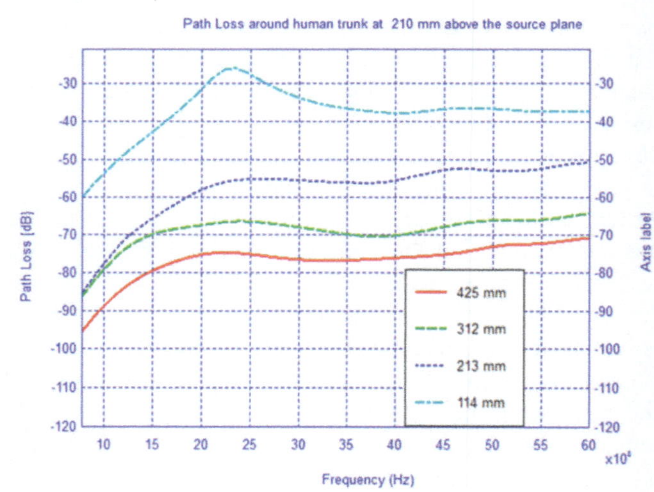

Figure 3. Path loss around the cylindrical human trunk model 210 mm above source plane.

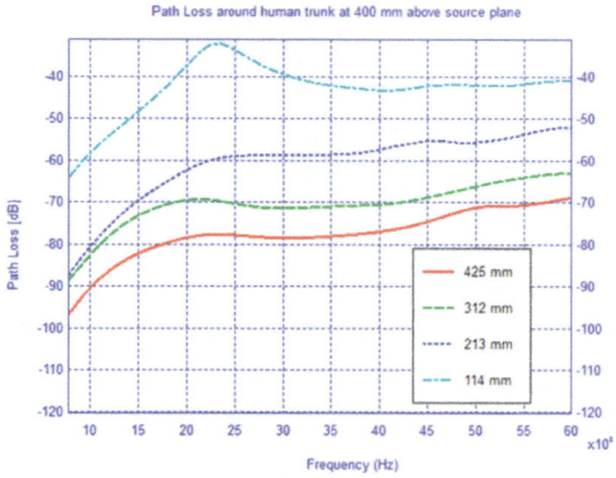

Figure 4. Path loss around the cylindrical human trunk model 400 mm above source plane.

For BANs it is typically assumed that the wave propagation through the body is so attenuated that its contribution can be ignored. To verify this we have plotted the fields for the cross-section of the ellipse in the source plane as well as a vertical cross-section along the extent of the ellipse model. The electric field plots are shown in Figures 5 and 6. From these figures it can be seen that the field magnitude inside the model is less than −100 dB and the main source of energy transfer is along the surface of the trunk in the form of a creeping wave.

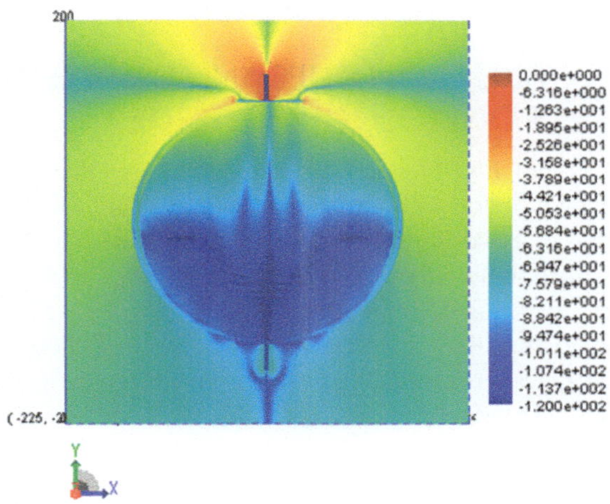

Figure 5. Electric field distribution in the source plane.

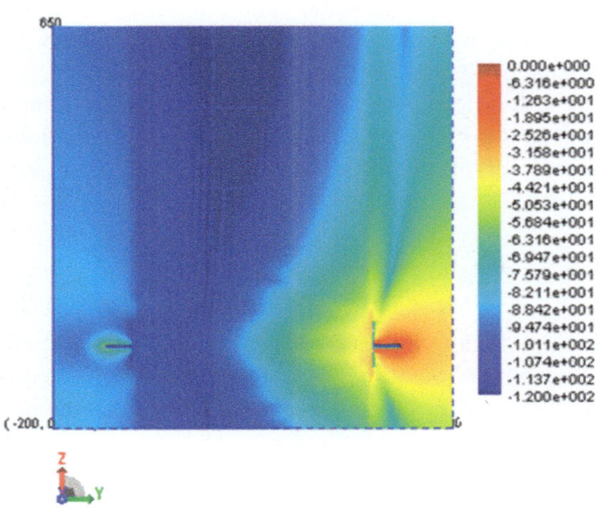

Figure 6. Electric field distribution on a vertical cut plane bisecting the cylindrical model.

In Figures 7–9 the path loss is plotted versus the distance around the elliptical model in each of the different observation planes. As shown in Figure 7, which depicts the observations made in the source plane, the attenuation at higher frequencies exhibit a clear monotonic behavior and is characteristic of a creeping wave. The results in Figures 7–9 have been found to agree closely with those in the literature for other body models [1,2,7,9–19]. These results can be useful as reference points when approximating the rate of attenuation along the trunk in body-centric network scenarios.

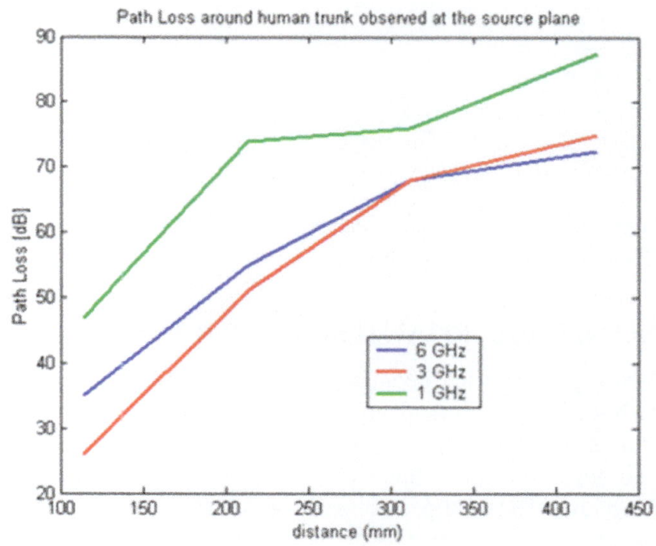

Figure 7. Path loss versus separation distance with receiving antenna at the source plane.

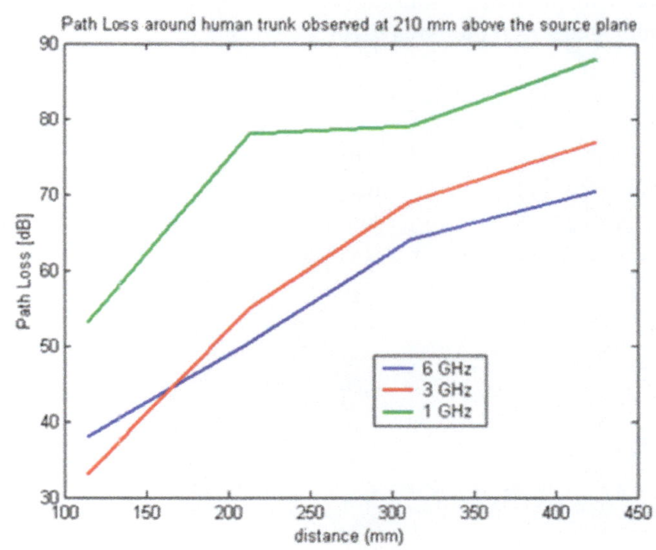

Figure 8. Path loss versus separation distance with receiving antenna 210 mm above the source plane.

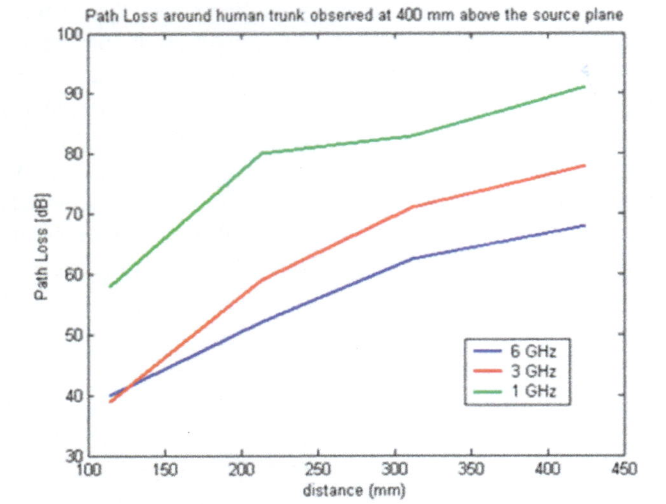

Figure 9. Path loss versus separation distance with receiving antenna 400 mm above the source plane.

NUMERICAL PHANTOMS FOR BANS

Although simplified geometries have their use in analyzing body-centric communication networks, the availability of realistic human body models enables us to use a more rigorous approach to understand how the irregular shape of the body affects the performance of BANs. One of the most commonly used body models is shown in Figure 10, where the original data set is comprised of voxels (i.e., 3D matrix data set). Furthermore, a hypothetical BAN configuration is shown with the transmitting and receiving antennas. To reduce the computational resources needed in the simulation, a lossy homogeneous model was used.

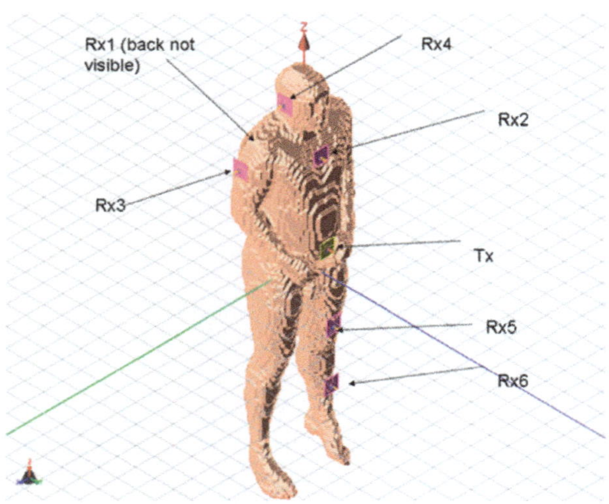

Figure 10. Numerical human body phantom based on 3D CT scan voxel set with transmitting and receiving antennas in typical BAN scenario.

Additionally, the resolution of the original voxel data set, which is 1 mm × 1 mm × 1 mm, was down-sampled to 3 mm × 3 mm × 3 mm. Even at this lower resolution the computational resources needed are high, specifically 8 hours on 16 Pentium 4 CPUs. Although this may seem too coarse for the conventional λ/20 cell size, the fields couple on or near the surface of the body, where the mesh can be coarser, and we are interested in separation distance well into the asymptotic region of the transmitting antenna. To demonstrate that the loss of accuracy is negligible in doing so, we have performed a simple experiment by computing the difference in field magnitude at 3 GHz—far away from the transmitting antenna on the same plane as shown in Figure 11. It is evident from the results that in the asymptotic region there is little loss in accuracy by utilizing a down-sampled phantom model.

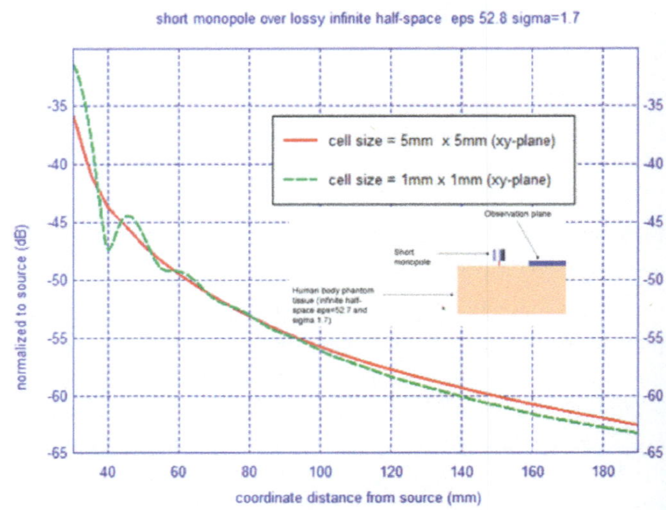

Figure 11. Experiment to determine if the down-sampled human body voxel set causes numerical inaccuracies (λ = 10cm).

DIELECTRIC PROPERTIES AND DISPERSION MODELS OF BIOLOGICAL TISSUE

It is commonly claimed in the literature that a good approximation for the body can be made by using a full muscle phantom or a two-thirds muscle equivalent phantom, the latter being more commonly used in EMC analysis for SAR [20,21]. However, realistic biological tissues exhibit dispersion characteristics that are not of the traditional Debye type used in the FDTD model, whose spectral behavior is usually represented by the form:

$$\hat{\varepsilon}(\omega)=\varepsilon_{\infty}+\sum_{n=1}^{N}\frac{\Delta\varepsilon_n}{1+j\omega\tau_n}+\frac{\sigma_n}{j\omega\varepsilon_0} \tag{1}$$

where τ_n is the relaxation time constant $\Delta\varepsilon_n$ is the spectral coefficient and σ_n is the static ionic conductivity. This model can be handled in FDTD through a simple convolution extension of the basic FDTD algorithm. In general, biological tissue exhibit a broadened spectrum which is empirically accounted for by the parameter $0 \le \alpha_n < 1$, (see Gabriel et al. [22]) and is typically referred to as the Cole-Cole model. It is given by:

$$\hat{\varepsilon}(\omega)=\varepsilon_{\infty}+\sum_{n=1}^{N}\frac{\Delta\varepsilon_n}{1+\left(j\omega\tau_n\right)^{1-\alpha_n}}+\frac{\sigma_n}{j\omega\varepsilon_0} \tag{2}$$

The difficulty in this model lies in the fact that there are not any simple closed-form expressions amenable to the FDTD algorithm. Specifically, this is due to the fact that the inverse Fourier transform is non-trivial.

Dielectic Spectrum Approximation

In this Section we will introduce a technique that can be used to implement the Cole-Cole model into the desired frequency band of interest for simple geometries relevant to BANs (e.g., planar human tissue phantoms) in FDTD. Since the static conductivity term can be handled easily in the FDTD update equations we turn our attention to the term containing α_n. First we postulate that each term in the Cole-Cole model can be rewritten as an integral representation of a Debye spectrum as follows:

$$\frac{\hat{\varepsilon}_i(\omega)}{\Delta\varepsilon_i}=\frac{1}{1+(j\omega)^{1-\alpha_i}}=\frac{1}{1+(j\omega)^{\beta_i}}=\int_0^\infty \frac{g_i(\lambda)}{\lambda+j\omega}d\lambda \tag{3}$$

Then the expression for the electric field density follows as:

$$D(\omega)=\hat{\varepsilon}_i(\omega)E(\omega) \tag{4}$$

$$=\Delta\varepsilon_i\int_0^\infty \frac{E(\omega)g_i(\lambda)}{\lambda+j\omega}d\lambda \tag{5}$$

where by the inverse Fourier transform yields:

$$D(t)=F^{-1}\{\hat{\varepsilon}_i(\omega)E(\omega)\}(t)=\int_0^\infty E(t)\otimes e^{-\lambda t}g_i(\lambda)d\lambda \tag{6}$$

$$D(t)=E(t)\otimes L\{g_i(\lambda)\}(t) \tag{7}$$

Therefore, if it is possible to find a time domain representation for the Laplace transform of the unknown spectrum $g_i(\lambda)$ and, hence, we may extend the basic convolutional FDTD to implement non-Debye types of materials.

The problem at hand is to determine a suitable function for which we can derive the Laplace transform. To do this we first rewrite the Debye integral representation as:

$$\frac{\hat{\varepsilon}(\omega)}{\Delta\varepsilon}=\int_0^\infty \frac{g(\lambda)}{\lambda+j\omega}d\lambda=\int_0^\infty g(\lambda)\int_{-\infty}^{\infty} e^{-j\omega\rho}e^{-\lambda\rho}u(\rho)d\rho d\lambda \tag{8}$$

$$=\int_0^\infty\int_0^\infty g(\lambda)e^{-\lambda\rho}e^{-j\omega\rho}d\lambda d\rho \tag{9}$$

where u(ρ) is the Heaviside step function. Now by Fubini›s Theorem:

$$\frac{\hat{\varepsilon}(\omega)}{\Delta\varepsilon}=\int_0^\infty e^{-j\omega\rho}\left[\int_0^\infty g(\lambda)e^{-\lambda\rho}d\lambda\right]d\rho \tag{10}$$

To find g(λ) we expand it in terms of some complete basis set for which we know the Laplace transform. Specifically, we chose Bessel functions of the first kind which form a complete set in $\Re:\lambda\in[0,\infty)$. Therefore, we have:

$$g(\lambda)=\sum_{n=0}^{\infty} C_n J_n(a\lambda) \tag{11}$$

where a is a fitting parameter. Obviously we must truncate this infinite series for numerical purposes; hence we write:

$$g(\lambda)\approx\sum_{n=0}^{N} C_n J_n(a\lambda) \tag{12}$$

Now we use the Laplace transform relation given by:

$$\int_0^\infty J_n(a\lambda)e^{-\lambda\rho}d\lambda=\frac{1}{\sqrt{\rho^2+a^2}}\frac{a^n}{\left(\rho+\sqrt{\rho^2+a^2}\right)^n} \tag{13}$$

Of course, we could have used Equation (1) directly in Equation (5); however, we have found that the integral converges much more slowly than that of the reformulation given in Equation (10). After inserting Equation (13) into the equation given in Equation (10) we have:

$$\frac{\hat{\varepsilon}(\omega)}{\Delta\varepsilon}=\int_0^\infty e^{-j\omega\rho}\left(\sum_{n=0}^{N} C_n\frac{1}{\sqrt{\rho^2+a^2}}\frac{a^n}{\left(\rho^2+\sqrt{\rho^2+a^2}\right)^n}\right)d\rho \tag{14}$$

Next, we can apply the method of least squares to find the coefficients C_n. However, from Equation (7) it is obvious that we must require the C_n's to be real. Therefore we have a constrained least squares system of equations given by the following:

$$\bar{b}^T=\left[\mathrm{Re}\int_0^\infty \frac{\hat{\varepsilon}(\omega)}{\Delta\varepsilon}I_0 d\omega \quad \mathrm{Im}\int_0^\infty \frac{\hat{\varepsilon}(\omega)}{\Delta\varepsilon}I_0 d\omega \quad \mathrm{Re}\int_0^\infty \frac{\hat{\varepsilon}(\omega)}{\Delta\varepsilon}I_1 d\omega \quad \mathrm{Im}\int_0^\infty \frac{\hat{\varepsilon}(\omega)}{\Delta\varepsilon}I_1 d\omega \quad \bullet \quad \bullet \quad \mathrm{Im}\int_0^\infty \frac{\hat{\varepsilon}(\omega)}{\Delta\varepsilon}I_N d\omega\right] \tag{15}$$

$$\bar{x}^T=[C_1 \quad C_2 \quad C_3 \quad \bullet \quad \bullet \quad \bullet \quad C_N] \tag{16}$$

$$\bar{\bar{A}}=\begin{bmatrix} \mathrm{Re}\int_0^\infty I_0I_0d\omega & \mathrm{Re}\int_0^\infty I_0I_1d\omega & \mathrm{Re}\int_0^\infty I_0I_2d\omega & \bullet & \bullet & \bullet & \mathrm{Re}\int_0^\infty I_0I_Nd\omega \\ \mathrm{Im}\int_0^\infty I_0I_0d\omega & \mathrm{Im}\int_0^\infty I_0I_1d\omega & \mathrm{Im}\int_0^\infty I_0I_2d\omega & \bullet & \bullet & \bullet & \mathrm{Im}\int_0^\infty I_0I_Nd\omega \\ \mathrm{Re}\int_0^\infty I_1I_0d\omega & \mathrm{Re}\int_0^\infty I_1I_1d\omega & \bullet & & & & \\ \mathrm{Im}\int_0^\infty I_1I_0d\omega & \mathrm{Im}\int_0^\infty I_1I_1d\omega & & \bullet & & & \\ \bullet & & & & \bullet & & \\ \mathrm{Re}\int_0^\infty I_NI_0d\omega & & & & & \bullet & \\ \mathrm{Im}\int_0^\infty I_NI_0d\omega & & & & & & \mathrm{Im}\int_0^\infty I_NI_Nd \end{bmatrix} \tag{17}$$

where $\bar{\bar{A}}$ is 2NxN , $\bar{b}$ is 2Nx1 , $\bar{x}$ is Nx1 , $\bar{x}=\bar{\bar{A}}^{+}\bar{b}$ and

$$I_n=\int_0^\infty \frac{e^{-j\omega\tau}a^n}{\sqrt{\tau^2+a^2}\cdot\left(\tau^2+\sqrt{\tau^2+a^2}\right)^n}d\tau \tag{18}$$

In common with most over-constrained inverse problems, we must use the Moore-Penrose pseudoinverse for $\bar{\bar{A}}^{+}$ to solve the system of equations. In practice, the integral cannot be evaluated over all possible frequencies; hence, it is typically replaced by a discrete summation over all frequencies of interest where the spectral match is desired.

As an illustrative example of this approach, the 4-term Cole-Cole model was taken for the tissue corresponding to the muscle. Although the preceding formulation corresponds to a single Cole-Cole term, the postulated Debye spectrum can equally well be applied to the entire summation of Cole-Cole terms. The parameters given by Gabriel [22] listed in Table 1.

Table 1. 4 term Cole-Cole model parameters for muscle.

$\Delta\varepsilon_1=50$	$\tau_1=7.23e-12$	$\alpha_1=0.1$
$\Delta\varepsilon_2=7000$	$\tau_2=353.68e-9$	$\alpha_2=0.1$
$\Delta\varepsilon_3=1.2e6$	$\tau_3=318.31e-6$	$\alpha_3=0.1$
$\Delta\varepsilon_4=2.5e7$	$\tau_4=2.27e-3$	$\alpha_4=0$

Using the proposed algorithm we have obtained the following coefficients for an 11th-order expansion that are listed in Table 2. In addition, we have found that a good match was achieved when the fitting parameter was chosen to be a = 0.5e − 9. Although initially the order of the expansion may seem to be large, the expansion is only 2.75 times that of the original 4-term Cole-Cole model. Figures 12–13 show the magnitude and phase spectrum match for the band of interest, specifically 1–10 GHz. Furthermore, Figure 14 shows that the relative error is less than 5 percent over the entire band, for both magnitude and phase, and this can be considered a very good match for simulation purposes.

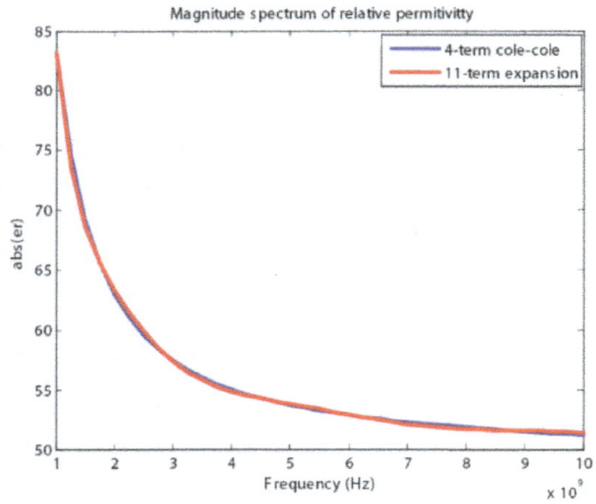

Figure 12. Magnitude spectrum of the relative permittivity for the 4 term Cole-Cole model and the spectrum approximation.

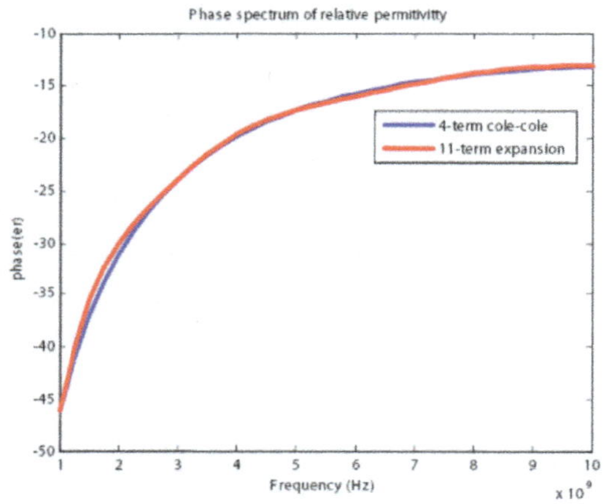

Figure 13. Phase spectrum of the relative permittivity for the 4 term Cole-Cole model and the spectrum approximation.

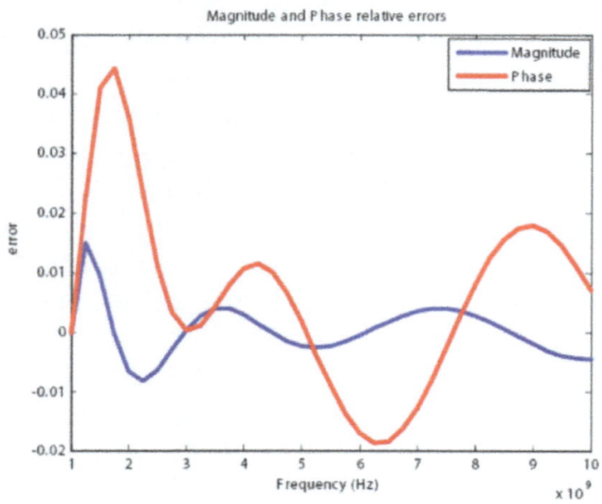

Figure 14. Magnitude and Phase relative errors of the spectrum approximation and the 4 Cole-Cole model.

Table 2. Calculated coefficients from the spectral approximation method.

Coefficients	Numerical Value
C1	0.0002e6
C2	−0.0032e6
C3	0.0394e6
C4	−0.2759e6
C5	1.1027e6
C6	−2.4578e6
C7	2.7137e6
C8	−0.5489e6
C9	−1.7652e6
C10	1.6288e6
C11	−0.4334e6

As an example, the calculated coefficients along with the fitting parameter were used to model the frequency response in a muscle medium. Using the convolution sum of Equation (7) for a Gaussian pulse excitation in the medium yields the result shown in Figure 15. It can be seen from Figure 15 that the electric flux density inside the muscle medium is influenced by its frequency domain characteristics as modeled by the proposed algorithm.

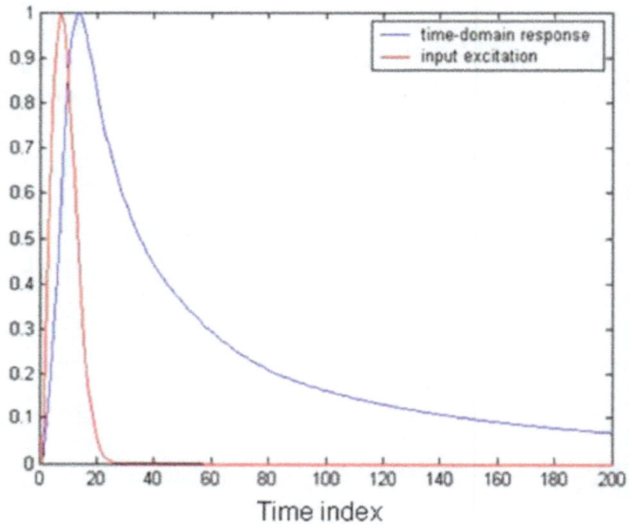

Figure 15. Normalized time-domain response of the electric flux density using the spectral approximation.

Recursive Convolution Method for Debye Materials

The problem with incorporating the spectral approximation algorithm proposed in the previous section is that FDTD requires a record-keeping of all past time values of the electric field components. If either the problem size is small, or reflection properties from an infinite planar slab are of interest then this approach may be feasible. However, for practical problems, the computational domain can become exceedingly large (e.g., upward of 1 billion unknowns for realistic human body phantoms), and other approximation methods must be used. Often times, the dispersion properties of biological tissue can be approximated by a Debye model over a finite frequency bandwidth. In this case it is possible to formulate an FDTD updating scheme for dispersive materials that does not require keeping all past histories of the fields.

The recursive convolution FDTD is one such method and is discussed in [5]. The essential features of this method are based on the Debye dispersion model given by:

$$\varepsilon(\omega) = \varepsilon_o \varepsilon_\infty + \varepsilon_o \chi(\omega) \tag{19}$$

where $\chi(\omega)$ is explicitly given as:

$$\chi(\omega) = \sum_{p=1}^{P} \frac{A_p}{\omega - W_p} \tag{20}$$

After some extensive manipulations the final form of the FDTD updating equations for the electric field become:

$$\bar{E}^{n+1} = \frac{\frac{\varepsilon_{eff}}{\Delta t} - \frac{\sigma_{eff}}{2}}{\frac{\varepsilon_{eff}}{\Delta t} + \frac{\sigma_{eff}}{2}} \bar{E}^n + \frac{1}{\frac{\varepsilon_{eff}}{\Delta t} + \frac{\sigma_{eff}}{2}} \bar{\nabla} \times \bar{H}^{n+1/2} + \frac{\frac{\varepsilon_o}{\Delta t}}{\frac{\varepsilon_{eff}}{\Delta t} + \frac{\sigma_{eff}}{2}} \sum_{p=1}^{P} \psi_p^n \tag{21}$$

and ψ_p^n is given by:

$$\psi_p^n = \bar{E}^n \Delta\chi_p^0 + e^{jW_p \Delta t} \psi_p^{n-1} \tag{22}$$

The formulation given in Equation (21) can be used to incorporate materials that exhibit Debye-type dispersive properties into the FDTD. If the geometry is simple (e.g., a dispersive sphere), this method has been shown to give improved accuracy. However, in the case of the human body, where the object has irregular features, and many dispersive dielectric layers, the conformal (CFDTD) version, combined with the recursive convolution method, can lead to unstable results. In fact, when using the parameters in Table 1 with $\alpha_i = 0$ to simulate the body shown in Figure 10, the results were found to diverge. This problem further demonstrates the difficulties when trying to rigorously simulate a realistic human body. Therefore, we have used a common approximation [21] for the human body as consisting of a non-dispersive muscle ($\varepsilon_r = 52.7$, $\sigma = 1.7$) or its two-thirds equivalent phantom in the band of 0.8–3 GHz.

SIMULATIONS AND RESULTS

Using the muscle and the two-thirds muscle equivalent human body phantom, the CFDTD algorithm was used to simulate the setup previously shown in Figure 10.

For each model, the network configuration was simulated and the S_{21} was calculated. Additionally, the network was also simulated in the absence of the body in order to understand the influence of its presence. The free-space case (body absent) is shown in Figure 16 and the results of the two body models are shown in Figures 17 and 18. To gain insight into the field distribution around the body the electric field is shown in Figure 19 at 3 GHz. It is evident from the results that the line-of-sight (LOS) or near LOS paths are not greatly influenced by the presence of the body for monopole antennas.

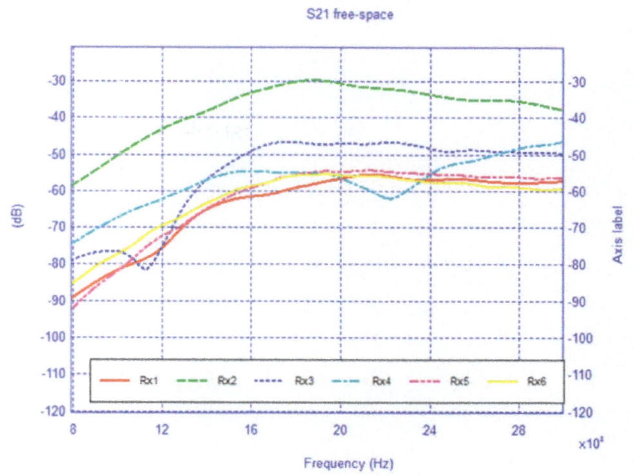

Figure 16. S_{21} of the BAN network scenario with body absent.

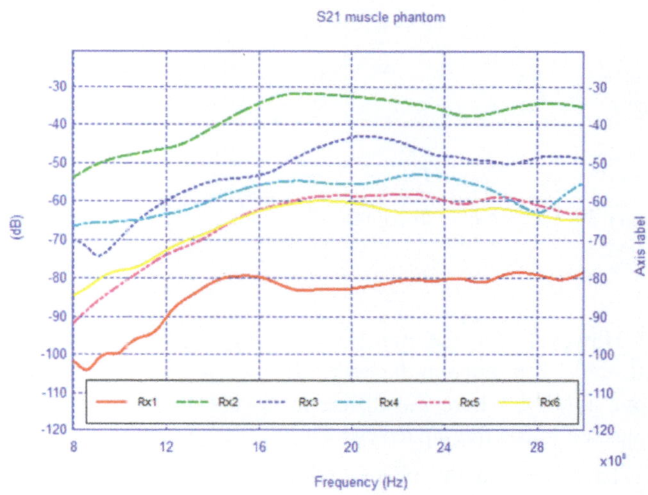

Figure 17. S_{21} of the BAN network scenario with muscle phantom.

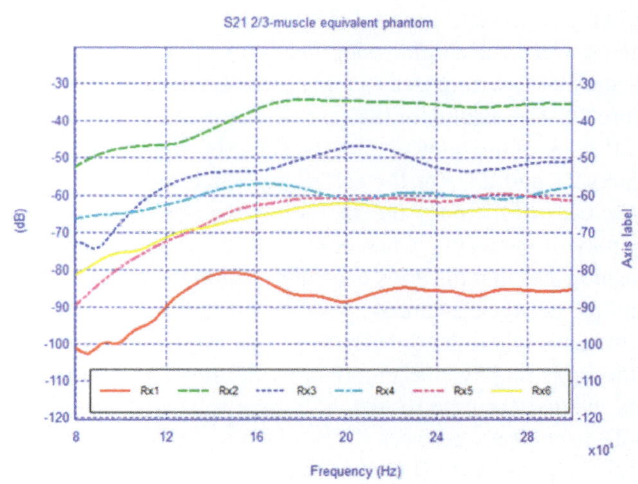

Figure 18. S_{21} of the BAN network scenario with 2/3-muscle phantom.

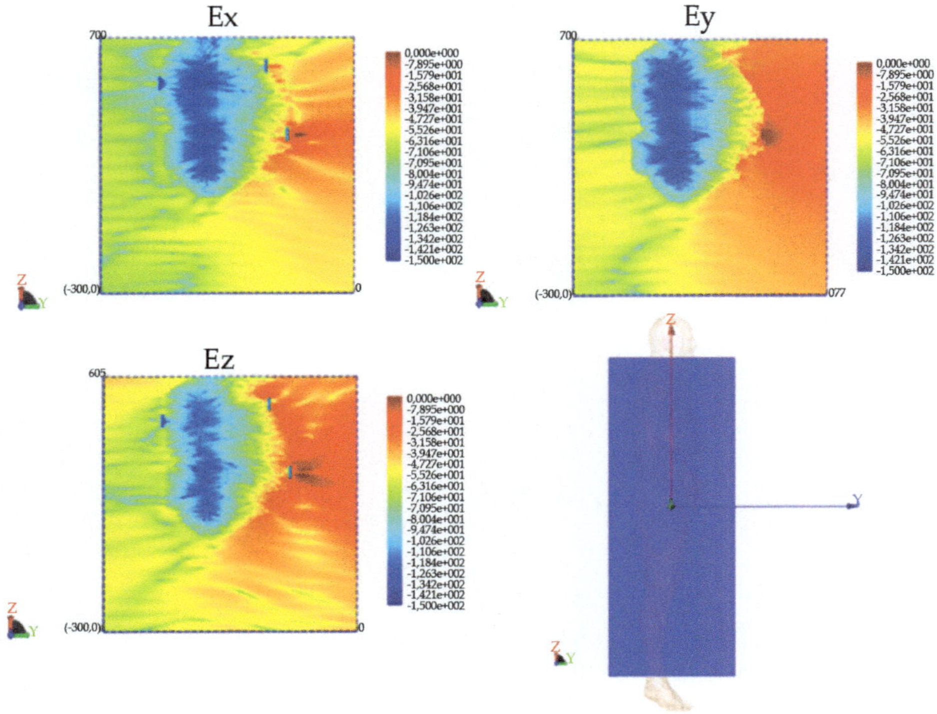

Figure 19. Electric field distributions (dB) on a vertical cut plane bisecting the body.

However, for the non-line-of-sight (NLOS) there is a noticeable difference in the path loss, especially at frequencies above 1 GHz, where the difference is approximately 35 dB compared to the free-space case.

To determine the accuracy of the elliptical model, we compare two cases, namely when the receiver is located on the back of the body and on the shoulder-side. Additionally, in order to improve the elliptical model from the previous section, two muscle tissue arms, composed of cylinders and spheres, were placed alongside the elliptical torso model and are shown superimposed on the actual physical model in Figure 20. The results for the case when the receiver is located on the back of the body and on the shoulder-side are shown in Figures 21 and 22. For the case when the receiver is located on the back of the body the addition of the arms improves the accuracy of the results when compared to the physical model. This indicates that the arms can have a significant effect on the coupling between antennas mounted on the body. However, for the case when the antenna is placed on the shoulder side of the body, neither the elliptical model nor the ellipse with arms yielded accurate results, and there was approximately a 10 dB discrepancy between the results obtained with these and the physical models. Therefore, we surmise that the addition of other scattering features such as the head, irregular arm position and shape—not exhibited in the simplified models—play a crucial role in accurately capturing the behaviors of the fields around the human body.

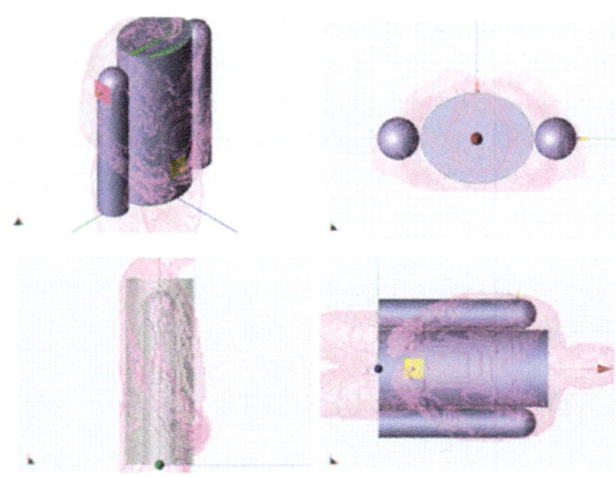

Figure 20. Simplified model superimposed on the human body numerical phantom.

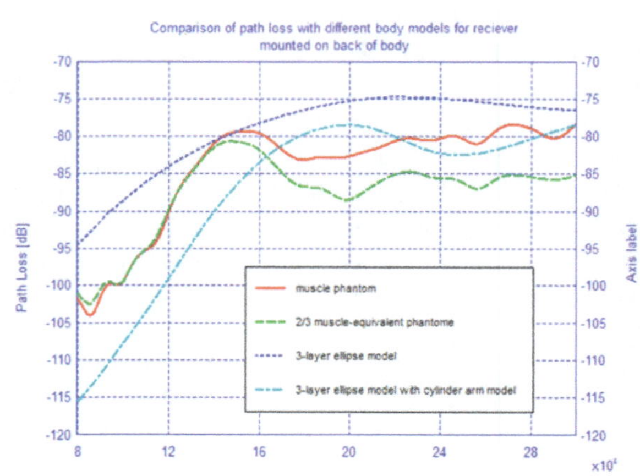

Figure 21. Comparison of the path loss for different phantom models with receiving antenna on the back of the body and transmitting antenna at the waist.

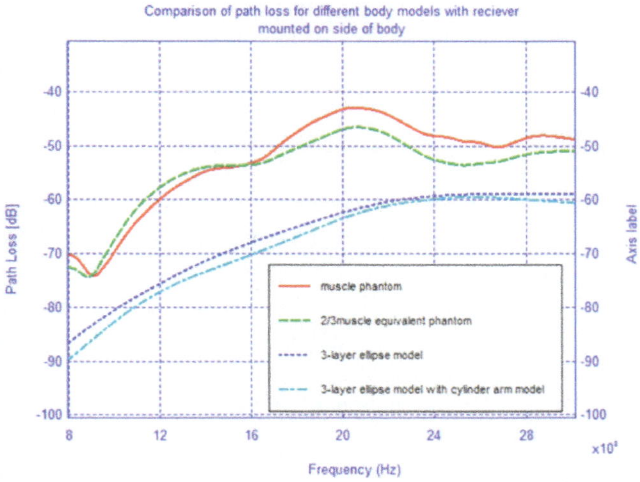

Figure 22. Comparison of the path loss for different phantom models with receiving antenna on the shoulder side of the body and transmitting antenna at the waist.

CONCLUSIONS

Despite the capability of FDTD to model the complex human body in its original form, it is often desirable to seek out good geometrical approximations to simply the computational cost. This has been documented in the literature for both measurement and simulation purposes. However, to-date a quantitative study that demonstrates the accuracy of the results using simplified models has not yet been reported. A detailed characterization of this comparison has been carried out and presented in this paper. It has been shown that the rigorous FDTD phantom model yielded results that are consistent with the measurements found in the literature, and they provide valuable information into potential cosite BAN antenna systems. In addition, we have shown simple models can give reasonably accurate results for some antenna configurations, but they can fail to reproduce the mutual coupling results when the antennas are not in line-of-sight. Therefore, it is highly desirable to carry out an in-depth study of these approximations and to see how well the results based on these approximations compare to the rigorous simulation of the human body.

REFERENCES

1. Alves, T.; Poussot, B.; Laheurte, J.-M. Analytical propagation modeling of BAN channels based on the creeping-wave theory. IEEE Trans. Antennas Propag. 2011, 59, 1269–1274.
2. Ma, D.; Zhang, W.X. Analytic propagation model for wireless body-area networks. IEEE Trans. Antennas Propag. 2011, 59, 4749–4756.
3. Dumanli, S.; Railton, C.J. Analysis of coupled tilted slot antennas in FDTD using a novel time domain huygens method with application to body area networks. Trans. Antennas Propag. 2012, 60, 1987–1994.
4. Eid, A.M.; Wallace, J.W. Accurate modeling of body area network channels using surface-based method of moments. Trans. Antennas Propag. 2011, 59, 3022–3030.
5. Yu, W.; Mittra, R.; Su, T.; Liu, Y.; Yang, X. Parallel Finite-Difference-Time-Domain Method; Artech House Inc.: Norwood, MA, USA, 2006.
6. Taflove, A.; Hagness, S.C. Computational Electrodynamic—The Finite Difference Time-Domain Method, 3rd ed.; Artech House Inc.: Norwood, MA, USA, 2005; pp. 415–417.
7. Zhao, Y.; Hao, Y.; Alomainy, A.; Parini, C. UWB on-body radio channel modeling using ray theory and subband FDTD method. IEEE Trans. Microwave Theory Tech. 2006, 54, 1827–1835.
8. Hall, P.S.; Hao, Y. Antennas and Propagation for Body Centric Communications Systems; Artech House: Norwood, MA, USA, 2006; pp. 241–255.
9. Hall, P.S.; Hao, Y.; Nechayev, Y.I.; Alomalny, A.; Constantinou, C.C.; Parini, C.; Kamarudin, M.R.; Salim, T.Z.; Hee, D.T.M.; Dubrovka, R.; et al. Antennas and propagation for on-body communication systems. IEEE Antennas Propag. Mag. 2007, 49, 41–58.
10. Hall, P.S.; Hao, Y. Wearable Antennas for Body Area Networks. In Microstrip and Printed Antennas: New Trends, Techniques and Applications; John Wiley & Sons: Hoboken, NJ, USA, 2011; pp. 183–210.
11. Gallo, M.; Hall, P.S.; Bai, Q.; Nechayev, Y.I.; Constantinou, C.C.; Bozzetti, M. Simulation and measurement of dynamic on-body communication channels. IEEE Trans. Antennas Propag. Feb. 2011, 59, 623–630.
12. Alomainy, A.; Hao, Y.; Hu, X.; Parini, C.G.; Hall, P.S. UWB on-body radio propagation and system modelling for wireless body-centric networks. IEE Proc. Commun. 2006, 153, 107–114.
13. Kamarudin, M.R.; Nechayev, Y.; Hall, P.S. Performance of Antennas in the on Body Environment. Proceedings of the IEEE International Symposium on Antennas and Propagation, Washington, DC, USA, 3– 8 July 2005.
14. Kim, J.; Rahmat-Samii, Y. Implanted antennas inside a human body: Simulations design, and characterization. IEEE Trans. Microw. Theory Tech. 2004, 52, 1934–1943.
15. Mittra, R.; Bringuier, J.; Wiart, J.; Yoo, K. Modeling of Interaction Between Body-Mounted Antennas. Proceedings of the First European Conference on Antennas and Propagation EuCAP 2006, Nice, France, 6– 10 November 2006.
16. Scanlon, W.G.; Evans, N.E. Body-Surface Mounted Antenna Modelling for Biotelemetry Using FDTD with Homogeneous, Two- and Three-Layer Phantoms. Proceedings of the 10th International Conference on Antennas and Propagation, Edinburgh, Scotland, 14– 17 April 1997.
17. Zasowski, T.; Althaus, F.; Stager, M.; Wittneben, A.; Troster, G. UWAB for Noninvasive Wireless Body Area Networks: Channel Measurements and Results. Proceedings of the IEEE Conference on Ultra Wideband Systems and Technologies, Reston, VA, USA, 16–19 November 2003.
18. Ryckaert, J.; de Doncker, P.; Meys, R.; de Le Hoye, A.; Donnay, S. Channel model for wireless communication around human body. Electron. Lett. 2004, 40, 543–544.
19. Alomainy, A.; Hao, Y.; Owadally, A.; Parini, C.I.; Nechayev, Y.; Hall, P.S.; Constantinou, C.C. Statistical

analysis and performance evaluation for on-body radio propagation with microstrip patch antennas. IEEE Trans. Antennas Propag. 2007, 55, 245–248.

20. Furse, C.M.; Chen, J.-Y.; Gandhi, O.P. The use of the frequency-dependent finite-difference time-domain method for induced current and SAR calculations for a heterogeneous model of the human body. IEEE Trans. Electromagn. Compat 1994, 36, 128–133.
21. Gandhi, O.P.; Gao, B.-Q.; Chen, J.-Y. A frequency-dependent finite-difference time-domain formulation for general dispersive media. IEEE Trans. Microw. Theory Tech. 1993, 41, 658–665.
22. Gabriel, S.; Lau, R.W.; Gabriel, C. The dielectric properties of biological tissues: III parametric models for the dielectric spectrum of tissues. Phys. Med. Biol. 1996, 41, 2271–2293.
23. Abbasi, Q.H.; Sani, A.; Alomainy, A.; Hao, Y. On-body radio channel characterization and system-level modeling for multiband OFDM ultra-wideband body-centric wireless network.IEEE Trans. Microw. Theory Tech. 2010, 58, 3485–3492.
24. Khan, M.M.; Abbasi, Q.H.; Alomainy, A.; Hao, Y. Radio Propagation Channel Characterisation Using Ultra Wideband Wireless Tags for Body-Centric Wireless Networks in Indoor Environment. Proceedings of International Workshop on Antenna Technology, Hong Kong, China, 7–9 March 2011.
25. Tuovinen, T.; Yazdandoost, K.Y.; Iinatti, J. Monopole Ultra Wideband antenna for on-body communication in Wireless Body Area Network. Proceedings of the Antennas and Propagation Conference (LAPC), Loughborough, UK, 14–15 November 2011.
26. Zhang, Y.P.; Li, Q. Performance of UWB impulse radio with planar monopoles over on-human-body propagation channel for wireless body area networks. IEEE Trans. Antennas Propag. 2007, 55, 2907–2914.
27. Fort, A.; Desset, C.; Ryckaert, J.; de Doncker, P.; van Biesen, L.; Donnay, S. Ultra Wide-band Body Area Channel Model. Proceedings of the IEEE International Conference Communication, Seoul, Korea, 16– 20 May 2005; Volume 4. pp. 2840–1844.
28. Hao, Y.; Alomainy, A.; Song, W.; Parini, C.G.; Hall, P.; Nechayev, Y.; Constantinou, C. Numerical Modeling of On-body Radio Propagation Channel. Presented at the IEEE AP-S International Symposium, Washington, DC, USA, 3– 8 July 2005.
29. Alomainy, A.; Hao, Y.; Parini, C.; Hall, P. Comparison between two different antennas for UWB on-body propagation measurements. IEEE Antennas Wirel. Propag. Lett. 2005, 4, 31–34.
30. Kamarudin, M.; Nechayev, Y.; Hall, P. Antennas for On-body Communication Systems. Proceedings of the IEEE International Workshop Antenna Technology, Singapore, 7– 9 March 2005; pp. 17–20.
31. Nechayev, Y.I.; Hall, P.S.; Constantinou, C.C.; Hao, Y.; Alomainy, A.; Dubrovka, R.; Parini, C. Antennas and Propagation for On-Body Communication Systems. Proceedings of the 11th International Symposium. on Antenna Technology and Applied Electromagnetics ANTEM, Saint Malo, France, 15–17 June 2005.
32. Nechayev, Y.I.; Hall, P.S.; Constantinou, C.C.; Hao, Y.; Alomainy, A.; Dubrovka, R.; Parini, C. On-Body Path Gain Variations with Changing Body Posture and Antenna Position. Proceedings of the IEEE International Symposium on Antennas and Propagation, Washington, DC, USA, 3– 8 July 2005.

CHAPTER 10

COMPUTATIONAL METHODS FOR COUPLED FLUID-STRUCTURE-ELECTROMAGNETIC INTERACTION MODELS WITH APPLICATIONS TO BIOMECHANICS

Felix Mihai, Inja Youn, Igor Griva and Padmanabhan Seshaiyer

INTRODUCTION

In the last decade, the rapid development of computational science has provided new methodologies to solve complex multiphysics applications involving fluid-structure interaction to a variety of fields. These include solving applications involving blood flow interactions with the arterial wall to computational aeroelasticity of flexible wing micro-air vehicles to magnetohydrodynamic of liquid-metal cooled nuclear reactor to ferromagnetics with biological applications. In these applications, the challenge is to understand and develop algorithms that allow the structural deformation, the flow field, and temperature variations to interact in a highly nonlinear fashion.

Coupling these multiphysics with electromagnetic effects makes the associated computational model too complex. Not only is the nonlinearity in the geometry challenging but in many of these applications the material is nonlinear as well, which makes the problem even more complex. Direct numerical solution of the highly nonlinear equations governing even the most simplified two-dimensional models of such multiphysics interaction requires that all the unknown fields, such as fluid velocity, pressure, the magnetic and the electric field, the temperature field, and the domain shape, be determined as part of the solution, since neither is known a priori.

The past few decades, however, have seen significant advances in the development of finite element and domain decomposition methods. These have provided new algorithms for solving such large scale multiphysics simulations. There have been several methods that have been introduced in this regard and their performance has been analyzed for a variety of problems. One such technique is the mortar finite element method which has been shown to be stable mathematically and has been successfully applied to a variety of applications and references therein. The basic idea is to replace the strong continuity condition at the interfaces between the different subdomains modeling different multiphysics by a weaker one to solve the problem in a coupled fashion. Such novel techniques provide hope for us to develop new faster and efficient algorithms to solve complex multiphysics applications. A variety of methods have been introduced including the level set methods [1], fictitious domain methods [2, 3], nonconforming hp finite element methods [4, 5], multilevel multigrid methods [6], and the immersed boundary methods [7]. While these methods help enhance our ability to understand complex processes, there is still a great need for efficient computational methods that cannot only help simulate physiologically realistic situations qualitatively but also analyze and study modeling of such processes quantitatively. Such multiphysics applications involve the interaction of various components, such as fluid with the structure, electromagnetics with the fluid, or fluid-structure interacting completely with electromagnetics.

Citation: Mihai, Inja Youn, Igor Griva, and Padmanabhan Seshaiyer, "Computational Methods for Coupled Fluid-Structure-Electromagnetic Interaction Models with Applications to Biomechanics," Mathematical Problems in Engineering, vol. 2015, Article ID 253179, 10 pages, 2015. doi:10.1155/2015/253179.

Electromagnetic-Fluid Interaction

An important application involving interaction of electromagnetics with fluid which describes the behavior of electrically conducting fluid is very complex under a magnetic field, since the additional Lorentz force is caused by the interaction between velocity field and electromagnetic field. Understanding such coupled behavior not only helps us to create efficient algorithms but also applies to a variety of magnetohydrodynamic (MHD) applications. Due to its multidisciplinary applications, a solid understanding of the MHD is required. In this regard, the Hartmann flow has been studied extensively. The Hartmann flow is the steady flow of an electrically conducting fluid between two parallel walls, under the effect of a normal magnetic and electric field. A thorough understanding of such models for electromagnetic fluid interaction can help us in developing new techniques for complex problems such as magnetic drug targeting in cancer therapy. Such a model would involve ferrohydrodynamics of blood that helps to study external magnetic field and its interaction with blood flow containing a magnetic carrier substance. The analytic models would involve solving Maxwell's equations in conjunction with Navier-Stokes equations. While new models in this area are just starting to evolve, these often consider the structure to be fixed. There is a need to extend these models to include fluid-structure interaction with electromagnetics, which would be another focus of this work.

Proposed New Models

In this paper, we will develop a computational infrastructure for solving coupled fluid-structure interaction with electromagnetic and temperature effects. The rest of the work is organized as follows. Section 2 presents the models, methods, and background required to develop and solve the coupled multiphysics systems. In Section 3, we consider the model of a blood vessel, a permanent magnet, and surrounding tissue and air in two dimensions. We will consider both a nonmoving structure and a moving structure. The deformed structure provides a new geometry, where the Navier-Stokes equations are solved for the velocity and pressure fields in the bloodstream. A magnetic vector potential generated by the permanent magnet is calculated, which in turn creates a magnetic volume force that impacts the flow in the blood vessel. The flow field changes the displacement of the structure, and the problem is solved once again for the new geometry. The proposed models are validated against benchmark applications numerically. Section 4 presents conclusion and a discussion of the results. Future work on the proposed problems is also presented.

A magnetically targeted drug delivery system [8] is based on magnetic particles under the action of an external magnetic field. This is becoming an increasingly effective approach in drug therapy. As this field has evolved in the last decade, lots of scientific interest led to this inquiry into efficient computational models that simulate this experimental process [9]. Our study indicates that the strong magnetic fields which aid in targeted drug delivery can impact not only fluid (blood) circulation but also the displacement of arterial walls. Thus, it is important to have a model, which includes the interactions between fluid, structure, and magnetic field in order to study and optimize drug delivery.

In this section, we will present a model that describes the interaction between these three components, which previously received little to no attention in the scientific community. To develop an electromagnetic fluid-structure interaction, we incorporate the effects of the electromagnetic field into a fluid-structure model. Gaining a thorough understanding of such a coupled model can help us to understand the efficacy of magnetic nanoparticle-based drug delivery for diseases such as cancer as has been proposed by various researchers [10, 11]. There is significant evidence that indicates a need for more promising models which overcome current limitations and improve magnetic targeting technique.

MATHEMATICAL MODEL AND GOVERNING EQUATIONS

The model we consider is a blood vessel with a permanent magnet near its surface, as illustrated in Figure 1. For simplicity of presentation, we consider a computational model that comprises three components. Let the computational domain $\Omega \subset \mathbf{R}^2$ be an open set with global system boundary Γ. Let Ω be decomposed into the four disjoint open sets, a fluid subdomain Ω_f denoted by blood flow, two solid subdomains $\Omega_s^i, i = 1,2$, (blood vessel walls) with respective boundaries Γ_f and Γ_s, and one electromagnetic domain Ω_m (permanent magnet). Let $\Gamma_I^j, i = 1,2,3,4$, be the interface between the solid, fluid, and electromagnetic domains. The structural domain consists of two symmetric arterial vessel walls denoted by Ω_s^1 and Ω_s^2. The electromagnetic domain consists of a permanent magnet of dimensions 10 μm×40 μm placed in free space. The arterial wall describes a structural mechanism that interacts with the flow dynamics of blood which in turn is impacted by a permanent magnet, which is described next.

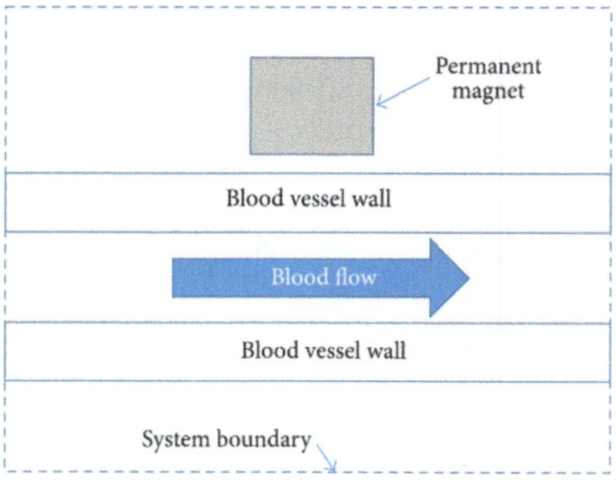

Figure 1: Electromagnetic fluid-structure interaction model.

For this, we use Maxwell's equation for the magnetostatic case (the field quantities do not vary with time) that relates the magnetic field intensity H and the electric current density J [12]:

$$\nabla \times \mathbf{H} = \mathbf{J},$$
$$\nabla \cdot \mathbf{J} = 0. \tag{1}$$

The constitutive relations between B and H depend on the domain [12, 13]:

$$\mathbf{B} = \begin{cases} \mu_0\mu_{r,\text{mag}}\mathbf{H} + \mathbf{B}_{\text{rem}} & \text{for the permanent magnet} \\ \mu_0\left(\mathbf{H} + \mathbf{M}_{ff}(\mathbf{H})\right) & \text{for the blood stream} \\ \mu_0\mathbf{H} & \text{for the tissue and air,} \end{cases} \tag{2}$$

where μ_0 is the magnetic permeability of vacuum (V·s/(A·m)), $\mu_{r,\text{mag}}$ is the relative magnetic permeability of the permanent magnet (dimensionless), B_{rem} is the remanent magnetic flux (A/m), and $\mathbf{M}_{ff}$ is the magnetization vector in the blood stream (A/m), which is a function of the magnetic field, H. By defining a magnetic vector potential A such that

$$\mathbf{B} = \nabla \times \mathbf{A}, \quad \text{with } \nabla \cdot \mathbf{A} = 0, \tag{3}$$

we get

$$\nabla \times \left(\frac{1}{\mu}\nabla \times \mathbf{A} - \mathbf{M}\right) = \mathbf{J}. \tag{4}$$

Assuming no perpendicular currents, we can simplify to a 2D problem and reduce this equation to

$$\nabla \times \left(\frac{1}{\mu_0}\nabla \times \mathbf{A} - \mathbf{M}\right) = \mathbf{0}. \tag{5}$$

This assumes that the magnetic vector potential has a nonzero component only perpendicularly to the plane, which is A = (0, 0, A_z). The induced magnetization $\mathbf{M}_{ff}(x, y) = (M_{ffx}, M_{ffy})$ is characterized by [14–17]

$$\mathbf{M}_x = \alpha \arctan\left(\frac{\beta}{\mu_0}\frac{\partial A_z}{\partial y}\right),$$
$$\mathbf{M}_y = \alpha \arctan\left(\frac{\beta}{\mu_0}\frac{\partial A_z}{\partial x}\right). \tag{6}$$

To capture the magnetic fields of interest we can linearize these expressions to obtain

$$\mathbf{M}_x = \frac{\chi}{\mu_0}\frac{\partial A_z}{\partial y}, \qquad \mathbf{M}_y = \frac{\chi}{\mu_0}\frac{\partial A_z}{\partial x}, \tag{7}$$

where $\chi = \alpha\beta$ is the magnetic susceptibility. This magnetic field induces a body force on the fluid. With the assumption that the magnetic nanoparticles in the fluid do not interact, the magnetic force F = (F_x, F_y) on the ferrofluid for relatively weak fields is given by [16]

$$F = |M|\, \nabla\, |H|\,. \tag{8}$$

Substituting (2) and (3) in (8) leads to the expression

$$F_x = k_{ff}\frac{\chi}{\mu_0\mu_r^2}\left(\frac{\partial A_z}{\partial x}\frac{\partial^2 A_z}{\partial x^2} + \frac{\partial A_z}{\partial y}\frac{\partial^2 A_z}{\partial x \partial y}\right),$$
$$F_y = k_{ff}\frac{\chi}{\mu_0\mu_r^2}\left(\frac{\partial A_z}{\partial x}\frac{\partial^2 A_z}{\partial x \partial y} + \frac{\partial A_z}{\partial y}\frac{\partial^2 A_z}{\partial y^2}\right), \tag{9}$$

where k_{ff} is the fraction of the fluid which is ferrofluid. The vector F_f = (F_x, F_y) is the volume force, which is input for the Navier-Stokes equations in the next subsection.

Modeling the Unsteady Blood Flow

We model the fluid domain for the blood flow via the unsteady Navier-Stokes equations for an incompressible, isothermal fluid flow written in nonconservative form as

$$\rho_f\frac{\partial u_f}{\partial t} + \rho_f\left(u_f \cdot \nabla\right)u_f + \nabla p = \nabla \cdot \tau_f + F_f,$$
$$\rho_f \nabla \cdot u_f = 0, \tag{10}$$

where u_f is the velocity, ρ_f is the density, p is the pressure, and F_f is the body forces. The viscous stress tensor is $(u_f) = 2\eta(u_f)$,, where η is the dynamic viscosity and the deformation tensor is

$$D(u_f) = \mu_s\left(\frac{\nabla u_f + (\nabla u_f)^T}{2}\right) \tag{11}$$

The fluid equations are subject to the boundary conditions:

$$u_f = u_{\text{wall}}, \quad x \in \Gamma_I^j, \quad j = 2, 3$$
$$\tau_f \cdot n = t \cdot n, \quad x \in \Gamma_N,$$
$$u_f = \frac{\partial d_s}{\partial t} \qquad x \in \Gamma_I^j, \quad j = 2, 3, \tag{12}$$

where $t = -pI + 2D(u_f)$ is the prescribed tractions on the Neumann part of the boundary with n being the outward unit normal vector to the boundary surface of the fluid. Conditions of displacement compatibility and force equilibrium along the structure-fluid interface are enforced. In order to solve a fluid-structure interaction problem in a coupled fashion we employ an arbitrary Lagrangian-Eulerian (ALE) formulation where the characterizing velocity is no longer the material velocity u_f, but a grid velocity $\widehat{u}_f$. This allows us to replace the material velocity u_f in (10) with the convective velocity $c = u_f - \widehat{u}_f$ [5]. The weak variational formulation of the fluid problem then becomes

$$\int_{\Omega_f} \tau_f \cdot \nabla\phi \, d\Omega_f$$
$$= \int_{\Omega_f} F \cdot \phi \, d\Omega_f + \int_{\Gamma_f} t \cdot \phi \, d\Gamma$$
$$+ \int_{\Omega_f} \rho_f \frac{\partial u}{\partial t} \cdot \phi \, d\Omega_f + \int_{\Omega_f} \rho_f (c \cdot \nabla) u_f \cdot \phi \, d\Omega_f,$$
$$\int_{\Omega_f} q\nabla \cdot u \, d\Omega_f = 0. \tag{13}$$

Modeling the Structure Equations

The structural domains for the blood vessel walls consist of the arterial vessel walls denoted by Ω_s^1, Ω_s^2. They are modeled via the following equation:

$$\rho_s\frac{\partial^2 d_s}{\partial t^2} = \nabla \cdot \tau_s + F_s, \tag{14}$$

where d_s is the structure displacement, ρ_s is the structure density, τ_s is the solid stress tensor, and $\partial^2 d_s/\partial t^2$ is the local acceleration of the structure. This is solved with the boundary conditions:

$$d_s = d_s^D \quad x \in \Gamma_S^D,$$

$$\tau_s \cdot n_s = t_s \quad x \in \Gamma_S^N,$$

$$\tau_s \cdot n_s = -\tau \cdot n + t_S^I \quad x \in \Gamma_I^j \quad j = 2, 3. \tag{15}$$

Here Γ_S^D and Γ_S^N are the respective parts of the structural boundary where the Dirichlet and Neumann boundary conditions are prescribed. Also, t_s are the applied tractions on Γ_S^N and t_S^I are the externally applied tractions to the interface boundaries $\Gamma_I^j,\ j = 1,2,3,4.$. The unit outward normal vector to the boundary surface of the structure is n_s. The stresses are computed using the constitutive relation described next. Equations (15) enforce the equilibrium of the traction between the fluid and the structure on the respective fluid-structure interfaces. The total strain tensor for a typical geometrically nonlinear model is written in terms of displacement gradients:

$$\varepsilon = \frac{1}{2}\left(\nabla d_s + \nabla d_s^T + \nabla d_s \nabla d_s^T\right). \tag{16}$$

For small deformations, the last term on the right hand side is omitted to obtain a geometrically linear model. Since the objective of this section is to investigate the influence of electromagnetic effects on fluid-structure interaction models, we will consider a geometrically linear model combined with a linear constitutive law. The solid stress tensor τ_s is given in terms of the second Piola-Kirchoff stress S:

$$\tau_s = \left(S \cdot (I + \nabla d_s)\right). \tag{17}$$

For the linear material model, we employ the following constitute law relating the stress tensor to the strain tensor:

$$S = S_0 + C : \varepsilon, \tag{18}$$

where C is the 4th order elasticity tensor and ":" stands for the double-dot tensor product. S_0 and ε_0 are initial stresses and strains, respectively. The weak variational form of the structural equations then becomes the following: find the structure displacement u_s such that

$$\int_{\Omega_f} \tau_s \cdot \varepsilon_s \, d\Omega_s$$

$$= \int_{\Omega_f} F_s \cdot \phi_s \, d\Omega_s + \int_{\Gamma_S^N} t_s \cdot \phi_s \, d\Gamma$$

$$- \int_{\Omega_f} \rho_s \frac{\partial^2 d_s}{\partial t^2} \cdot \phi_s \, d\Omega_s - \int_{\Gamma_I} \left(t_S^I - \tau_f \cdot n\right) \cdot \phi_s \, d\Gamma. \tag{19}$$

NUMERICAL RESULTS

In this section, we present the numerical results for the electromagnetic-fluid-structure interaction model problem presented in this section. To understand the effects of the coupling between electromagnetic field and fluid-structure interaction models better, we first consider the interaction with a rigid structure, which is often employed in the most research problems that are only interested in studying the electromagnetic-fluid interaction. The computational domain (see Figure 2) represents a blood vessel that is 300 micrometers long and 100 micrometers in diameter, with walls 20 micrometers in thickness. All the results presented are for three magnetic fields: 0T (no magnetic field), 0.5 T, and 1T. The structure model we consider is linear (MLGL), which was introduced in Section 2.

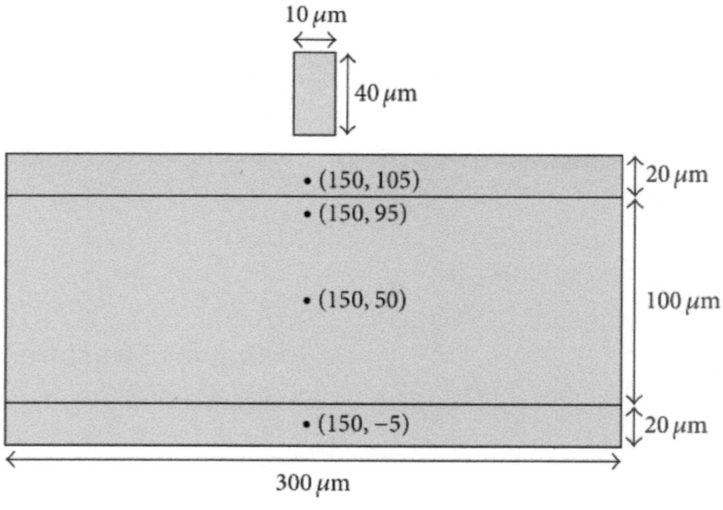

Figure 2: Domain and points of interest.

Coupled Interaction with Rigid Structure

Figures 3(a), 4(a), and 5(a) illustrate the influence of the magnetic field on the interaction. These figures show the surface von Mises stress along with streamlines of spatial velocity field and the z-component of the magnetic vector potential. While there is no significant impact of increasing the magnetic field on the velocity profile in each of the graphs in Figures 3(a), 4(a), and 5(a), the impact on the magnetic vector potential is as expected. As it can be seen, the z-component of the magnetic potential doubles when magnetic field doubles.

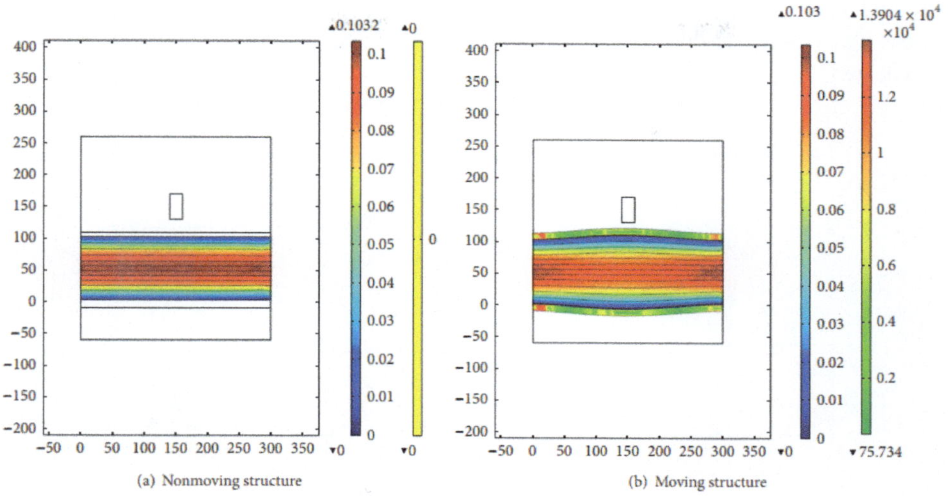

Figure 3: Surface von Mises stress with streamlines of spatial velocity field and magnetic field for B_{rem} =0T at t=0.215. Time = 0.215, surface: von Mises stress (N/m^2), surface: velocity magnitude (m/s), and streamline: velocity field (spatial).

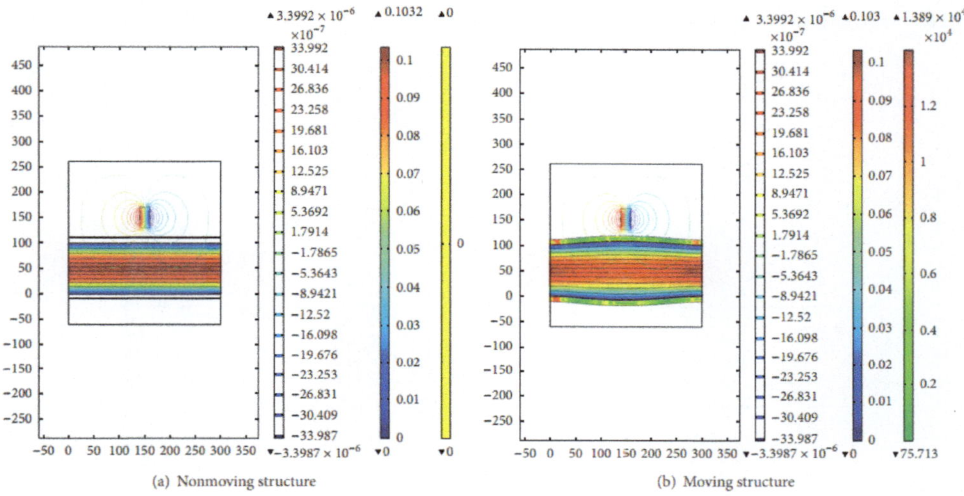

Figure 4: Surface von Mises stress with streamlines of spatial velocity field and magnetic field for B_{rem} = 0.5 T at t = 0.215. Time = 0.215, surface: von Mises stress (N/m^2), surface: velocity magnitude (m/s), contour: magnetic vector potential, z component (Wb/m), and streamline: velocity field (spatial).

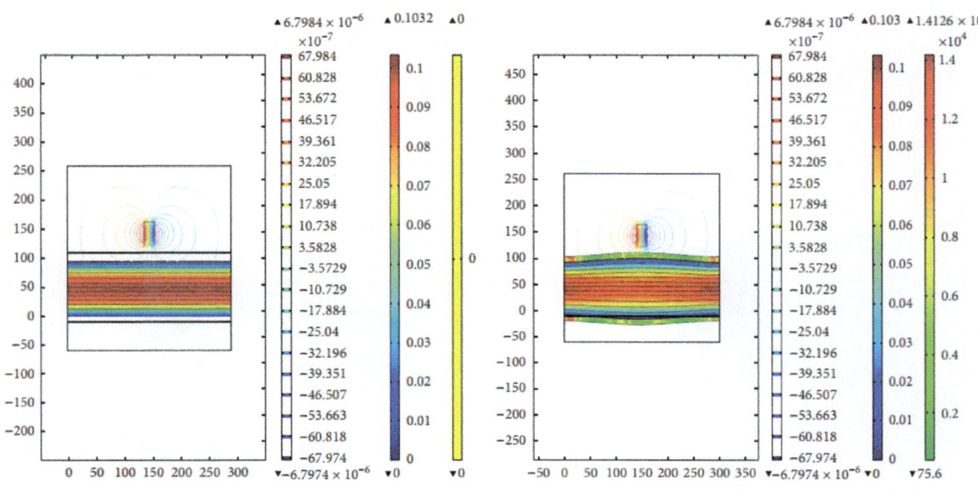

Figure 5: Surface von Mises stress with streamlines of spatial velocity field and magnetic field for B_{rem} = 1 T at t = 0.215. Time = 0.215, surface: von Mises stress (N/m^2), surface: velocity magnitude (m/s), contour: magnetic vector potential, z component (Wb/m), and streamline: velocity field (spatial).

Figures 6(a), 7(a), and 8(a) compare the effect of varying the magnetic field on the surface pressure. Unlike the impact on the velocity profile, these figures suggest that the surface pressure is impacted by increasing the magnetic field and the doubling effect is also seen as expected.

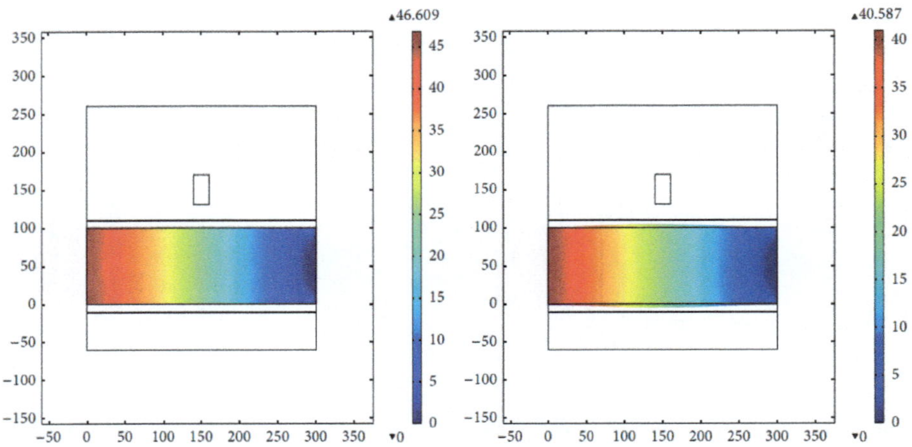

(a) Nonmoving structure (b) Moving structure

Figure 6: Pressure for B_{rem} = 0 T shown at t=4. Time = 4; surface: pressure (Pa).

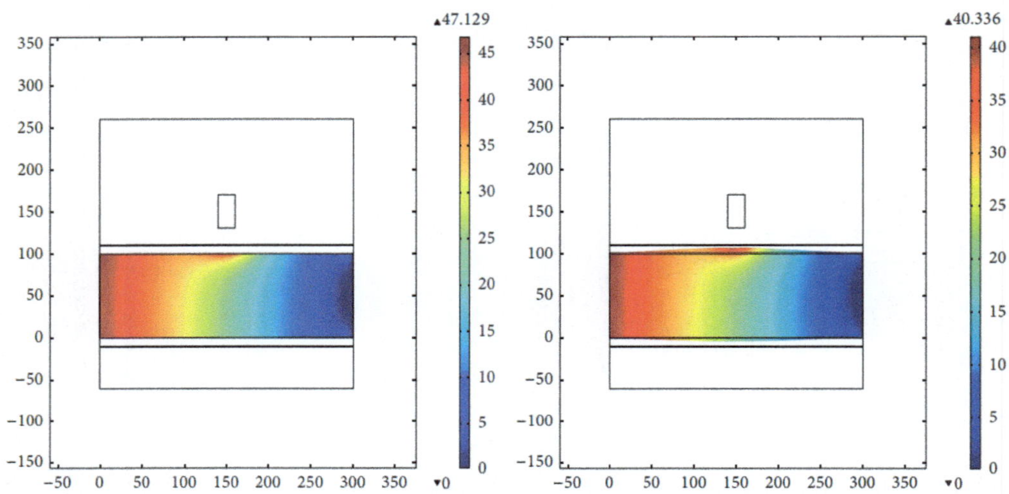

(a) Nonmoving structure (b) Moving structure

Figure 7: Pressure for B_{rem} = 0.5 T shown at t=4. Time = 4; surface: pressure (Pa).

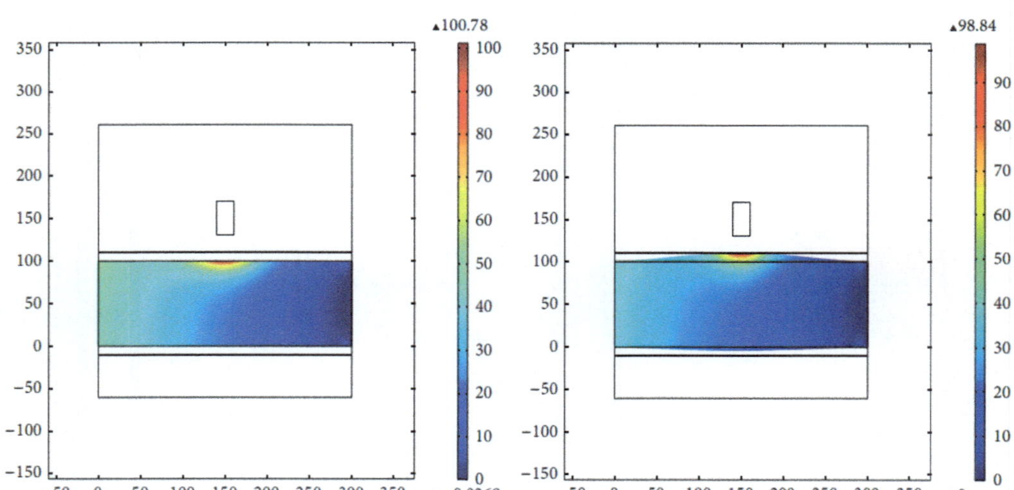

(a) Nonmoving structure (b) Moving structure

Figure 8: Pressure for B_{rem} = 1 T shown at t=4.. Time = 4; surface: pressure (Pa).

Coupled Interaction with Moving Structure

Next, we consider the benchmark problem presented with the structure moving. For this, we employ the ALE formulation for the fluid-structure interaction as described in Section 2. We notice from Figures 3(b),4(b), and 5(b) that, at t=0.215 (when the fluid velocity has maximum value), the structure and the flow pattern are not very much impacted by the magnet. For the maximum studied magnetic field of 1 T, the arterial wall is slightly bent towards the magnet. For even larger magnetic fields not shown in the picture (the order of magnetic field of 5 T), the magnet intersects with the arterial wall.

Even though we have not seen a big difference in structural deformation and fluid flow for our study case, the fluid pressure is entirely different between two considered magnetic fields (see Figures 6(b), 7(b), and 8(b)). If for B_{rem}=0 T the pressure is completely symmetric with respect to the x-axis, the pressure around the magnet increases when magnetization is 0.5 T and becomes more than double the maximum pressure in the rest of the fluid when B_{rem}=1T.

Another experiment we perform is to measure the velocity profile and displacement of two specific points. From Figures 9 and 10, we notice that, as expected, the velocity and pressure decrease at the center and increase around the boundaries when the structure is moving, mainly because of the dilatation of the structure. While the pressure in the center is not affected much by the presence or absence of magnetic field, near the magnet the pressure is steadily increasing with the time.

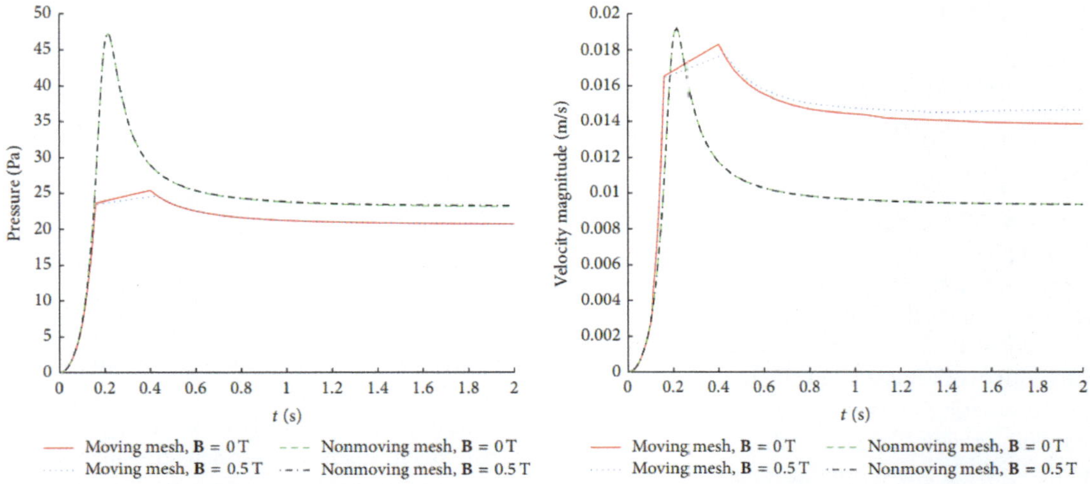

(a) Velocity profile for coordinates (150, 50)

(b) Velocity profile for coordinates (150, 95)

Figure 9: Velocity for a center and edge point inside the fluid.

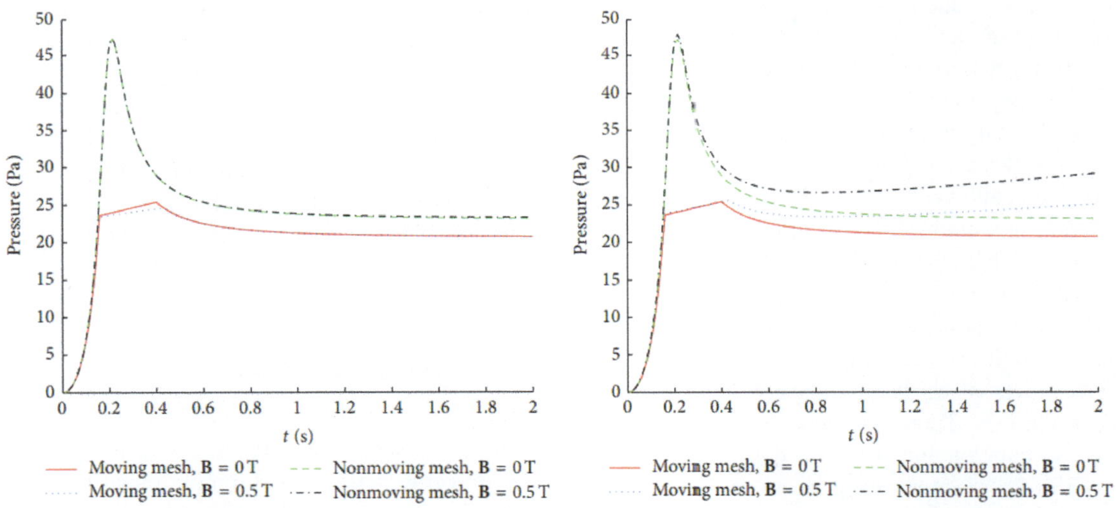

(a) Pressure profile for coordinates (150, 50)

(b) Pressure profile for coordinates (150, 95)

Figure 10: Pressure for a center and edge point inside the fluid.

For the measured displacement, we notice in Figure 11 that the wall towards the magnet is getting closer to the magnet because of the increasing pressure, while the other wall is virtually unaffected by the presence of the magnetic field.

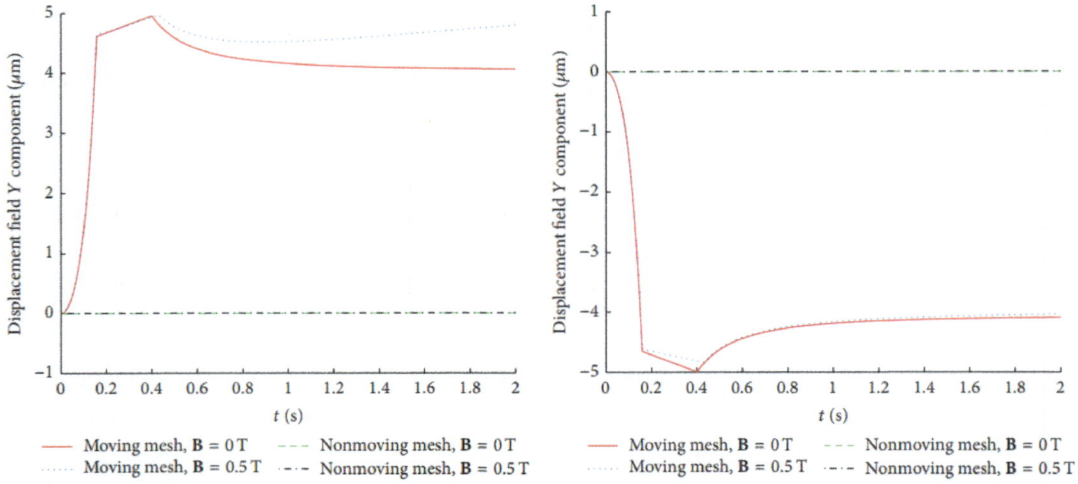

(a) y-displacement for coordinates (150, 105)

(b) y-displacement for coordinates (150, −5)

Figure 11: y-displacement of two points.

CONCLUSION

In this work, we presented the computational modeling and simulation of coupled multiphysics applications. These included a variety of processes such as fluid dynamics, structural mechanics, and electromagnetic interaction that impacted the behavior of the physical system in a coupled way. Specifically, this work considered the research question of "how does incorporating electromagnetic field into fluid-structure interaction models influence the fluid flow and structural deformation?" In answering this question, this work led to the development of a two-dimensional multi-physics problem involving electromagnetics coupled with fluid-structure interaction.

In order to answer this research question, we first presented the mathematical background and simulation of the interaction between fluid, structure, and magnetic field. The motivation of this came from researching models for targeted drug delivery for delivering drugs in human body, to increase the concentration of the drug in the target area. For example, the chemotherapy drug dosage is limited by the negative impact on the drugs on the healthy cells. By delivering the drugs with high accuracy and maximum concentration to specific areas of the body, it is possible to increase local dosage of the drug on the tumor, with lower concentration in the rest of the body. The drug effectiveness is increased while the side effects are reduced. Other examples of the applications of magnetic drug targeting are treatment of cardiovascular conditions, such as stenosis and thrombosis. Thus, it is important to model not only the blood circulation but also the deformation of the blood vessel, in order to improve the accuracy which is the focus of the second problem in this thesis. In particular, it is important to have an accurate model of the interaction between the three components for optimizing the shape, size, and magnetic power, in order to deliver the drugs efficiently in the desired place and minimize the side effects. Our results from this work clearly indicate the importance of the magnetic field to be coupled with a fluid-structure interaction model. More importantly, the results suggest the importance of using moving walls versus nonmoving walls in this coupled electromagnetic fluid-structure interaction.

While this work provided a lot of insight into the importance of electromagnetic effects in fluid-structure interaction, there is scope to enhance this work by considering effects of non-Newtonian rheological properties incorporated along with the extension to materially and geometrically nonlinear models. In the last two decades, collagenous soft tissues have been found to exhibit viscoelastic behavior, which includes time-dependent creep and stress relaxation, rate-dependence, and hysteresis in a loading cycle. As suggested in [18], this hysteresis is less sensitive than the stiffness to the loading rate, and this phenomenon is generally found in soft tissues and elastomers [18]. One of the future directions would be to extend the structural mechanics module to incorporate viscoelasticity and then study the influence of this on our models. The computational models in this work included two-dimensional models for simplicity, but our models can be naturally extended to three dimensions. With increasing the size of the problem comes the need for more computational resources. There is intensive work that is evolving in the area of domain decomposition that helps to address how to solve coupled multiphysics problems efficiently. So as the problem dimension becomes bigger, one must also resort to domain decomposition type approaches which can then open up more venues on parallelization of the algorithms that have been developed.

ACKNOWLEDGMENT

The authors wish to thank the STEM Accelerator Program in the College of Science at George Mason University for supporting this project in part to help promote scientific development for high school students and undergraduate and graduate students to learn about multidisciplinary research in STEM.

REFERENCES

1. Y. C. Chang, T. Y. Hou, B. Merriman, and S. Osher, "A level set formulation of Eulerian interface capturing methods for incompressible fluid flows," Journal of Computational Physics, vol. 124, no. 2, pp. 449–464, 1996.

2. F. P. Baaijens, "A fictitious domain/mortar element method for fluid-structure interaction,"International Journal for Numerical Methods in Fluids, vol. 35, no. 7, pp. 743–761, 2001.

3. R. Glowinski, T. W. Pan, T. I. Hesla, D. D. Joseph, and J. Périaux, "A fictitious domain approach to the direct numerical simulation of incompressible viscous flow past moving rigid bodies: application to particulate flow," Journal of Computational Physics, vol. 169, no. 2, pp. 363–426, 2001.

4. P. Seshaiyer, "Stability and convergence of nonconforming hp finite-element methods," Computers & Mathematics with Applications, vol. 46, no. 1, pp. 165–182, 2003.

5. E. W. Swim and P. Seshaiyer, "A nonconforming finite element method for fluid-structure interaction problems," Computer Methods in Applied Mechanics and Engineering, vol. 195, no. 17, pp. 2088–2099, 2006.

6. E. Aulisa, A. Cervone, S. Manservisi, and P. Seshaiyer, "A multilevel domain decomposition approach for studying coupled flow applications," Communications in Computational Physics, vol. 6, no. 2, pp. 319–341, 2009.

7. C. S. Peskin, "Numerical analysis of blood flow in the heart," Journal of Computational Physics, vol. 25, no. 3, pp. 220–252, 1977.

8. A. O. Mulyar, Magnetically Targeted Drug Delivery System, GRIN Verlag, 2010.

9. A. D. Grief and G. Richardson, "Mathematical modelling of magnetically targeted drug delivery,"Journal of Magnetism and Magnetic Materials, vol. 293, no. 1, pp. 455–463, 2005.

10. J. Dobson, "Magnetic nanoparticles for drug delivery," Drug Development Research, vol. 67, no. 1, pp. 55–60, 2006.

11. S.-I. Takeda, F. Mishima, S. Fujimoto, Y. Izumi, and S. Nishijima, "Development of magnetically targeted drug delivery system using superconducting magnet," Journal of Magnetism and Magnetic Materials, vol. 311, no. 1, pp. 367–371, 2007.

12. J. M. Jin, The Finite Element Method in Electromagnetics, John Wiley & Sons, 2002.

13. J. D. Jackson, Classical Electrodynamics, John Wiley & Sons, 1998.

14. D. Strauss, Magnetic Drug Targeting in Cancer Therapy, 2005, http://www.comsol.com/showroom.

15. R. E. Rosensweig, "An introduction to ferrohydrodynamics," Chemical Engineering Communications, vol. 67, no. 1, pp. 1–18, 1988.

16. R. E. Rosensweig, Ferrohydrodynamics, Dover, 1997.

17. P. A. Voltairas, D. I. Fotiadis, and L. K. Michalis, "Hydrodynamics of magnetic drug targeting," Journal of Biomechanics, vol. 35, no. 6, pp. 813–821, 2002.

18. Y. C. Fung, Biomechanics: Mechanical Properties of Living Tissues, Springer, 1993.

CHAPTER 11

APPLICATION OF AN ULTRA-HIGH-FREQUENCY FRACTAL ANTENNA TO PARTIAL DISCHARGE DETECTION IN SWITCHGEARS

Chenguo Yao, Pan Chen, Congjian Huang, Yu Chen and Panpan Qiao

Keywords: UHF, PD, switchgear, UHF-sensor, FDTD, propagation characteristics, installation position of sensor

INTRODUCTION

Insulation defects, such as protrusions, bad contacts and free metallic particles, may trigger partial discharges (PDs) in high-voltage equipment, and in most occasions, the existence of PD in equipment is treated as the indicator of insulation integrity degeneration [1]. Although PD has little influence on the insulation strength, without appropriate corrective measures it can come to be the cause of insulation flashovers or breakdowns [2]. Therefore, the detection of PD has become one of the most significant insulation condition assessment methods [3]. High-voltage switchgears which directly provide power supply to distribution networks and consumers have been widely used in the power system. Great economic costs have resulted from blackouts due to switchgear equipment faults. Furthermore, switchgears have a quite complex internal structure, many parts of differing physical qualities and small internal space. As a result, these devices are prone to be affected by partial discharges.

Currently, the main means of PD detection for switchgears involve ultrasonic detection [4–6], radio frequency (RF) detection [7], transient earth voltage detection [8] and UHF detection [9,10]. In comparison, the UHF method features a great anti-interference capability, high sensitivity and broad application. As part of UHF detection system, the performance of the UHF sensor directly conditions the whole detection system performance in signal extraction and recognition. Basically, UHF antenna sensors can be broken down into two main categories: internal and external. The electromagnetic waves generated by PD are shielded by the steel casing of the switchgear, but a small amount of electromagnetic waves still radiates outside the box from joints and outgoing line ports. These radio waves decay severely in the air. For this reason, the sensitivity of the UHF method is not very high when the UHF sensor is installed outside the cabinet. However, if an inner sensor is adopted, not only will the detection sensitivity be increased, but the outer electromagnetic interference will also be avoided. Most UHF antenna sensors described in the related research are external ones. Furthermore, existing UHF inner sensors are designed for some specific devices such as transformers and gas insulated switchgear (GIS) [11,12]. A special inner sensor designed for switchgear hasn't been developed yet. Besides, there is a particularly close relationship between the location of the UHF sensor and the sensitivity of this method. The object of study in the UHF technique is electromagnetic waves. Because of the limited space and the complex structure of the switchgear, its components affect the propagation of the radio waves differently, which makes it hard to determine the best place to install the sensor [8]. Intensive study of UHF is very necessary to analyze the propagation characteristics of UHF signals in switchgears. The GIS simulation model has been developed by the FDTD method

Citation: Chenguo Yao, Pan Chen, Congjian Huang, Yu Chen and Panpan Qiao, Ultra-High-Frequency Fractal Antenna to Partial Discharge Detection in Switchgears, doi:10.3390/s131217362.

for the research on propagation characteristics [13–16]. With a coaxial cavity structure, however, GIS has an entirely different structure from switchgears. The GIS conclusions may not apply to switchgear because their different structures. The transformer simulation model has also been developed for the study of propagation characteristics [17,18]. Transformer cores and windings play a decisive role in the propagation characteristics of transformers while the internal structure of switchgears is more complex and the two kinds of equipment can't be lumped together.

In this paper, an inner UHF antenna sensor based on the fractal technique has been developed. The sensor is designed in accordance with the features of UHF monitoring in switchgear. In addition, a switchgear simulation model has been developed by the FDTD method for the research into the propagation characteristics. The detection results of different monitoring points under various discharges have been analyzed by simulation and test methods, which offer the theoretical basis of the location of the inner sensor.

INNER UHF ANTENNA SENSOR

The antenna is developed to be installed in the switchgear for the UHF monitoring method. Since the frequency of signals in air used in the UHF method are commonly distributed from 300 MHz to 1 GHz [9], and wide band information is necessary for such a method, the mentioned antenna should work over 300 MHz with wide band capability. However, as too much low-frequency noise exists, the antenna working frequency should be a little higher than 300 MHz to achieve better SNR. Besides, a fast data acquisition card would be used to transform the signal from the antenna into digital data for further analysis. With consideration of the UHF method requirements and economic factors, an antenna with a working band from about 500 MHz to 1 GHz would be developed.

The inner structure of switchgear is complicated, with a number of electric components, including the cabinet, partition boards, current transformers (CTs), circuit breakers (CBs), bushings, insulators, bus-bars, etc. CTs, CBs, insulators and bus-bars are the basic components installed in all types of switchgears. Due to the components within switchgears, there is not much space for an inner antenna. The space is further limited by the insulation requirements due to the air insulation used in switchgears. As a result, this places demands on the size and shape of the inner antenna. In addition, the inner antenna should also work with low standing wave ratio (VSWR), high gain and suitable directivity to collect PD signals inside the switchgear.

To satisfy the above requirements, a micro-strip with a Koch structure will be developed as the inner antenna. The Koch curve is a type of fractal curve. This kind of fractal curve could make up the micro-strip antenna with Koch structure shown in Figure 1.

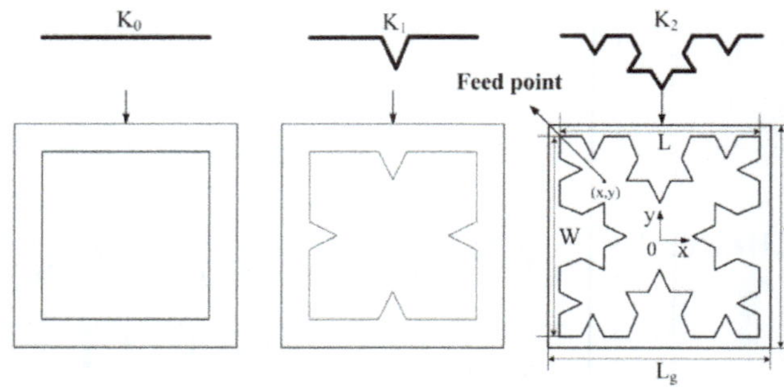

Figure 1. Koch fractal patch antennas with fractal dimensions of 0, 1 and 2.

The general micro-strip antenna commonly has a bandwidth of a few of percent, while the fractal structure could contribute to extend the frequency band [19]. Moreover, the adoption of a substrate with low permittivity is beneficial to spectral band extension. For this reason, as shown in Figure 2, the substrate and ground plate are separated by an air layer because of its low permittivity.

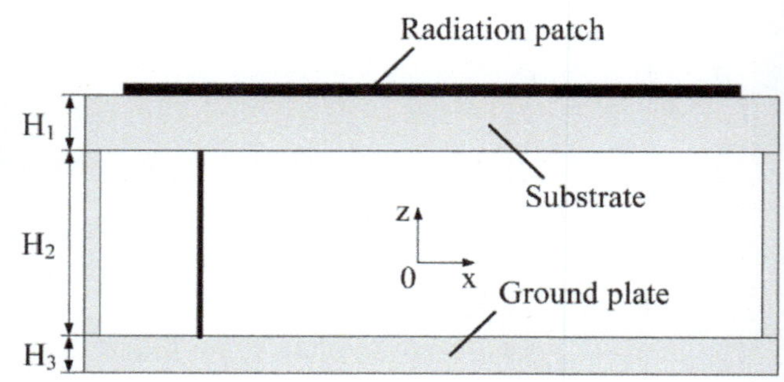

Figure 2. Patch antenna with an air layer structure.

As shown in Figures 1 and 2, the structural parameters of the designed antenna consist of the dimensions of the radiation patch W and L, the feeding point coordinates (x,y), the relative permittivity of the substrate, the thickness of substrate H_1, the dimensions of substrate W_g and L_g, the thickness of the air layer H_2, the thickness of the ground plate H_3 and the fractal dimension. The simulation model of the Koch fractal patch antenna has been developed using the High Frequency Structure Simulator (HFSS) software. Furthermore, the parameters above have been optimized with the optimization of the coordinated of the feeding point as an example shown in Figure 3. The feed point is located in the diagonal of the micro-strip, with its coordinate origin in the center of the micro-strip as seen in Figure 1.

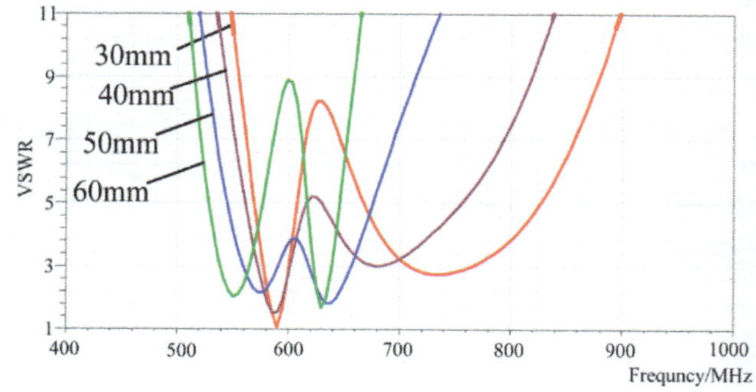

Figure 3. The VSWR curve of antennas with different feed point positions.

The general parameters of the sensor are listed in Table 1. A picture of the designed antenna with a flat-plate structure is shown in Figure 4. Since the ground plate of antenna isn't the radiating element, the performance of this antenna is less affected by the switchgear cabinet when the sensor is installed on the inner side plate of the switchgear body. The VSWR curve, gain and directivity of the sensor have been measured by an AV3629 vector network analyzer, AV1487B signal source, HP8592L spectrum analyzer, and HP8720D vector network analyzer, respectively. The measured VSWR curve with two resonance frequencies is illustrated in Figure 5b. The first resonance frequency is 583 MHz with the VSWR of 1.3, the frequency band from 565 MHz to 623 MHz, the relative bandwidth of 10%, while the second one is 732 MHz with the VSWR of 1.7, the frequency band from 668 MHz to 850 MHz, the relative bandwidth of 24%. Compared with the VSWR curve in HFSS, the two curves are almost the same. Due to the construction, the peak between two resonance points of the measured curve is higher than the simulated one, while the overall bandwidth of the measured curve is better than the simulated one.

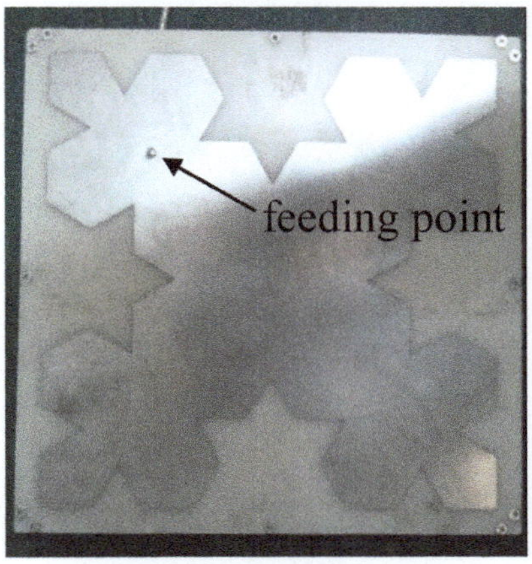

Figure 4. Picture of the designed antenna.

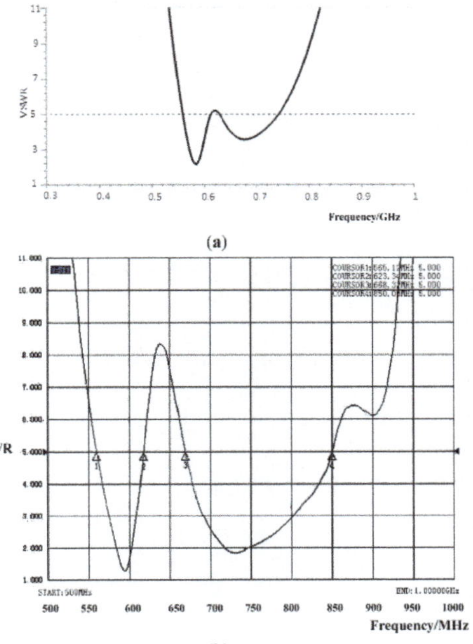

Figure 5. VSWR curve of the designed antenna. (**a**) VSWR curve in HFSS, (**b**) Measured VSWR curve.

Table 1. Dimensions of the microstrip antenna (in mm).

Fractal Dimension	W	L	Wg	Lg	H1	H2	H3	Relative Permittivity	Feeding Point
2	170	170	180	180	5	20	1	4.2	(40.5,40.5)

The gain of the sensor is listed in Table 2. The gain within the working band is over 1.4 dB, with which the sensor can detect weak PD signals. However, the gain is only −1.5 dB for 700 MHz because it's out of the working band. This indicates that signals outside the antenna working band will be attenuated too severely to be collected. Therefore, the low-frequency noise will be avoided efficiently.

Table 2. Measured gain of the microstrip antenna.

Frequency (MHz)	600	700	800	900
Gain (dB)	3	−1.5	1.4	1.7

Compared to ultra-wide-band (UWB) antennas, the size of the developed antenna is smaller, which is more suitable for installation in switchgears. Its gain within the working band would be relatively higher due to deep VSWR curve in the two resonance frequency point. Last but not the least, as shown in Figure 6, the excellent vertical directivity of the antenna ensures that the signals above it can be collected best. In summary, the antenna satisfies the above requirements well.

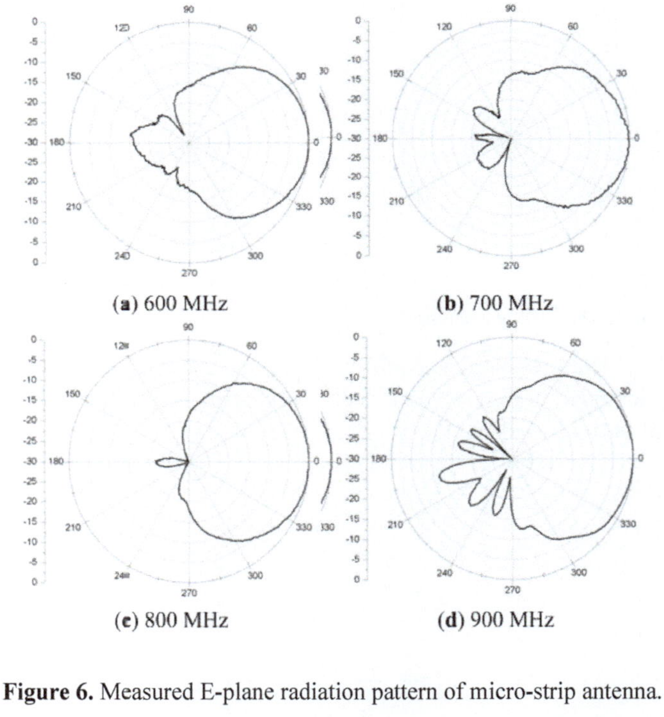

Figure 6. Measured E-plane radiation pattern of micro-strip antenna.

Four kinds of insulation defect models have been developed in this paper, to simulate four main kinds of PDs by experiment. These insulation defect models are an air-gap insulation defect, surface flashover insulation defect, metallic outshoot insulation defect, and free metal particles insulation defect. The proposed sensor was investigated by experiment using the circuit illustrated in Figure 7. In the test, the detection impedance of the impulse current method was used as the reference sensor. The output signal was recorded using a Tektronix TDS 7104 oscilloscope with the band width of 1 GHz and a maximum sampling rate of 20 GS/s. The insulation defects and sensor were placed in a real switchgear cabinet as shown in Figure 8.

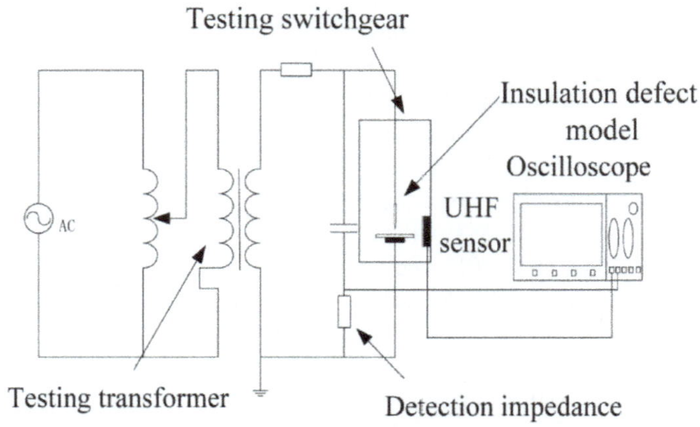

Figure 7. Experimental circuit.

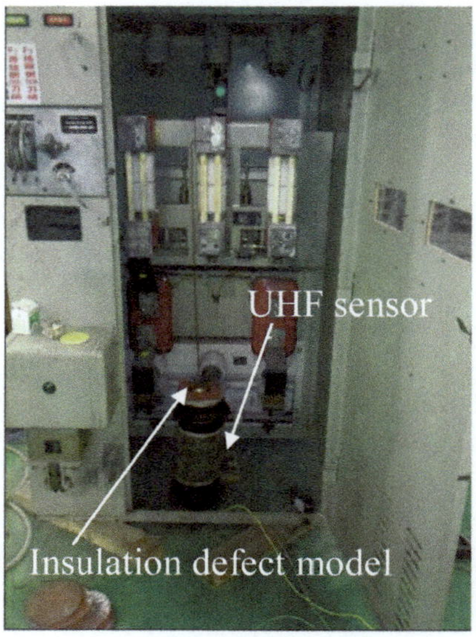

Figure 8. Testing switchgear.

The measured waveforms of the air-gap, surface flashover, metallic outshoot, and metal particles insulation defects are illustrated in Figure 9a–d respectively. They show that the UHF sensor could reproduce the waveforms of the partial discharges of various kinds well.

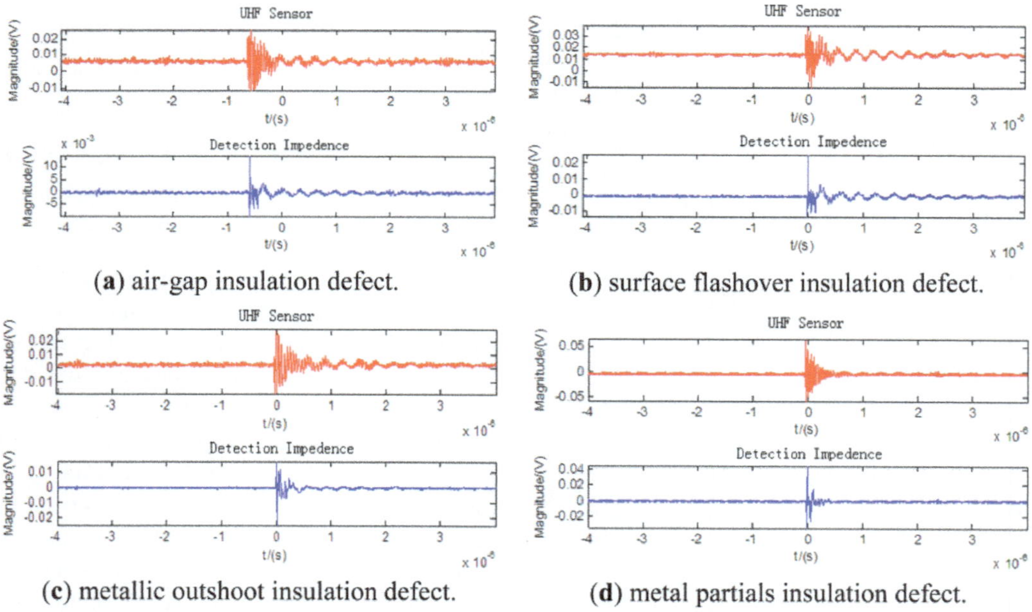

Figure 9. The measured waveforms of different kinds of discharges.

INFLUENCE OF EACH COMPONENT IN THE SWITCHGEAR ON THE PROPAGATION OF ELECTROMAGNETIC WAVES

A switchgear has a quite complex internal structure, and each component of a switchgear has different effects on the propagation of UHF signals. Hence, the emphasis of this section is placed on the influence of the actual size and structure of the switchgear on the propagation of radio waves.

Figure 10 shows an FDTD simulation model of a switchgear with the dimensions of 1,110 mm × 1,250 mm × 2,620 mm that consists of a cabinet, partition boards, CTs, CBs, bushings, insulators and bus-bars. The material parameters of each component are listed in Table 3. The FDTD method is based on mesh generation, and the meshing size closely relates to the accuracy of the result and the amount of calculation. To balance calculation time and precision, the meshing size should be appropriate. According to the Courant condition, the maximum mesh size L_{max} should be calculated as [20]:

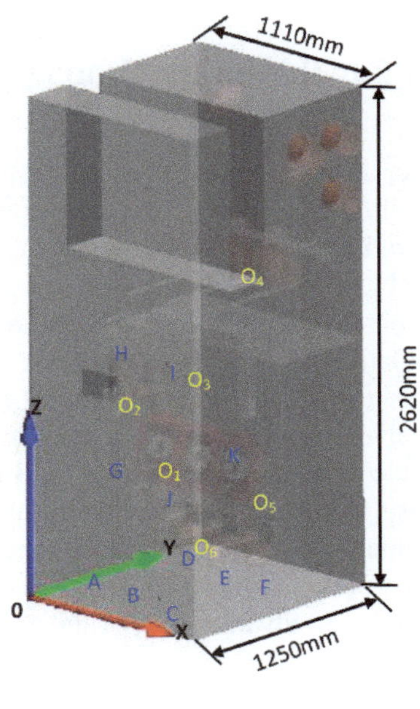

Figure 10. FDTD switchgear simulation model.

Table 3. The material parameters of each component.

Each Component of the Switchgear	Relative Permittivity (ε_r)	Relative Permeability (μ)	Electrical Conductivity σ (S/m)
Cabinet	1	-	1.1×10^6
Copper bus-bar	1	-	5.7×10^7
Umbrella skirt of insulator	5.7	-	1×10^{-8}
Pouring material of CT	3.6	-	1×10^{-8}
Iron core of CT	-	2,000	-
Porcelain vase of break	1	-	6.5×10^7

$$L_{max} = \frac{c}{10 \times f} \tag{1}$$

where c is the speed of light and f is the maximum frequency of the excitation source. The maximum frequency of our simulation analysis was set at 3 GHz due to the UHF band being used. Therefore, according to Equation (1), the maximum mesh size L_{max} was 10 mm. In this article, the simulation model was meshed uniformly in a three-dimensional system. Then the corresponding time step Δt was calculated to be 19.26 ps. It must be mentioned that the simulation boundary condition is determined to be a perfectly matched layer (PML). The partial discharge source used in the simulation was a Gaussian pulse with broad frequency characteristics as defined in Equation (2):

$$I(t) = I_0 \exp(-\frac{4\pi(t-t_0)^2}{\tau^2}) \tag{2}$$

where τ is a constant that determines the width of the pulse waveform, t_0 is the time when the peak waveform occurs and I_0 is the amplitude of the waveform. The waveform was determined to have τ of 616 ps, t_0 of 616 ps and I_0 of 1 A. The discharge capacity of the Gaussian pulse was calculated to be 359.6 pC by waveform integration.

Influence of CT on the Propagation of EM Waves

The size of CT with the iron core structure and windings is relatively larger than the other components, so the influence of the component on the propagation of EM waves can't be ignored. The proposed discharge source was placed near the B-phase CT. Two detection points were located on the near and far sides of the CT relative to the discharge source, respectively. Figure 11a,b is the simulation results of the near and far sides, respectively.

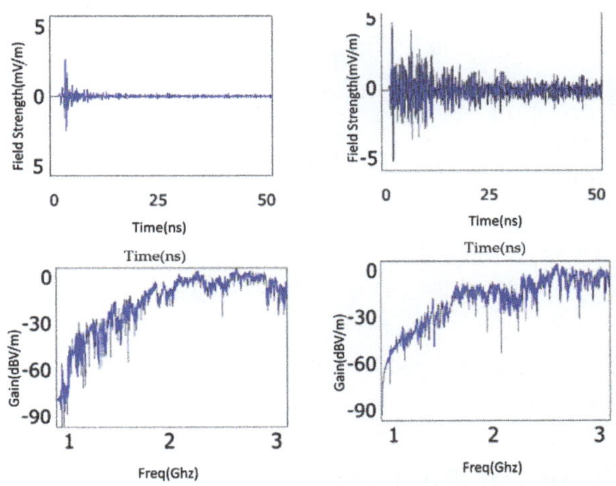

(a) far side (b) near side

Figure 11. Simulation results of the influence of CT on the propagation of EM waves.

It can be seen from the comparison of the time domain waveforms that the waveform amplitude reduces by as much as 6 dB when the EM waves pass through the CT. In addition, the amplitude dwindles rapidly by more than 20 dB after the first peak appears. The Fast Fourier Transform (FFT) waveforms show that the CT makes the low frequency components (the frequency band below 1.5 GHz, the same below) attenuate greatly while the high frequency ones (the frequency band above 1.5 GHz, the same below) hardly change. According to these results, the sensor should not be positioned near CTs due to their influence of attenuation of EM waves.

Influence of CB on the Propagation of EM Waves

CB is the most important part of switchgear, and is normally located in the breaker chamber. The CB of the simulation model is a vacuum circuit breaker with a porcelain vase structure outside and a metal shield shell inside. The pulsed discharge signal was placed near the B-phase CB. Two detection points were located on the near and far sides of the CB relative to the discharge source, respectively. Simulation results are shown in Figure 12a,b. From the comparison of the time domain waveforms, the radio waves spread through the CB with little attenuation. It is likely that although the EM waves pass into the porcelain vase with a little dielectric attenuation, they diffract with little decay when meeting the metal shield shell as the size of the shell is rather smaller than that of electric waves. It's worth noting that the waveform will distort thanks to the existence of CB. The FFT waveforms also indicate that CB has a limited effect on the spectrum distribution. From these simulation results, it is concluded that the installation position of the sensor does not have to avoid the CB areas under the condition that the small waveform distortion can be neglected.

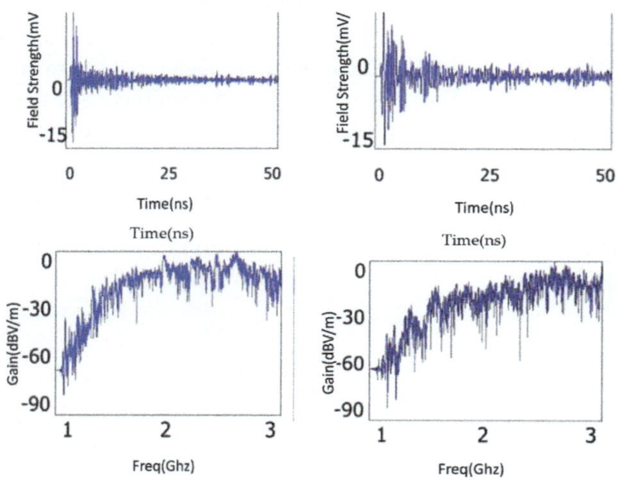

(a) far side (b) near side

Figure 12. Simulation results of the influence of CB on the propagation of EM waves.

Influence of the Insulator on the Propagation of EM Waves

Since there is a large quantity of insulators in switchgear, EM waves inevitably spread into the regions of these insulators. The artificial discharge source was placed near an insulator. Then two detection points were placed on the near and far sides of the insulator relative to the discharge source, respectively. By comparing the measured waveforms illustrated in Figure 13a,b, we see that the insulator affects the waveform amplitude slightly, but with great wave shape distortion. It is shown from the FFT waveforms, the low frequency components are weakened by the insulator. It can be concluded that the UHF antenna should not be located next to insulators because the severe wave distortion would affect pattern recognition.

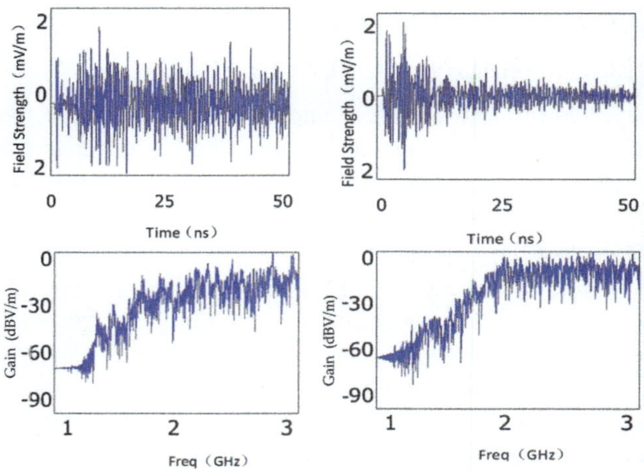

(a) far side (b) near side

Figure 13. Simulation results of the influence of insulator on the propagation of EM waves.

Influence of Bus-Bar on the Propagation of EM Waves

The metal-material bus-bars are widely distributed in the cabinet, so the influence of bus-bars on the propagation of EM waves cannot be ignored. The above discharge source was located near the bus-bar connected with the B-phase CT. Then two detection points were placed on the near and far sides of the insulator relative to the discharge source, respectively. Figure 14 a,b is the simulation results of the near and far sides, respectively. Through the time domain waveforms, it is shown that the electric wave decays by 3.5 dB because of the reflection losses. Then the FFT waveforms indicate that the high frequency components are restrained by the bus-bar. From these results, the sensor should not be positioned near bus-bars because they will be disadvantageous to signal reception.

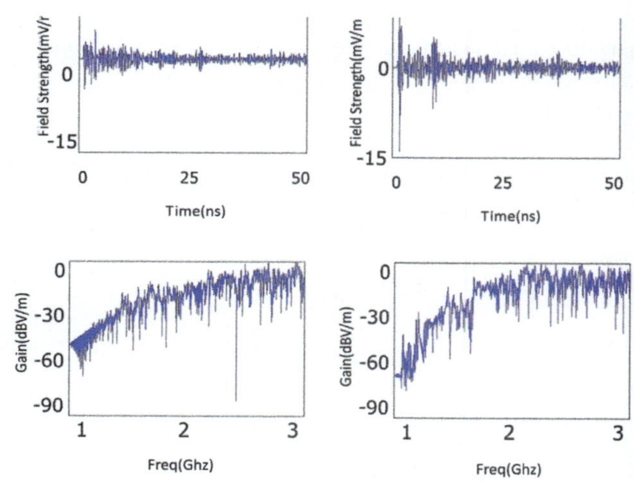

(a) far side (b) near side

Figure 14. Simulation results of the influence of bus-bar on the propagation of EM waves.

THE OPTIMAL INSTALLATION POSITION OF THE INNER SENSOR

According to the results of Section 3, it can be found that the components in the switchgear affect the electromagnetic waves differently and they affect electromagnetic waves as a whole. As one of the most important components in the UHF monitoring system, the UHF sensor should be installed in an appropriate place to optimize detection results. In this section, the best place to position the inner sensor will be studied.

Simulation Analysis

According to the simulation analyses of Section 3, in this article, several detection points were set in the simulation model. Then, referring to the common types and locations of discharges that occur in switchgear in [21], various discharge sources were placed. The waveform amplitudes would be used to determine the best detection point, namely, the optimal installation position of the inner sensor. Moreover, the mentioned discharge source is the same as in Section 3.1. The locations of discharge sources and detection points are shown in Figure 10 and the coordinates are listed in Table 4.

Table 4. Coordinates of discharge sources and detection points.

Locations			Coordinates
Discharge sources	Inside the B-phase CT (O_1)		(440,510,590)
	On the surface of the supporting insulator of the A-phase CB (O_2)		(440,440,1060)
	Around the C-phase CB (O_3)		(770,330,1210)
	On the surface of the bushing at the upside of back cabinet (O_4)		(390,855,1910)
	Inside cable joint at the downside of back cabinet (O_5)		(580,780,290)
	On the surface of the bushing at the downside of front cabinet (O_6)		(630,520,280)
Detection points	Bottom plate of the front cabinet	Left side (A)	(150,340,0)
		Middle side (B)	(520,340,0)
		Right side (C)	(770,340,0)
	Bottom plate of the back cabinet	Left side (D)	(150,940,0)
		Middle side (E)	(520,940,0)
		Right side (F)	(770,940,0)
	Front cupboard door	Bottom-left side (G)	(520,0,580)
		Upper-left side (H)	(520,0,1500)
		Bottom-right side (I)	(770,0,580)
		Upper-right side (J)	(770,0,580)
	Side plate of the front cabinet	Middle side (K)	(1110,340,600)

The amplitudes of electric field strength in each detection point are listed in Table 5. In order to facilitate comparison of results, in Table 5 each value is divided by the maximum value in each row for normalization. Then the processed values of each column were added up to a total value, which is related to the overall detection effect of the corresponding detection point. Furthermore, the larger the value is, the better the overall detection effect is. The variance of values in each column was also figured out to determine the volatility of the detection for different discharge sources. The smaller the variance is, the less the volatility is. All the processed data was listed in Table 6. In general, the detection point C located on the right side of the bottom plate and the detection point I located on the bottom-right side of the front cupboard door have good satisfied detection performance with overall consideration. From the viewpoint of safety, only the detection point I is close to CTs, which probably brings a new insulation fault hazard. Moreover, if the detection point I is selected for the sensor, the sensor need to be anchored vertically with more difficulty than point C, so in conclusion, the inner sensor should be positioned in the point C for the best detection performance.

Table 5. Amplitude field strength of each detection point.

Point	Amplitude Field Strength of Each Detection Point (mV/m)					
	O_1	O_2	O_3	O_4	O_5	O_6
A	2.61	4.87	1.98	2.81	1.87	1.47
B	2.65	4.62	2.49	2.34	1.27	6.73
C	2.37	5.53	2.37	3.12	1.37	6.24
D	0.72	1.34	1.21	0.72	7.12	1.84
E	0.89	1.12	1.22	0.56	15.75	1.62
F	0.83	1.47	1.2	0.62	17.74	1.86
G	17.46	4.12	1.72	1.86	0.81	3.14
H	2.22	4.46	2.49	4.45	0.42	1.55
I	10.16	4.46	3.45	3.08	0.32	3.14
J	2.34	4.23	4.79	3.87	0.38	1.43
K	3.52	8.34	4.12	3.12	1.23	1.67

Table 6. Amplitude field strength of each detection point (normalized).

Point	Amplitude Field Strength of Each Detection Point (Normalized)							
	O_1	O_2	O_3	O_4	O_5	O_6	Total	Variance
A	0.149	0.12	0.413	0.631	0.105	0.218	1.636	0.044
B	0.146	0.554	0.52	0.526	0.072	1	2.818	0.111
C	0.136	0.663	0.495	0.701	0.077	0.927	2.999	0.112
D	0.041	0.161	0.253	0.72	0.401	0.273	1.849	0.055
E	0.051	0.134	0.255	0.56	0.888	0.241	2.129	0.098
F	0.048	0.176	0.251	0.62	1	0.276	2.371	0.124
G	1	0.494	0.359	0.418	0.046	0.467	2.784	0.095
H	0.127	0.511	0.52	1	0.024	0.23	2.412	0.126
I	0.582	0.535	0.72	0.692	0.046	0.23	2.805	0.073
J	0.134	0.507	1	0.87	0.021	0.212	2.744	0.164
K	0.202	1	0.86	0.701	0.069	0.248	3.08	0.151

Experimental Verification

According to the results in Table 6, the detection performance of each detection point located on the same mounting surface obviously don't vary from each other, so there is no necessity to verify the detection performance of every point. As a result, the four detection points with the best detection performance in each mounting surface were chosen to be the experimental verification points. They are point C, point F, point I and point K (marked in gray).The wiring diagram is illustrated in Figure 7, and the components of this circuit are the same as those in Section 2. It must be mentioned that in each test, the values of the voltage on the same insulation defect model and the discharge capacity of discharge (i.e., waveform amplitudes got by the detection impedance) in the same defect model should be equal to each other. The test data can be collected ten minutes after the voltage was applied. Furthermore, each case contained 10 tests, and the average of these results was taken as the final result in order to reduce the random error as much as possible.

The comparison between the results of test verification and simulation computation is shown in Figure 15. According to these Figures, it can be seen that, on the condition of the discharge O_3, the error of the two methods is relatively larger than that under other conditions, because the discharge O_3 is a kind of free metal particles PD discharge. It's a sort of non-stationary discharge. The metal particles would move irregularly in an electric field, which makes the strength of the discharge change greatly throughout the experiment. In general, errors occurred under most conditions. However, the simulated curves agree well with those of experiments. In other words, the simulation analysis method of is correct and feasible. Besides, the results indicate that the antenna with vertical directivity is best suitable for the location point C. Since the antenna would be installed at point C, the PD signal above it can be directly collected by the antenna, which best avoids the distortion and attenuation due to catadioptric effects and it ensures the integrity of the signals for better analysis.

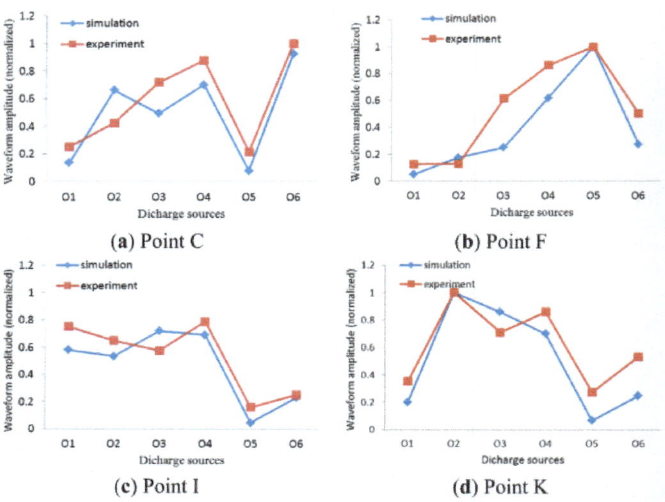

Figure 15. Comparison between the results of test verification and simulation computation.

CONCLUSIONS

An inner UHF fractal antenna sensor which was specially designed for PD detection in switchgear was developed. The UHF sensor had two frequency bands. The first resonance frequency is 583 MHz with the VSWR of 1.3, the frequency band from 565 MHz to 623 MHz, the relative bandwidth of 10%, while the second one is 732 MHz with the VSWR of 1.7, the frequency band from 668 MHz to 850 MHz, the relative bandwidth of 24%. The waveforms of different kind of PDs in the switchgear can be measured with a high sensitivity.

In order to determine the optimal installation position of the inner senor, the FDTD simulation model was developed to study the propagation characteristics of UHF signals in switchgears. The simulation indicated that CTs and bus-bars would make the EM waves attenuate greatly and insulators would affect the EM waves with obvious distortion of the wave shape. The sensor should not be positioned near these three kinds of components. CBs had little influence on the radio waves, so the installation position of sensor may not avoid their areas. According to simulation analyses and experimental verification, the UHF sensor should be positioned at the right side of the bottom plate in the front cabinet (i.e., breaker chamber) to achieve the best detection performance.

ACKNOWLEDGMENTS

This work was supported by the National Natural Science Foundation of China (No:51021005), Science and Technology Tackle Key Problem Program of Chongqing (No:CSCT2010BC3003) and State Key Laboratory Autonomic Research Foundation(No:2007DA10512709104).

REFERENCES

1. Portugues, I.E.; Moore, P.J.; Glover, I.A.; Johnstone, C.; McKosky, R.H.; Goff, M.B.; van der Zel, L. RF-based partial discharge early warning system for air-insulated substations. IEEE Trans. Power Del. 2009, 24, 20–29.
2. Coenen, S.; Tenbohlen, S.; Markalous, S.M.; Strehl, T. Sensitivity of UHF PD measurements in power transformers. IEEE Trans. Dielectr. Electr. Insul. 2008, 15, 1553–1558.
3. Hikita, M.; Okabe, S.; Murase, H. Cross-equipment evaluation of partial discharge measurement and diagnosis techniques in electric power apparatus for transmission and distribution. IEEE Trans. Dielectr. Electr. Insul. 2008, 14, 505–518.
4. Kweon, D.J.; Chin, S.B.; Kwak, H.R. The analysis of ultrasonic signals by partial discharge and noise from thetransformer. IEEE Trans. Power Del. 2005, 20, 1976–1983.
5. Chen, L.J.; Tsao, T.P.; Lin, Y.H. New diagnosis approach to epoxy resin transformer partial discharge using acoustic technology. IEEE Trans. Power Del. 2005, 20, 2501–2508.
6. Li, D.J.; Liang, J.Z.; Bu, K.W.; Yang, J.G.; Li, Y.M. Ultrasonic detection of partial discharge on typical defects in GIS (in Chinese). High Volt. Appar. 2009, 45, 72–75.
7. Guan, Y.G.; Qian, J.L. Practical study of radio frequency signal used in on-line PD monitoring of hv switchboard (in Chinese). High Volt. Appar. 2007, 37, 1–3.
8. Ren, M.; Dong, M.; Ren, Z.; Peng, H.D.; Qiu, A.C. Transient earth voltage measurement in PD detection of artificial defect models in SF6. IEEE Trans. Plasma Sci. 2012, 40, 2002–2008.
9. Tenbohlen, S.; Denissov, D.; Hoek, S.M.; Markalous, S.M. Partial discharge measurement in the ultra high frequency (UHF) range. IEEE Trans. Dielectr. Electr. Insul. 2008, 15, 1544–1552.
10. Li, T.H.; Rong, M.Z.; Zheng, C.; Wang, X.H. Development simulation and experiment study on UHF partial discharge sensor in GIS. IEEE Trans. Dielectr. Electr. Insul. 2012, 19, 1421–1430.
11. Hoshino, T.; Kato, K.; Hayakawwa, N. A novel technique for detecting electromagnetic wave caused by partial discharge in GIS. IEEE Trans. Power Del. 2001, 16, 545–551.
12. Tang, Z.G.; Li, C.R.; Cheng, X.; Wang, W.; Li, J.Z.; Li, J. Partial discharge location in power transformers using wideband RF detection. IEEE Trans. Dielectr. Electr. Insul. 2006, 13, 1193–1199.
13. Hikita, M.; Ohtsuka, S.; Teshima, T.; Okabe, S.; Kaneko, S. Examination of electromagnetic mode propagation characteristics in straight and L-section GIS model using FD-TD analysis.IEEE Trans. Dielectr. Electr. Insul. 2007, 14, 1477–1483.
14. Hoshino, T.; Maruyama, S.; Sakakibara, T. Simulation of propagating electromagnetic wave due to partial discharge in GIS using FDTD. IEEE Trans. Power Del. 2009, 24, 153–159.
15. Hu, X.; Judd, M.D.; Siew, W.H. A Study of PD Location Issues in GIS using FDTD Simulation. Proceedings of 45th International Universities' Power Engineering Conference, Cardiff, UK, 31 August– 3 September 2010.
16. Li, X.; Li, C.R.; Li, Y.S.; Wang, W.; Li, H.L. Analysis on partial discharge in GIS by FDTD method. Proc. CSEE 2005, 25, 150–155.
17. Judd, M.D.; Li, Y.; Hunter, B.B.I. Partial discharge monitoring for power transformers using UHF sensors part I: Sensors and signal interpretation. IEEE Electr. Insul. Mag. 2005, 21, 5–12.
18. Aung, M.T.; Milanvic, J.V. The Influence of transformer winding connections on the propagation of voltage sags. IEEE Trans. Power Del. 2006, 21, 262–269.
19. Puente-Baliarda, C.; Romeu, J.; Pous, R.; Cardama, A. On the bahavior of the Sierpinski multiband fractal antenna. IEEE Trans. Antennas Propag. 1998, 46, 517–524.
20. Lee, J.H.; Kalluri, D.K. Three-dimensional FDTD simulation of electromagnetic wave transformation in a dynamic inhomogeneous magnetized plasma. IEEE Trans. Antennas Propag. 1999, 47, 1146–1151.
21. Wang, F.L. Well-Chosen Cases of Applications of Condition Monitoring in Electrical Equipment (in Chinese).; China Electric Power Press: Beijing, China, 2009

CHAPTER 12

ELECTROMAGNETIC FIELD ANALYSIS AND MODELING OF A RELATIVE POSITION DETECTION SENSOR FOR HIGH SPEED MAGLEV TRAINS

Song Xue, Ning He and Zhiqiang Long

Keywords: Relative position detection sensor, electromagnetic modeling, multipole theory, new equivalent source method, second-order vector potential and difference coil structure

INTRODUCTION

High speed maglev trains are driven by a linear synchronous motor [1,2]. The rotor is the suspension electromagnet, and the long stator inlaid with 3-phase windings has a tooth-slot structure [1,3], shown in Figure 1.

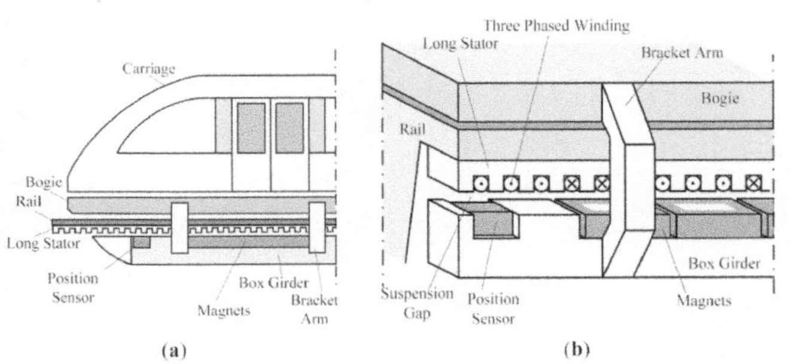

Figure 1. (**a**) Diagram of a high speed maglev train; (**b**) Diagram of the substructure of a high speed maglev train.

The operating principle of the relative position detection sensor is based on nondestructive examination technology [4–7]. There are detecting coils arranged on one side of the sensor facing the tooth-slot structure, as shown in Figure 2. When the train is moving, the coil inductance varies periodically because of the tooth-slot structure. Thus, the relative position can be obtained by detecting the changing coil inductance [8,9].

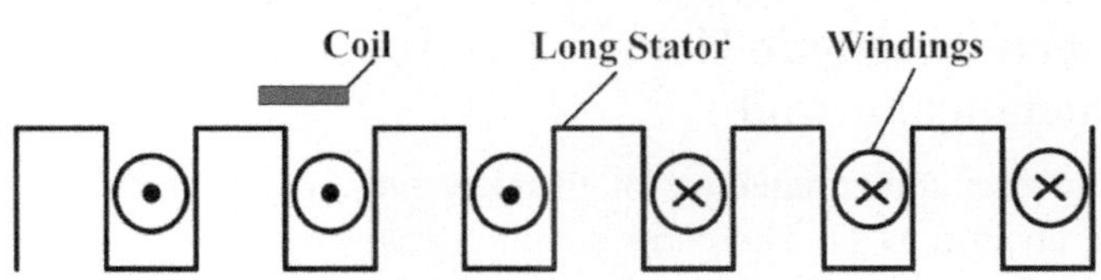

Figure 2. Diagram of the sensor coil and the long stator.

This paper studies the electromagnetic modeling of the sensor analytically, which lays an essential foundation for the sensor design work. Nowadays, with the development of PC technology, complex electromagnetic modeling problems can be realized through numerical methods precisely and conveniently, whereas, analytical methods are complicated, time consuming, and may even have no solution. Usually, in order to get a solution, model simplification is needed, which decreases the precision of the solution. However, analytical methods can get the expression of the field distribution function, represent less of a calculation load for computers, and are beneficial for the development of theories to some extent. They can be used to do some preliminary analysis and get a rough estimate for a problem to guide the principle design work. Numerical methods can be introduced later to guide the detailed design and parameter optimization.

Usually, it's difficult to solve 3-D electromagnetic problems with complex boundaries directly through separation

Citation: Song Xue, Ning He and Zhiqiang Long, Electromagnetic Field Analysis and Modeling of a Relative Position Detection Sensor for High Speed Maglev Trains, doi:10.3390/s120506447.

of variables or mirror methods, etc. So, many researches are focused on 3-D field problems which can be converted to 2-D ones. Besides, second-order vector potential (SOVP) [10,11] can be introduced in to simplify vector field problems to scalar field ones. Reference [10] studies the electromagnetic field in an electrically conductive right-angled wedge due to an inductive excitation by a coil in air. The problem is simplified to a 2-D one. The magnetic induction intensity is expressed by a second-order vector potential (SOVP). The scalar components of the SOVP are expressed in terms of series expansions through Fourier transform. Reference [12] studies the 3-D Eddy-Current modeling problem with a spherical boundary in a spherical coordinate system. References [13–15] propose a semi-analytical method called Multipole Theory (MT). Its solution procedure and expressions are similar to those in reference [12], but this method is based on refined theoretical derivations, and can be applied to electromagnetic modeling problems with complicated boundaries by introducing New Equivalent Sources (NES).

The electromagnetic modeling of the relative position detection sensor is studied analytically in this paper. Firstly, the electromagnetic field in air due to an inductive excitation by a rectangular coil is calculated and analyzed. Based on this, a simplified model of the sensor is established. Then the magnetic induction density distribution of the simplified model is calculated semi-analytically according to the MT and NES methods, and a corresponding numerical solution is obtained through Finite Element Analysis (FEA). The comparison results show that the semi-analytical solution approximately agrees with the numerical solution. Finally, based on the modeling, a difference coil structure is designed to improve the sensitivity and precision of the sensor.

SENSOR MODEL SIMPLIFICATION

Electromagnetic Field in Free Space Due to a Rectangular Coil

The excitation frequency of the detecting coil is chosen to be about 2 MHz. The corresponding wavelength is λ = v/f = 150 m, which is much larger than the length of a tooth-slot period (86 mm). Thus, it can be considered a static magnetic field. Considering an ideal rectangular coil with current I and turn number N shown in Figure 3, according to Biot-Savart Law, the magnetic induction intensity at an arbitrary position P(x,y,z) due to each side of the coil is calculated respectively as follows [16–18]:

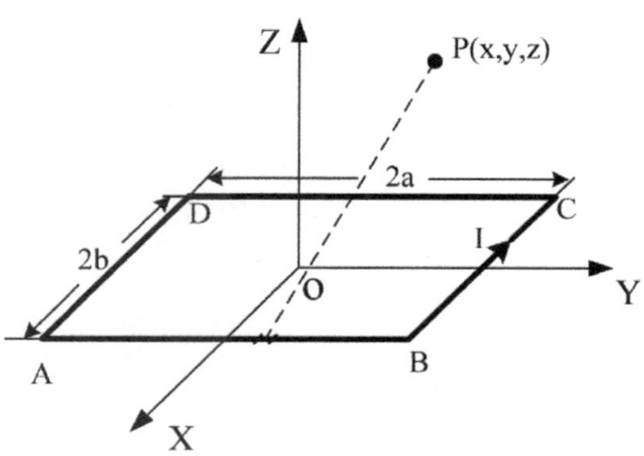

Figure 3. A rectangular coil in a Cartesian coordinate system.

$$\bar{B}_{AB} = \frac{\mu_0 NI\left[z\bar{j} - (y+b)\bar{k}\right]}{4\pi\left[(y+b)^2 + z^2\right]}(\sin\beta_1 - \sin\alpha_1) \tag{1}$$

where: $\alpha_1 = tg^{-1}\frac{-(a+x)}{\sqrt{(y+b)^2+z^2}}$, $\beta_1 = tg^{-1}\frac{a-x}{\sqrt{(y+b)^2+z^2}}$;

$$\bar{B}_{BC} = \frac{\mu_0 NI\left[(x-a)\bar{k} - z\bar{i}\right]}{4\pi\left[(x-a)^2 + z^2\right]}(\sin\beta_2 - \sin\alpha_2) \tag{2}$$

where: $\alpha_2 = tg^{-1}\frac{-(b+y)}{\sqrt{(x-a)^2+z^2}}$, $\beta_2 = tg^{-1}\frac{b-y}{\sqrt{(x-a)^2+z^2}}$;

$$\bar{B}_{DC} = \frac{\mu_0 NI\left[z\bar{j} - (y-b)\bar{k}\right]}{4\pi\left[(y-b)^2 + z^2\right]}(\sin\beta_3 - \sin\alpha_3) \tag{3}$$

where: $\alpha_3 = tg^{-1}\frac{a-x}{\sqrt{(y-b)^2+z^2}}$, $\beta_3 = tg^{-1}\frac{-(a+x)}{\sqrt{(y-b)^2+z^2}}$;

$$\bar{B}_{AD} = \frac{\mu_0 NI\left[(x+a)\bar{k} - z\bar{i}\right]}{4\pi\left[(x+a)^2 + z^2\right]}(\sin\beta_4 - \sin\alpha_4) \tag{4}$$

where: $\alpha_4 = tg^{-1}\frac{b-y}{\sqrt{(x+a)^2+z^2}}$, $\beta_4 = tg^{-1}\frac{-(b+y)}{\sqrt{(x+a)^2+z^2}}$, and $\bar{i}$, $\bar{j}$ and $\bar{k}$ denote the unit vectors of coordinate axes X, Y and Z respectively.

According to the electromagnetic field superposition principle, the magnetic induction intensity in free space due to the rectangular coil is obtained as follows:

$$\overline{B_c} = \overline{B_{AB}} + \overline{B_{BC}} + \overline{B_{CD}} + \overline{B_{DA}} \tag{5}$$

Sensor Model Simplification

The coil-stator system model is shown in Figure 4.

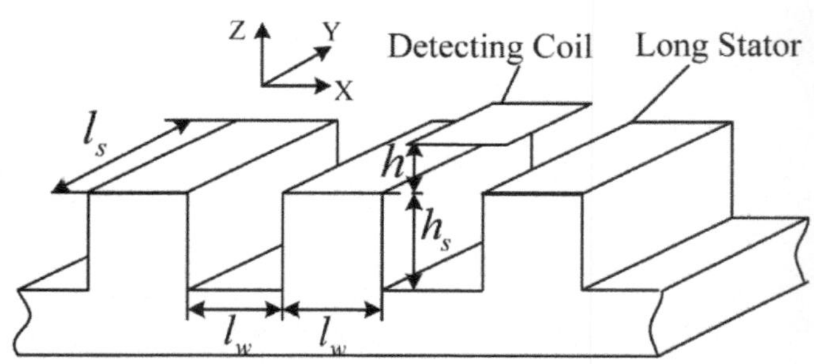

Figure 4. The coil-stator system.

The length and width of the coil and the width of a slot or a tooth denoted by l_w are all 43 mm, the depth of a slot is about h_s = 50 mm, and the width of the long stator is about l_s = 100 mm. Under normal working conditions, the distance between the coil plane and tooth plane is h = 8 mm.

The model is too complex for an analytical solution of the field, so some reasonable simplification is needed. Supposing I = 1 A and N = 1, according to Equations (1) to (5), the Z-direction component of the magnetic induction intensity at the tooth plane and the slot plane due to the rectangular coil excitation are simulated as shown in Figure 5.

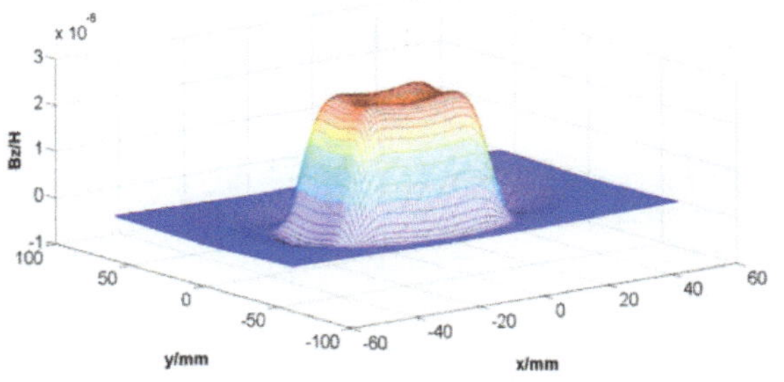

(a)

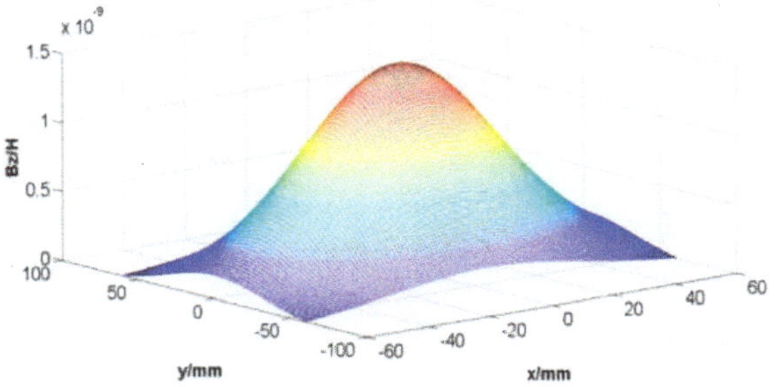

(b)

Figure 5. The Z-direction component of the magnetic induction intensity due to the rectangular coil excitation at (**a**) the tooth plane and (**b**) the slot plane.

According to Figure 5, the magnetic field is mainly distributed near the area facing the coil, and the magnetic induction intensity decreases rapidly with the increasing distance between the field plane and the coil plane, so the long stator shown in Figure 4 can be simplified to a series of cubes with an edge length l_w = 43 mm shown in Figure 6.

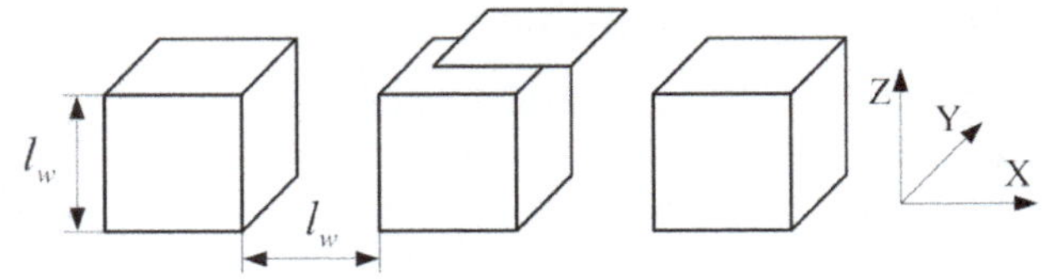

Figure 6. Simplified model of the coil-stator system.

Furthermore, considering the periodicity, the coil inductance is figured out only when $0<l_d<l_w$ shown in Figure 7. That's to say, the simplified model only consists of a cube and a square coil. In addition, the magnetic field due to the excitation coils is relatively weak, so the cube made of laminated silicon steel is regarded to be working in the linear region.

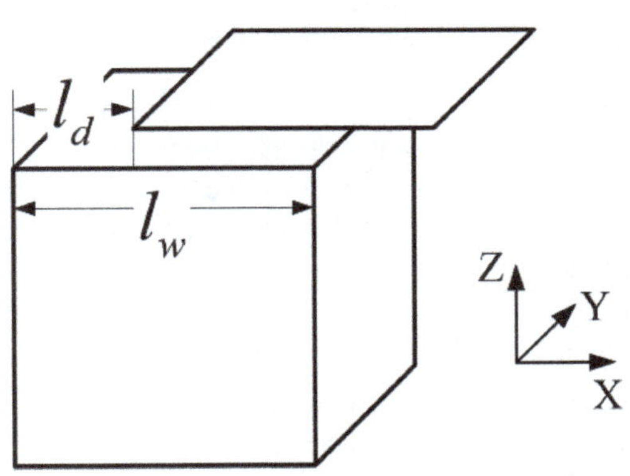

Figure 7. The final simplified model of the sensor.

ELECTROMAGNETIC MODELING OF THE SENSOR

Multipole Theory for 3-D Static Magnetic Field

In a 3-D static field comprising several domains separated by medium boundaries, there exists a certain scalar W, such as electric potential or scalar magnetic potential, satisfying the Laplace equation:

$$\nabla^2 W=2 \tag{6}$$

The domain in which W is to be solved is called the valid domain. In order to get the solution of W in this domain, auxiliary source points should be introduced into the field according to MT. An auxiliary source point set in the valid domain is called an inside pole. Otherwise, the pole is called an outside pole. For an open valid domain, there is no inside pole. For a finite valid domain, at most one inside pole can be set in it, and the pole had better be set at the geometrical center of the domain to make its corresponding series expansion converge fast [10].

The outside poles are set according to the boundaries. If all the inner angles between adjacent boundaries of the valid domain are less than 180°, each boundary should be assigned to an outside pole. For a spherical boundary surface, the pole should be set at the centre of the spherical surface to make the series expansion converge fast. For a plane boundary, it can be treated as a part of a spherical surface with an infinite radius. In this case, the series expansion corresponding to the plane boundary is reduced to a constant. Furthermore, if there exist inner angles greater than 180°,W can't be solved based on MT. In this case, the NES method [10] is introduced to solve the field approximately. The NES method will be discussed in the following section. The requirement of the inner angles of boundaries is called the expansion condition in MT.

Consider the situation shown in Figure 8. The inside pole is set at the centre of the valid domain denoted by O_1. The outside pole is set at O_2, the centre of spherical surface Γ_2. The solution to the Laplace Equation (6) in the valid domain is given as follows [4,7,8]:

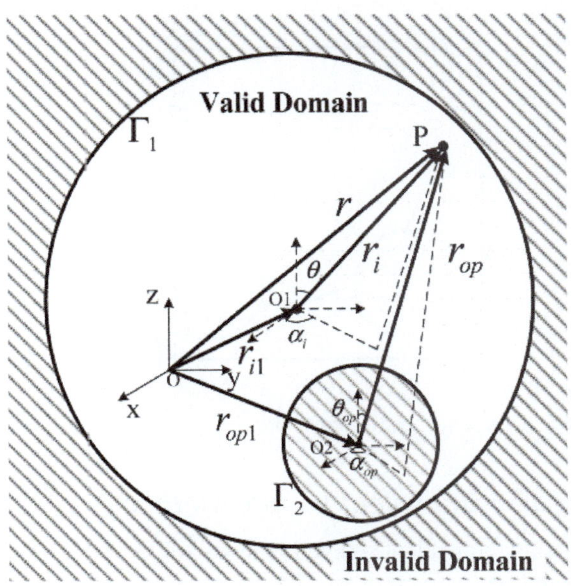

Figure 8. A field with two spherical boundaries.

$$W = A + \sum_{l=1}^{N_i} r_i^l (A_i^l P_l(\cos\theta_i) + \sum_{m=1}^{l} P_l^m(\cos\theta_i)(A_i^{lm}\cos m\alpha_i + B_i^{lm}\sin m\alpha_i))$$
$$+\sum_{p=1}^{N_p}[\frac{A_{op}}{r_{op}} + \sum_{l=1}^{N_{op}} r_{op}^{-(l+1)}(A_{op}^l P_l(\cos\theta_{op}) + \sum_{m=1}^{l} P_l^m(\cos\theta_i)(A_{op}^{lm}\cos m\alpha_{op} + B_{op}^{lm}\sin m\alpha_{op}))$$
$$=\sum_{k=1}^{N}\alpha_k U_k \qquad (7)$$

where $r_i = \left|\vec{r} - \overrightarrow{r_{i1}}\right|$, $\overrightarrow{r_{i1}}$ is the position vector of the inside pole; $r_{op} = \left|\vec{r} - \overrightarrow{r_{opn}}\right|$, $\overrightarrow{r_{opn}}$ is the position vector of the P th outside pole; N_i is the order of the inside pole ; N_p is the number of the outside poles (in Figure 8, $N_p = 1$); N_{op} is the order of the P th outside pole; N is the total number of the terms of the series expression; $A, A_i^l, A_i^{lm}, B_i^{lm}, A_{op}, A_{op}^l, A_{op}^{lm}$ are coefficients determined by boundary conditions; $P_l(\cos\theta_i)$ is Legendre polynomial and $P_l^m(\cos\theta_i)$ is associated Legendre function. a_i and θ_i are coordinates of field point P in spherical coordinate system O_1. a_{op} and θ_{op} are coordinates of field point P in spherical coordinate system of the pth outside pole (in Figure 8, it is O_2).

Experientially, to get a precise enough solution, the order of a pole corresponding to a spherical surface boundary can be set to 6, and the order of an inside pole corresponding to a cubic domain can be set to 10.

New Equivalent Source Method

Spherical Equivalent Source Method

Suppose the domain to be solved is the space outside the cube shown in Figure 9. The inner angle between adjacent boundaries is 270° dissatisfying the expansion conditions. In order to apply the MT method, we fill the cube with auxiliary spheres and treat these spherical surfaces as the boundaries instead of the surface of the cube. After this approximate substitution, MT can be applied to the problem. The substituted boundaries could be more approximate to the cube, if the cube is filled with more spheres. But increasing the number of the spheres will complicate the solution procedure.

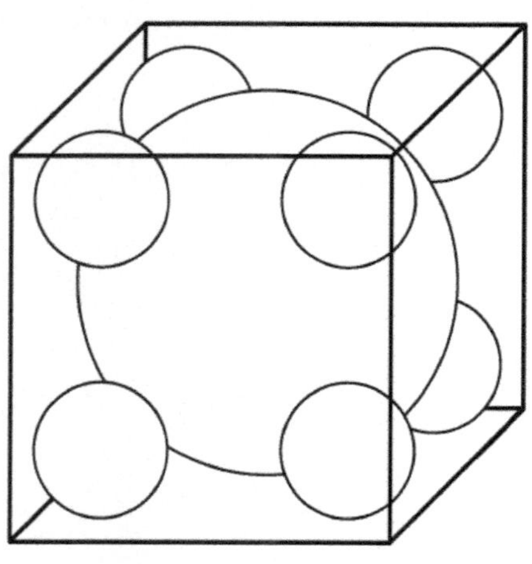

Figure 9. Spherical equivalent source.

Circular Equivalent Source Method

A circular equivalent source is shown in Figure 10. The scalar magnetic potential Ù at an arbitrary position P(x,y,z) due to the source satisfies Laplace equation, and has a solution given as follows [10]:

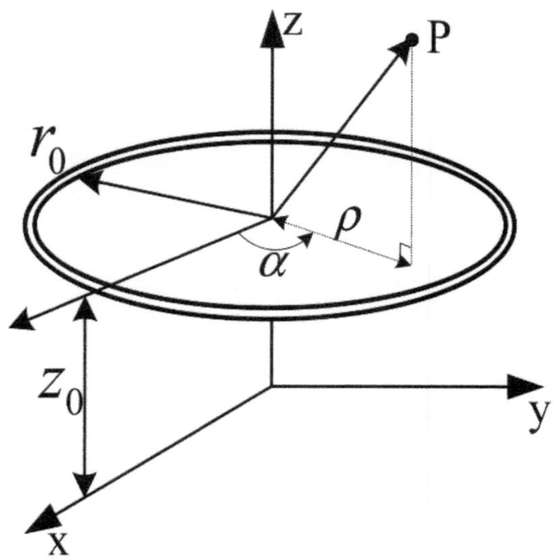

Figure 10. Circular equivalent source.

$$\Omega = \frac{1}{2\pi\mu_0}\sqrt{\frac{r_0}{\rho}}\sum_{k=0}^{m_c} Q_{k-1/2}(Y)(c_k\cos k\alpha + d_k\sin k\alpha) \qquad (8)$$

where c_k and d_k are undetermined coefficients; $Y=((z-z_0)^2+r_0^2+\rho^2)/2r_0\rho$; and $Q_{k-1/2}$ is half-integer degree Legendre function [19], which can be expressed by the hypergeometric function F(a,b,c,z) as $Q_l(t)= F(-l,l+1,1,(1-t)/2)$ [14].

Thus, similar to the spherical equivalent sources, the invalid domain can be filled with auxiliary circular sources to replace the actual boundaries.

Semi-Analytical Solution of the Electromagnetic Field of the Simplified Sensor Model

Usually, it is hard to get the solution of a 3-D electromagnetic field problem by solving the Maxwell's equations directly, so Smythe [20] proposed a second-order vector potential (SOVP) $\vec{W}$, which can be expressed as the sum of two components normal to each other:

$$\vec{W} = \vec{e}W_1 + \vec{e}\times\nabla W_2 \qquad (9)$$

where $\vec{e}$ denotes $\vec{i},\vec{j},\vec{k}$ or $\vec{r} = r\vec{e_r}$;; W_1 and W_2 are scalar components. The vector potential $\vec{A}$ can be expressed as the curl of $\vec{W}$:

$$\vec{A} = \nabla\times\vec{W} \qquad (10)$$

In 3-D passive static electromagnetic field, the vector potential $\vec{A}$ satisfies the Helmholtz equation:

$$\nabla^2\vec{A} + k^2\vec{A} = 0 \qquad (11)$$

where $k^2 = -j\omega\mu\sigma$, ω denotes the angular frequency of the field, σ denotes the electric conductivity and μ denotes the magnetic permeability. Considering the long stator of the high speed maglev train is made of laminated silicon steel, its eddy current is weak enough to be ignored, so σ can be regarded to be 0, and Equation (11) is reduced to a Laplace equation.

According to Equations (9–11), it can be concluded that W_1 and W_2 satisfy the Laplace equations:

$$\nabla^2 W_1 = 0 \qquad (12)$$

$$\nabla^2 W_2 = 0 \qquad (13)$$

The magnetic induction intensity $\vec{B}$ satisfies:

$$\vec{B}=\nabla\times\vec{A}=\nabla\times\nabla\times\vec{W} \quad (14)$$

Formula (14) results in the following expressions in spherical coordinate system:

$$\begin{cases} B_r = r\dfrac{\partial^2 W_1}{\partial r^2}+2\dfrac{\partial W_1}{\partial r} \\ B_\theta = \dfrac{\partial^2 W_1}{\partial r\partial\theta}+\dfrac{1}{r}\dfrac{\partial W_1}{\partial\theta} \\ B_\varphi = \dfrac{1}{\sin\theta}\dfrac{\partial^2 W_1}{\partial r\partial\varphi}+\dfrac{1}{r\sin\theta}\dfrac{\partial W_1}{\partial\varphi} \end{cases} \quad (15)$$

According to Equation (15), $\vec{B}$ is only a function of W_1 which satisfies the Laplace Equation (12) and can be solved through MT and the NES method discussed above.

For the simplified problem model shown in Figure 7, there are two domains: the air domain (denoted by (I) and the silicon steel domain (denoted by II). The scalar components of SOVP in domain I and II denoted by W_{I1} and W_{II1} are calculated as follows, respectively:

Domain I

In domain I, there exists an excitation coil, so the vector potential $\vec{A_1}$ satisfies the Possion equation:

$$\nabla^2\vec{A_1}=-\mu_0\vec{J} \quad (16)$$

where $\vec{J}$ is the current density of the coil. $\vec{A_1}$ is the sum of a particular solution $\vec{A_{1c}}$ of Equation (16) and the solution of the Laplace equation about $\vec{A_{10}}$, so the magnetic induction intensity $\vec{B_1}$ can be expressed as:

$$\vec{B_1}=\vec{B_{10}}+\vec{B_{1c}} \quad (17)$$

The induction intensity due to the excitation coil in the free space can serve as a particular solution as $\vec{B_{1c}}$, which is achieved in Equation (5).

Considering domain I as the valid domain and according to the analysis above, it does not satisfy the expansion condition, so equivalent sources are introduced to replace the cubic boundary. The sectional view of the arrangement of equivalent sources in the cube is shown in Figure 11.

The cube is filled with six equivalent sources. Source no. 1 is a spherical equivalent source and sources no. 2, 3, 4, 5 and 6 are circular equivalent sources. To calculate the inductance of the coil above the cube, the magnetic induction intensity near the upper surface of the cube should be precise enough, so equivalent sources are mainly set at the upper part of the cube to make the upper surface of equivalent sources more approximate to the upper plane of the cube. Because domain I is an open domain, there is no inside pole.

Let W_{I1i} denote the SOVP scalar component due to the i th source. According to the discussion above, we have:

$$W_{I11}=\frac{A_{op}}{r_{op}}+\sum_{l=1}^{N_{op}} r_{op}^{-(l+1)}(A_{op}^{l}P_l(\cos\theta_{op})+\sum_{m=1}^{l}P_l^m(\cos\theta_i)(A_{op}^{lm}\cos m\alpha_{op}+B_{op}^{lm}\sin m\alpha_{op})) \quad (18)$$

$$W_{I12}+W_{I13}+W_{I14}+W_{I15}+W_{I16}=\sum_{i=1}^{5}\frac{1}{2\pi\mu_0}\sqrt{\frac{r_{0i}}{\rho}}\sum_{k=0}^{m_c}Q_{k-1/2}(Y_i)(c_k\cos k\alpha_i+d_k\sin k\alpha_i) \quad (19)$$

As shown in Figure 7, the coil is moving along the X-axis, and the model is symmetrical referring to the X-O-Z plane, so the undetermined coefficients of the sine terms about α in Equations (18) and (19) are 0. Thus, W_{I1} can be expressed as follows:

$$\begin{aligned} W_{I1} &= W_{I11}+W_{I12}+W_{I13}+W_{I14}+W_{I15}+W_{I16} \\ &= \frac{A_{op}}{r_{op}}+\sum_{l=1}^{N_{op}} r_{op}^{-(l+1)}(A_{op}^{l}P_l(\cos\theta_{op})+\sum_{m=1}^{l}P_l^m(\cos\theta_i)A_{op}^{lm}\cos m\alpha_{op}) \\ &+\sum_{i=1}^{5}\frac{1}{2\pi\mu_C}\sqrt{\frac{r_{0i}}{\rho}}\sum_{k=0}^{m_c}Q_{k-1/2}(Y_i)c_k\cos k\alpha_i=\sum_{i=1}^{N_1}a_iU_{1i} \end{aligned} \quad (20)$$

where a_i ($i = 1\ldots N_1$) are undetermined coefficients, and N_1 is the term number. The meanings of other symbols are the same as in Equations (7) and (8). By substituting Equation (20) into Equation (15), the series expansion of $\vec{B_{10}}$ can be obtained.

Domain II

There is no excitation source in domain II, so the vector potential $\vec{A_{II}}$ satisfies the Laplace equation. The inner angles of the boundaries of domain II are 90° satisfying the expansion condition, and according to Section 3.1, the series expansions of the plane boundaries are reduced to constants, so there is only a series expansion of the inside pole set at the centre of the cube in this paper. Thus, W_{II1} can be expressed according to Equation (7) as follows:

$$W_{II1}=A+\sum_{l=1}^{N_i}r_i^l[A_i^lP_l(\cos\theta_i)+\sum_{m=1}^{l}P_l^m(\cos\theta_i)A_i^{lm}\cos m\alpha_i]=\sum_{i=1}^{N_2}b_iU_{2i} \quad (21)$$

Where b_i ($i = 1\ldots N_2$) are undetermined coefficients, N_2 is the term number. By substituting Equation (21) into Equation (15), the series expression of BII−→ can be obtained.

Coefficients Determination

The undetermined coefficients of Equations (20) and (21) can be determined by the Least Square (LS) method based on the boundary conditions. The boundary conditions between the air domain and the silicon steel domain are given as follows:

$$\begin{cases} B_{In}=B_{IIn} \\ B_{It1}/\mu_0=B_{IIt1}/\mu \\ B_{It2}/\mu_0=B_{IIt2}/\mu \end{cases} \quad (22)$$

where μ (about 5,600 μ_0) denotes the permeability of silicon steel, B_{In}, B_{IIn} and B_{It1}, B_{IIt1}, B_{It2}, B_{IIt2} denote the normal components and tangential components of the magnetic induction intensity, respectively. The LS method is adopted to minimize the error between the solutions of the two domains on the boundary. The target function is given as follows:

$$I_{LS}=\int_\Gamma[(B_{In}-B_{IIn})^2+(B_{It1}-\frac{\mu_0}{\mu}B_{IIt1})^2+(B_{It2}-\frac{\mu_0}{\mu}B_{IIt2})^2]ds \quad (23)$$

Where I_{LS} is the function of the undetermined coefficients a_i ($i = 1\ldots N_1$) and b_i ($i = 1\ldots N_2$). In order to minimize I_{LS}, let the partial derivatives of I_{LS} about the undetermined coefficients equal to 0:

$$\begin{cases} \frac{\partial I_{LS}}{\partial a_i} = 0 & i = 1,2...N_1 \\ \frac{\partial I_{LS}}{\partial b_j} = 0 & j = 1,2...N_2 \end{cases} \tag{24}$$

There are ($N_1 + N_2$) equations in Equation (24), so the ($N_1 + N_2$) undetermined coefficients can be determined. After substituting the coefficients a_i ($i = 1...N_1$) in the expression of BI—→, the coil inductance L can be obtained as:

$$L = \frac{\Phi}{I} = \frac{1}{I}\int_S \overrightarrow{B_1} \cdot d\vec{S} \tag{25}$$

where S denotes the area of the coil.

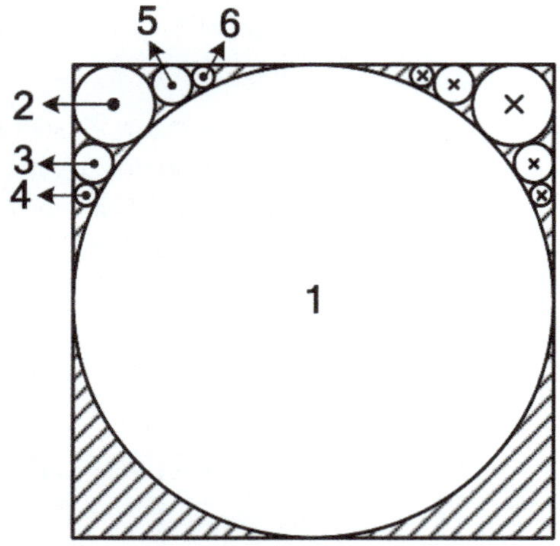

Figure 11. Configuration of the equivalent sources.

COMPARISON BETWEEN THE SOLUTIONS OF SEMI-ANALYTICAL METHOD AND NUMERICAL METHOD

Simulation Model of the Numerical Method

The Finite Element Analysis (FEA) software Ansoft Maxwell is used to simulate the electromagnetic field distribution of the relative position detection sensor. The simulation model is shown in Figure 12.

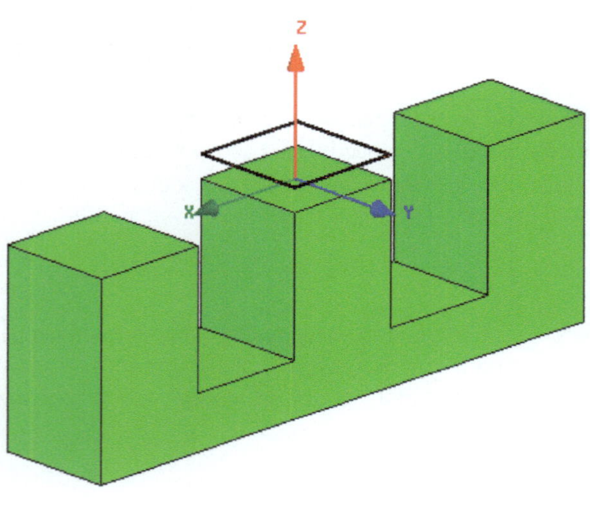

Figure 12. Simulation model of the numerical method.

Results of the Semi-Analytical Method and the Numerical Method

The calculation results of the coil inductance at different positions along the x-axis are shown in Figure 13, where the analytical modeling algorithm is implemented through MATLAB. Some special functions such as the Legendre polynomial and the hypergeometric function can be called straightforwardly from the MATLAB function library. According to Figure 13, there are errors between the numerical solution and the semi-analytical one. The maximal error is about 4.314 nH, which is much less than the coil inductance. The errors are mainly due to the model simplification and the discretization to get the undetermined coefficients by the LS method. In general, the coil inductance calculated through the semi-analytical method tends to approximately agree with that obtained by the numerical method, and the errors are within a tolerable limit.

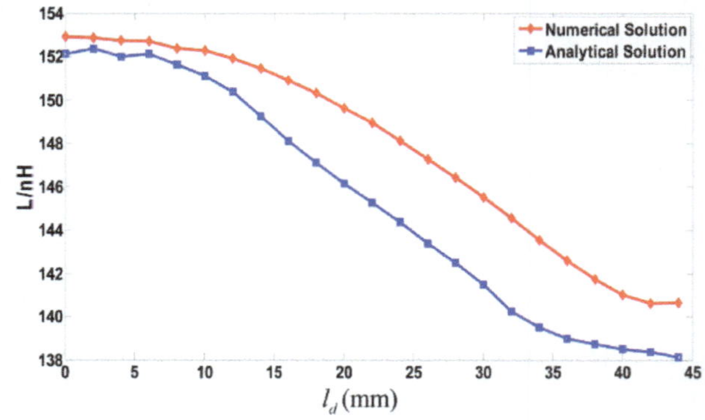

Figure 13. Relationship between coil inductance and relative position.

COIL DESIGN AND IMPROVEMENT

According the electromagnetic modeling results, the relationship between the coil inductance and the tooth-slot structure is shown in Figure 14. The relative position information can be gotten through table-lookup method according to this relationship.

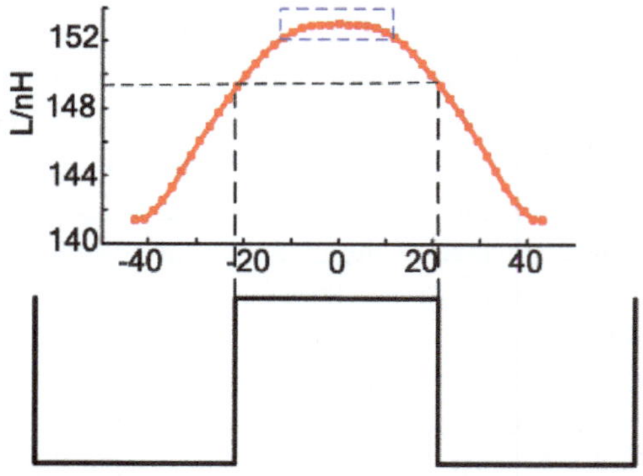

Figure 14. The relationship between coil inductance and the tooth-slot structure.

The curve shown in Figure 14 is similar to a sine wave but contains a large amount of harmonic components. When the coil is above a tooth, the slope ratio of the cure is small, which will reduce the sensitivity and accuracy of the table-lookup method. To improve this situation, a difference coil structure is proposed as shown in Figure 15. "8"-shaped Coils can counteract exterior electromagnetic disturbance. When the sensor is moving along the long stator at a certain constant speed, the relationship between the inductance of the two coils and the relevant displacement is shown in Figure 16(a). The phase difference of the two inductance curve is 180° because of the arrangement and dimension of the coils shown Figure 15. The difference inductance of the two coils has a better waveform as shown in Figure 16(b).

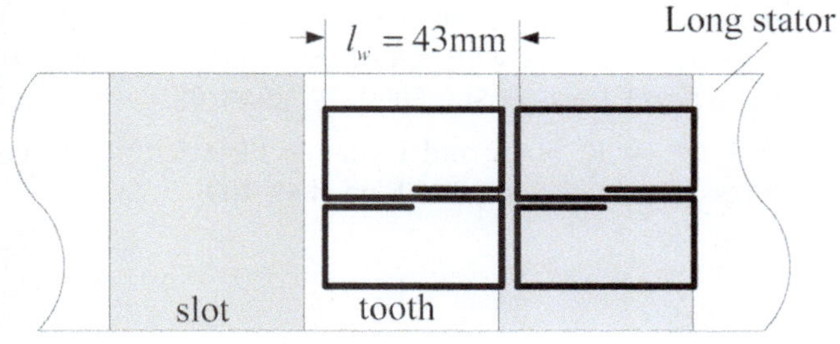

Figure 15. Difference coil structure.

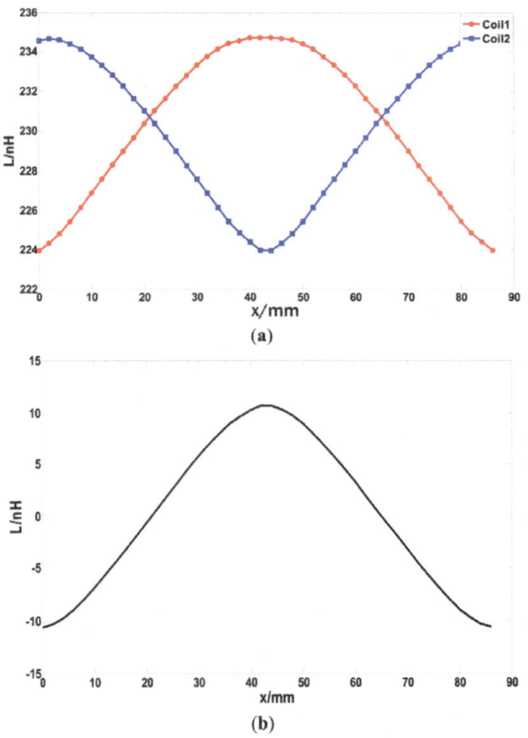

Figure 16. (**a**) Relative position and Inductance curve of coil 1 and 2; and (**b**) Inductance difference and relative position curve.

Thus, the precision of the sensor can be improved through the table-lookup method based on the relationship between the inductance difference and the relevant displacement. In addition, the difference operation can also counteract common mode disturbances such as temperature drift and enhance the performance of the sensor.

CONCLUSIONS

The relative position detection sensor obtains the relative displacement information between the train and the long stator by detecting the tooth-slot structure of the long stator based on nondestructive technology. This paper researches the electromagnetic field modeling of the sensor to guide the coil design and improvement. The electromagnetic field model of the coil-stator system is a complex three-dimensional (3-D) electromagnetic problem, so at first, the problem is simplified based on the analysis of the electromagnetic field distribution of a rectangular coil in free space. And then, MT and the NES method are adopted to get the series expansion of the magnetic induction intensity of the simplified model. The relationship between the coil inductance and the relative position is figured out afterwards. The inductance is also simulated through FEA. The comparison results show that the semi-analytical solution approximately agrees with the numerical solution. Finally, based on the modeling work, a difference coil structure is designed to improve the sensitivity and precision of the relative position detection sensor.

ACKNOWLEDGMENTS

This work was performed at the Engineering Research Center of Maglev Technology at National University of Defense Technology with funding from National Natural Science Foundation of China under grant no. 60874015.

REFERENCES

1. Liu, H.Q. Transrapid; University of Electronic Since and Technology of China Press: Chengdu, China, 1995; pp. 74–83.
2. Wu, X.M. Maglev Train; Shanghai Science and Technology Press: Shanghai, China, 2003.
3. Li, L.; Wu, J. Research and design of a position detection sensor for high speed maglev train.Chin. J. Sens. Actuat. 2007, 7, 1510–1513.
4. Deng, X.; Liu, Y. Electromagnetic imaging methods for nondestructive evaluation applications. Sensors 2011, 11, 11774–11808.
5. García-Martín, J.; Gómez-Gil, E.; Vázquez-Sánchez, J. Non-destructive techniques based on eddy current testing. Sensors 2011, 11, 2525–2565.
6. Zhu, Y.-K.; Tian, G.-Y.; Lu, R.-S.; Zhang, H. A review of optical NDT technologies. Sensors2011, 11, 7773–7798.
7. Zaoui, A.; Menana, H.; Feliachi, M.; Berthiau, G. Inverse problem in nondestructive testing using arrayed eddy current sensors. Sensors 2010, 10, 8696–8704.
8. Xue, S.; Dai, C.; Long, Z. Research on Location and Speed Detection for High Speed Maglev Train Based on Long Stator. Proceedings of the 8th World Congress on Intelligent Control and Automation (WCICA), Jinan, China, 7–9 July 2010; Volume 7, pp. 6953–6958.
9. Dai, C.; Long, Z.; Xie, Y.; Xue, S. Research on the filtering algorithm in speed and position detection of maglev trains. Sensors 2011, 11, 7204–7218.
10. Theodoulidis, T.P.; Bowler, J.R. Eddy current coil interaction with a right-angled conductive wedge. R. Soc. 2005, 461, 3123–3139.
11. Ma, Q.S.; Gao, G.; Sheng, J.N.; Jiang, K.X. Introducing a general analytical method for solving electromagnetic fields with complicated boundary. IEEE Trans. Magn. 1994, 30, 2920–2923.
12. Theodoulidis, T.P.; Tsiboukis, T.D. Analytical and numerical solution of eddy-current problem in spherical

coordinates based on the second-order vector potential formulation.IEEE Trans. Magn. 1997, 33, 2461–2472.

13. Shen, J.N.; Ma, Q.S.; Yuan, B.; Zheng, Q.H.; Yan, Z.W. Theoretical Methods and Applications of Semi-Analytical Method in Analysis of Electromagnetic Field and Wave; Science Press: Beijing, China, 2006.
14. Zheng, Q.H.; Xie, F. Solution of three-dimensional helmholtz equation by multipole theory method. Electromagnet Waves Appl. 1999, 13, 339–357.
15. Zheng, Q.H.; Xie, F.; Yang, Y.; Lin, W. Solution of three-dimensional laplace equation by multipole theory method. Electromagnet Waves Appl. 1999, 13, 1153–1171.
16. Lu, L. The study and design of suspensive gap detecting sensor for high speed maglev train.J. Natl. Univ. Def. Technol. 2007, 10, 14–17.
17. Li, T. Study and Realization of High-Speed Maglev Vehicle Relative Location Sensor. Master Thesis, Southwest Jiaotong University, Chengdu, China, 2006.
18. Hayt, W.H., Jr.; Buck, J.A. Engineering Electromagnetics; McGraw-Hill Companics: New York, NY, USA, 2001; pp. 212–215.
19. Peng, F.Q; Yu, J.H. The characters and computation of legendre function with non-integral degree in SCHA. Acta Geod. Cartogr. Sin. 2000, 29, 204–208.
20. Smythe, W.R. Static and Dynamic Electricity; Science Press: Beijing, China, 1982; pp. 404–409.

CHAPTER 13

MOLDING ACOUSTIC, ELECTROMAGNETIC AND WATER WAVES WITH A SINGLE CLOAK

Jun Xu, Xu Jiang, Nicholas Fang, Elodie Georget, Redha Abdeddaim, Jean-Michel Geffrin, Mohamed Farhat, Pierre Sabouroux, Stefan Enoch & Sébastien Guenneau

INTRODUCTION

Eight years ago, two seminal papers by Pendry, Schurig and Smith[1] and Leonhardt[2] have revolutionized the way we think about control of electromagnetic waves. These authors proposed to mold the flow of light through space coordinate transformations. As an illustration of this transformational optics (TO) tool, a design of a spherical invisibility cloak was proposed in[1] by mapping a sphere onto a hollow sphere. The shell surrounding the hole consists of an anisotropic heterogeneous metamaterial described by spatially varying, matrix valued, permittivity and permeability deduced from a formula based on the invariance of Maxwell's equations under geometric transforms. Such a route to cloaking can be viewed as analogous to what Einstein's equations teach us in relativity[3] light trajectories follow geodesics associated with the space (-time) metric. Interestingly, in[2] conformal mappings were used, which preserve right angles of the metric and thus avoid the anisotropic nature of coefficients in the transformed equations, but this conformal optics (CO) approach to cloaking is constrained by wavelengths small compared to the object to hide. Both TO and CO approaches have fueled metamaterials' research, with notably the experimental proof of concept of a cylindrical cloak consisting of concentric layers of split ring resonators of varying size molding the trajectory of transverse magnetic waves around a copper cylinder at 8.5 GHz via artificial anisotropic permeability[4]. The constraint on extreme material parameters has been since then relaxed with the advent of carpet cloaks[5] which is somewhat a mitigation of TO and CO. Their design is based upon quasi-conformal maps, which considerably reduce anisotropy of the transformed medium and have thus made possible fabrication and characterization of broadband three-dimensional ground cloaks at optical wavelengths[6,7] and even a nearly perfect two-dimensional omnidirectional cloak in the microwave regime[8].

In the tracks of electromagnetic cloaks, control of acoustic and water wave trajectories has been proposed[9,10,11,12,13] with acoustic metamaterials. It is also possible to detour elastodynamic waves[14,15,16], although this requires a fairly exotic (possibly asymmetric) elasticity tensor for fully coupled pressure and shear waves. This is due to the fact that when one moves into the area of elastodynamic waves, governing equations become much less tractable as pressure and shear waves are inherently coupled in structured solids. Another difficulty intrinsic to the tensor nature of the Navier equations is that they do not behave nicely upon a change of coordinates[14].

In contrast to Maxwell's equations, the form of Navier equations is not preserved under geometric transforms such as space folding, whereby the image plane of a flat lens with a negative refractive index is expected to map onto its source plane[3], and this explains perhaps why Pendry's perfect lens[17] can only be extended to specific elastodynamic wave polarizations, such as subwavelength focusing of Lamb waves in thin plates[18]. Let us note in passing that some form of external cloaking can be achieved via space folding for a finite set of dipoles located in the close neighborhood of a perfect lens[19].

Notwithstanding the fact that cloaking displays some universal features such as enhanced control of wave wavefronts (one might say a wave is a wave), some word of caution should be made regarding invisibility versus protection in various physical contexts. Indeed, cloaking for diffusion of heat[20,21,22] shows one can make a hole in the transformed space and achieve invisibility (i.e. control of isothermal curves in the case of heat flux management). However, waves and heat behave in different ways in the central region of invisibility cloaks: the amplitude of waves should vanish inside the core

Citation: Xu, J. et al. Molding acoustic, electromagnetic and water waves with a single cloak. Sci. Rep. 5, 10678; doi: 10.1038/srep10678 (2015).

of acoustic and electromagnetic cloaks, while the temperature might still undergo some elevation inside the core of thermal cloaks. In the present report, we would like to look at the field inside an electromagnetic cloak. We recall in Fig. 1(a–c) the experimental setup and earlier proof of cloaking via reduction of backscattering for the waterwave cloak introduced in10, see photo inFig. 1(e), for an acoustic source placed in its close neighborhood at 10 Hz. We would like to investigate the electromagnetic response of this cloak when it is placed in an anechoic chamber for microwaves, as schematically shown in Fig. 1(d), as well as for pressure waves, see Fig. 1(f). In order to assess the cloaking efficiency of the cloak in the microwave frequency range [2.6,6.0] GHz, we numerically computed the total radar cross section (RCS) of the cloak, a metal obstacle, and the cloak surrounding the metal obstacle using the CST Microwave Studio software, see Fig. 2. We further computed the counterpart of the total RCS of the cloak for pressure waves using the COMSOL Multiphysics software that indicate cloaking within the frequency range [1,5.8] KHz, see Fig. 3. In the microwave experiment, a ridged horn antenna placed a distance of 24 cm behind the cloak generates transverse electromagnetic waves between 0.8 and 18 GHz which are measured with a magnetic probe (mostly) sensitive to vertical magnetic fields. This allows us to measure one component of the magnetic field in the plane 5 mm above the cloak, outside and inside the cloak, both in magnitude and phase, see Figs. 4&5. One of our findings is that the field inside the core of the cloak is considerably reduced at the cloaking frequencies, as it transpires in the magnetic field's intensity profiles inFigs. 6&7, and the 2D plots of magnetic field's phase and intensity inFigs. 5,8&9. This suggests some form of protection for the object placed inside the cloak (a square metallic bar in the microwave experiment, seeFig. 1, and a glass bottle in the acoustic experiment). We also show in this paper that the cloak works equally well for pressure waves in air, see Figs. 3&4. This is due to the fact that the same mathematical model (Helmholtz equation with Neumann boundary conditions set at the structural elements of the aluminium cloak) is valid for transverse electric waves (structural elements behave like infinite conducting inclusions in microwaves), pressure waves (they behave like rigid inclusions in air) and waterwaves (no flow conditions). We note that the microwave cloak of Schurig et al.4 works for Transverse Magnetic (Ez) polarization, but that of Kanté & de Lustrac works for Transverse Electric (Hz) polarization23, simply by changing the orientation of split ring resonators. However, resonances of SRRs make these cloaks narrowband. We stress that our cloak works via homogenization24·25·26·27·28·29·30 and cloaking is here simply a consequence of a specific geometry (perforated concentric domain) as we shall see in the sequel. If one were to design an acoustic cloak based on acoustic SRRs, a good way forward would be to implement the design in31. Note that our approach to cloaking is also notably different from that in32, which is based on a reverse engineering algorithm allowing for considerable reduction of acoustic scattering at a given frequency, but our work is somewhat reminiscent to the approach of33for the design of ground plane acoustic cloaks via effective anisotropy. On a larger scale, our work might also inspire designs of seismic cloaks34.

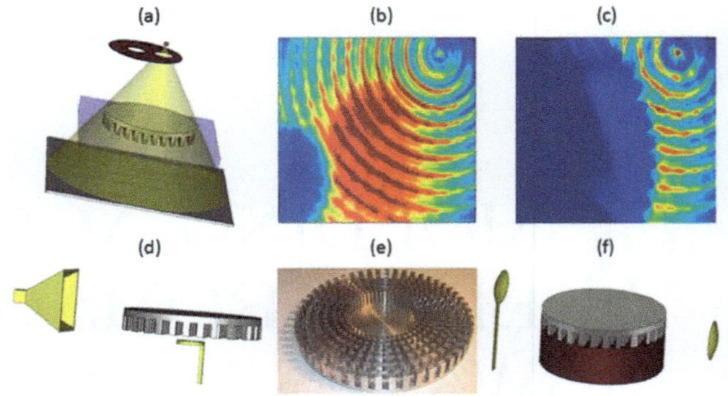

Figure 1: Overview of earlier results on waterwave cloak and new acoustic and electromagnetic experimental setups.

(**a**) artistic view of the experimental setup for linear surface water waves showing a source of light modulated by a perforated rotating disc, which illuminates a transparent vessel containing the liquid and the cloak, providing snapshots of wave pattern on a screen; (b-c) experimental results of reduced backscattering adapted from10 for a concentric liquid surface wave of frequency 10 Hz interacting with a rigid cylinder (7.6 cm in diameter) on its own (b) and the structured waterwave cloak (20 cm in diameter) (c); (d) artistic view of the experimental setup of the waterwave cloak tested in Fresnel Institute's anechoic chamber, with a rigded horn antenna generating transverse electric microwaves with frequencies ranging from 1 to 6 GHz; the longitudinal magnetic field is measured with a magnetic probe (underneath the cloak); (e) photo of the cloak; (f) artistic view of the experimental setup in MIT for the waterwave cloak tested for pressure acoustic waves generated by a speaker on the left, which are measured with a microphone mounted on a horizontal translational stage on the right.

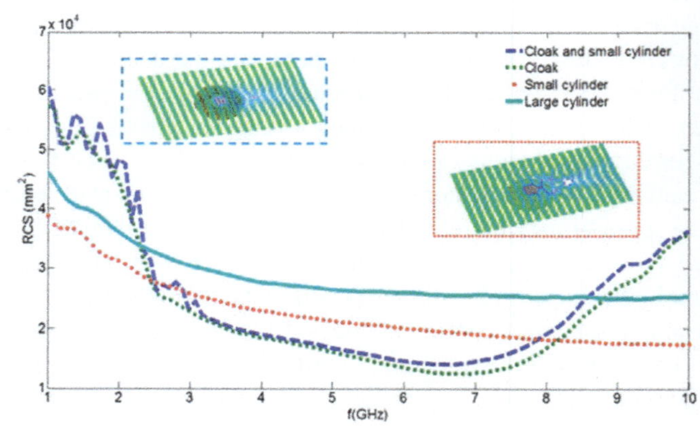

Figure 2: Numerical computation of radar cross section (RCS) for cloak and cylindrical metal obstacles versus frequency (three-dimensional electromagnetic simulations with Computer Simulation Technology Microwave Studio).

A small metallic obstacle surrounded by cloak (dashed dark blue curve) is nearly superimposed with RCS of cloak on its own (dotted green curve), and lower than RCS of small metallic obstacle (dotted red curve), which is itself smaller than RCS of large metallic obstacle of same diameter as the cloak (solid light blue curve) from 3 GHz to 7.8 GHz. Two insets further show isovalues of the real part of the longitudinal magnetic field for a plane wave incident at 3.5 GHz from the left on the small obstacle surrounded by the cloak –see inset marked with dashed blue lines—and on the small obstacle on its own –see inset with red dotted lines—, the latter displaying stronger forward scattering.

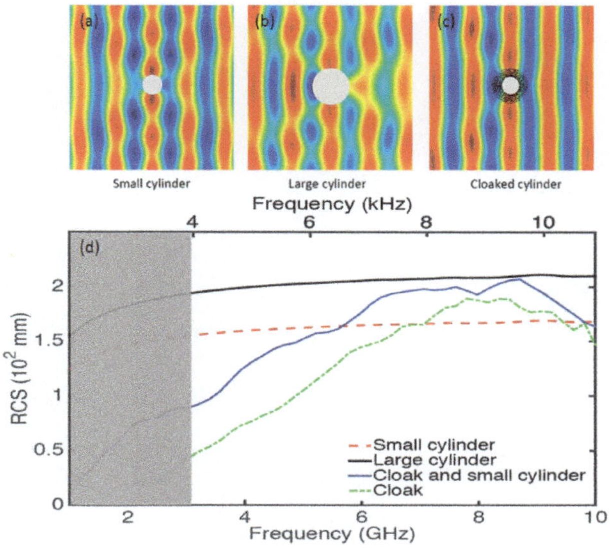

Figure 3: Numerical computation of radar cross section (RCS) for cloak and cylindrical rigid obstacles versus frequency (two-dimensional simulations with Comsol Multiphysics).

A small infinite conducting (resp. rigid for pressure waves) obstacle surrounded by cloak (solid dark blue curve) is slightly above RCS of cloak on its own (dotted green curve), but lower than RCS of small infinite conducting (resp. rigid) obstacle (dotted red curve), which is itself smaller than RCS of large infinite conducting (resp. rigid) obstacle of same diameter as the cloak (solid black curve) up to 7 GHz (resp. 8 KHz for pressure waves). Panels (**a**), (**b**), (**c**) further show the real part of the transverse magnetic (resp. pressure) field for a plane wave incident at 4.5 GHz (resp. 5 KHz) from the left on the small obstacle surrounded by the cloak on the small obstacle on its own (**a**), on the large obstacle (**b**), and on the cloaked obstacle (**c**).

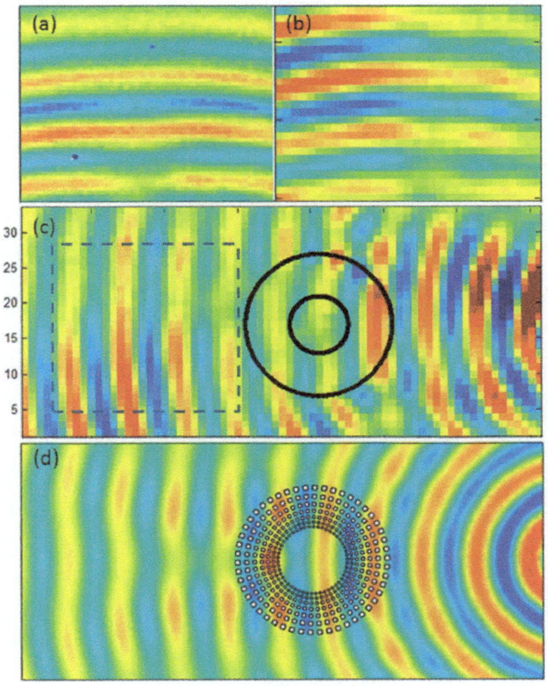

Figure 4: Acoustic and electromagnetic wave experiments versus finite element simulation.

(a, b) Experimental results in forward scattering for pressure waves (**a**) and transverse electric microwaves (**b**) for the same cloak containing a large glass bottle obstacle (**a**) and a small square metallic obstacle (**b**); The source is located 24 cm from the center of the cloak and its frequency lies inside the cloaking zone (5 KHz for pressure waves in (**a**) and 4.350 GHz for microwaves in (**b**); The real part of the pressure field in (**a**) and of the longitudinal magnetic field in (**b**) is measured within the inset of panel (**c**)-- see area marked with dashed lines--, which shows experimental results in backward and forward scattering for microwaves; (**d**) Comparison with finite element simulation, where Neumann (rigid for pressure field and infinite conducting for longitudinal magnetic field) boundary conditions are set on each inclusion. Cloaking is achieved as it transpires from the nearly unperturbed phase in both scenarios.

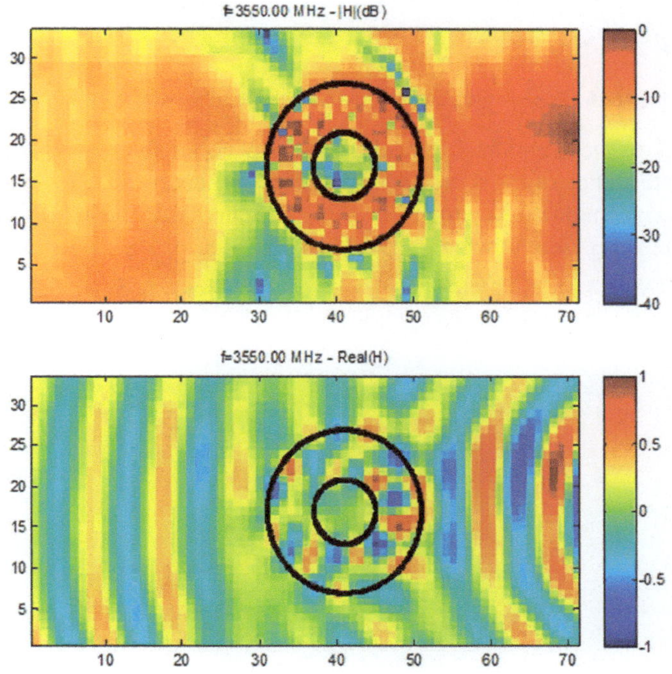

Figure 5: Experimental results in backward and forward scattering for microwaves.

The normalized modulus (upper panel) and the normalized real part (middle panel) of the magnetic field is measured within the rectangular domain shown in lower panel. The frequency of the source is 3.550 GHz. One notes that the magnetic field intensity is reduced by 18 dB at the center of the cloak with a reduction varying in the range [13.5,38]dB inside the inner disc. The field is enhanced inside the corona and both wavefront and field amplitude are nicely reconstructed behind the cloak. Notably the measured intensity is −11 dB and −1 dB at the left and right sides of the panel, respectively. See also Figure 5 for the profile of the field along the median passing through the center of the cloak. The square metallic obstacle inside the invisibility region is both invisible and protected from microwaves. Similar results hold in the cloaking interval [2.6,6]GHz, see Figs. 7&8.

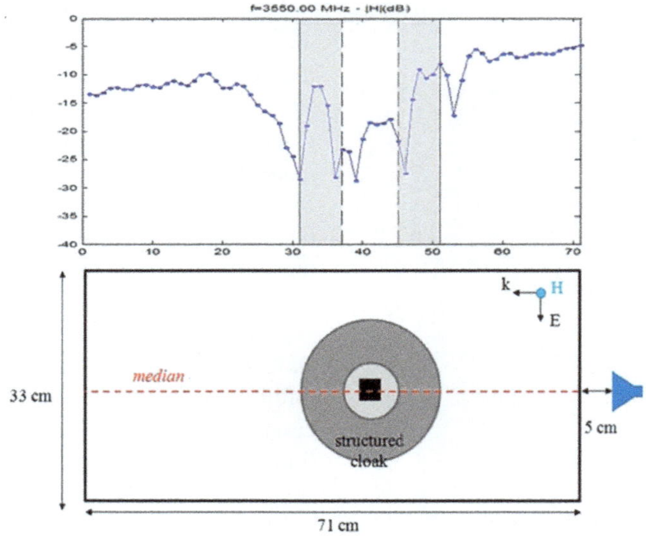

Figure 6: Experimental results in backward and forward scattering for microwaves.

The normalized modulus (upper panel)of the magnetic field is measured along the median shown in lower panel. The frequency of the source is 3.550 GHz. One notes that the magnetic field intensity is reduced by 18 dB at the center of the cloak with a reduction up to 30 dB inside the inner disc. The field is enhanced inside the corona and its intensity in front and behind the cloak is close to what it would be in free space. Similar results hold in the cloaking interval [2.6,6]GHz, see Fig. 6.

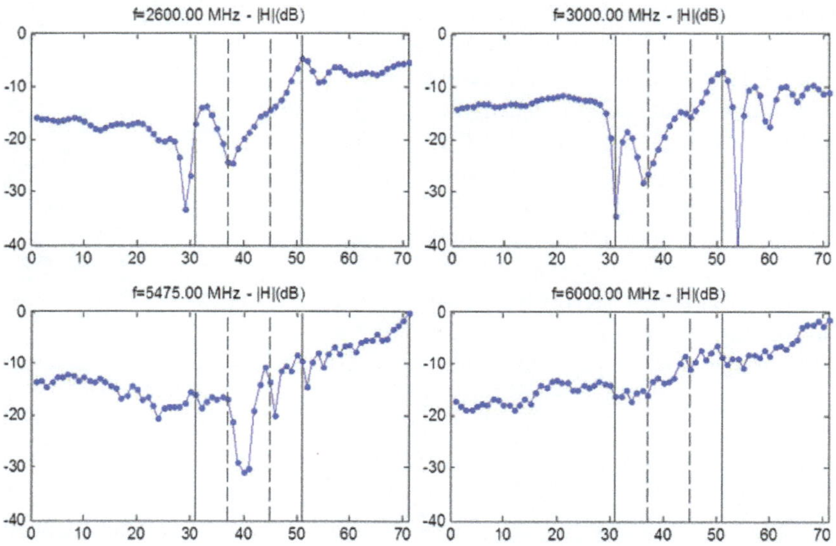

Figure 7: Experimental results for microwaves at four representative frequencies within the cloaking interval [2.6,6]GHz

The normalized modulus (upper panel) of the magnetic field is measured along the median passing through the center of the cloak as in Fig. 4 for the source at 2.6, 3, 5.475 and 6 GHz. One notes that the magnetic field intensity is reduced by up to 30 dB at the center of the cloak (whose inner and outer boundaries are marked by the solid and dashed vertical lines respectively). Protection is clearly more pronounced at 2.6 GHz than 6 GHz. Only invisibility is achieved for the latter frequency.

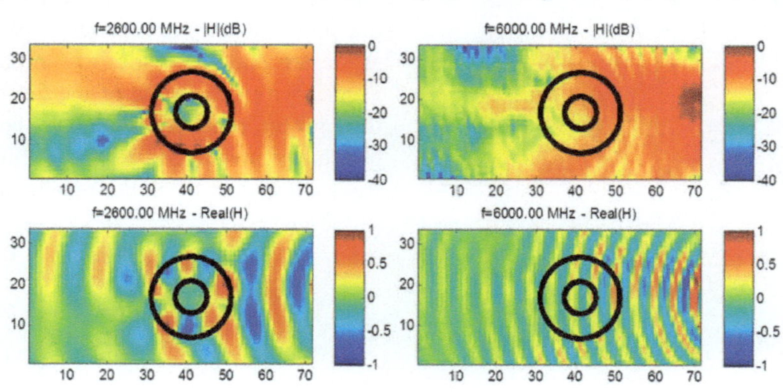

Figure 8: Experimental results for microwaves at the edges of the cloaking interval [2.6,6]GHz.

The normalized modulus (upper panel) and the normalized real part (middle panel) of the magnetic field is measured for the source at 2.6 and 6 GHz. One notes that the magnetic field intensity is reduced by 18.5 dB and 13 dB at the center of the cloak with a reduction varying in the range [11.5,35] and dB [8.5,17] dB inside the inner disc, respectively at 2.6 and 6 GHz. The field is enhanced inside the corona and both wavefront and field amplitude is better reconstructed behind the cloak at 6 GHz. Notably the measured intensity is −18 dB and −17 dB at the left side, respectively at 2.6 and 6 GHz and −1 dB at the right side, of the intensity panels. Protection is clearly more pronounced at 2.6 GHz than 6 GHz.

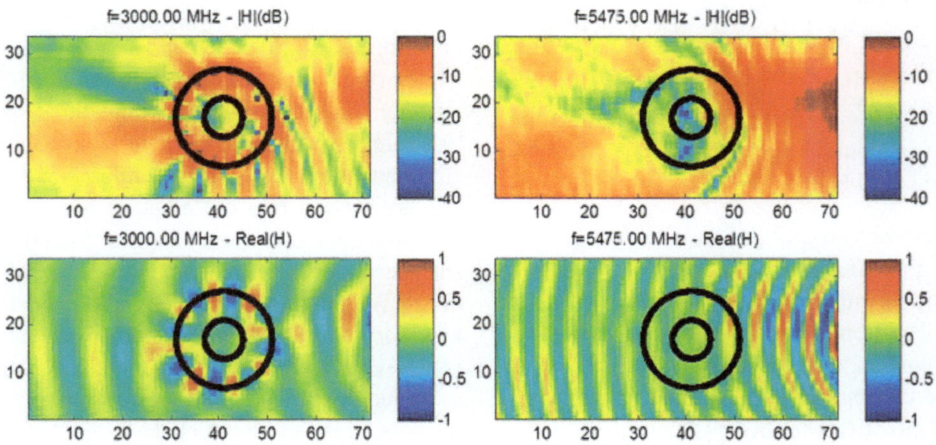

Figure 9: Experimental results for microwaves at two representative frequencies within the cloaking interval [2.6,6]GHz.

The normalized modulus (upper panel) and the normalized real part (middle panel) of the magnetic field is measured for the source at 3 and 5.475 GHz. One notes that the magnetic field intensity is reduced by 17 dB and 30 dB at the center of the cloak with a reduction varying in the range [12.5,24]dB and[11,38]dB inside the inner disc, respectively at 3.5 and 5.475 GHz. The field is enhanced inside the corona at 3 GHz and both wavefront and field amplitude is better reconstructed behind the cloak at 5.475 GHz. Notably the measured intensity is −14 dB and −12 dB at the left side, respectively at 3 and 5.475 GHz and −1 dB at the right side, of the intensity panels. Protection is clearly more pronounced at 5.475 GHz than 3 GHz.

RESULTS

Theory and numerics

We proceed with the mathematical approach of homogenization29·30, which amounts to replacing a structured material by an effective medium that captures the essential wave phenomena for wave wavelengths large compared to the typical heterogeneity size.

It can be shown, that sound waves with a wavelength much larger (in practice three times larger) than the largest structural elements of the cloak are governed by the following homogenized wave equation (written in polar coordinates)

$$-\nabla.\left(\begin{bmatrix}\rho_{rr}^{-1} & 0\\ 0 & \rho_{\theta\theta}^{-1}\end{bmatrix}(r)\,\nabla p(r,\ \theta)\right)=\frac{\omega^2}{\kappa_{eff}}p(r,\ \theta) \tag{1}$$

where the effective tensor of density is given by

$$\begin{bmatrix}\rho_{rr}^{-1} & 0\\ 0 & \rho_{\theta\theta}^{-1}\end{bmatrix}=\frac{1}{area\,(Y)}\int_0^{2\pi}\int_0^1 \rho(r,r',\theta')\begin{bmatrix}area\,(Y)-\Psi_{rr} & \Psi_{r\theta}\\ \Psi_{\theta r} & area\,(Y)-\Psi_{\theta\theta}\end{bmatrix}r'dr'd\theta' \tag{2}$$

Here, rr are periodic potentials given by

$$\Psi_{ij}=-\int_{\delta S}\psi_i n_j dS\,,\quad \Delta\psi_{ij}=0\,,\quad \frac{\partial\psi_i}{\partial n}=-\,n.\,e_i \tag{3}$$

where n is the outer unit normal to the boundary δS of the inclusion S in a unit cell Y.

The inverse of the effective bulk modulus is such that

$$\kappa_{eff}^{-1}(r)=\frac{1}{area\,(Y)}\int_0^{2\pi}\int_0^1 \kappa^{-1}(r,\theta,r')\,r'dr'd\theta' \tag{4}$$

Here r' and θ' are radial and azimuthal microscopic variables, and it should be noted that κ_{eff}^{-1} varies with the macroscopic variable r.

In a similar way, transverse electric waves are solution of the following homogenized Maxwell' equation in the cloak (see also supplemental material):

$$-\nabla.\left(\begin{bmatrix}\varepsilon_{rr}^{-1} & 0\\ 0 & \varepsilon_{\theta\theta}^{-1}\end{bmatrix}(r)\,\nabla H_z(r,\ \theta)\right) = \omega^2\mu_{eff}H_z(r,\ \theta) \tag{5}$$

where ε_{rr}^{-1} and $\varepsilon_{\theta\theta}^{-1}$ are the coefficients of the effective tensor of permittivity which are also given by (2) (replacing ρ by μ in the equation) and μ_{eff} is the effective permeability which is such that $\mu_{eff}(r) = \frac{1}{area(Y)}\int_0^{2\pi}\int_0^1 \mu(r,\ \theta,\ r')r'dr'd\theta'$. Note here that (4) is simply derived from the vector Maxwell equation by considering a magnetic field $H=(0,0,H_z)$.

It transpires from the integral expression of μ_{eff} above that $\mu_{eff}<\mu_0$ i.e. that the effective permeability has slightly lower, yet strictly positive, values than permeability of vacuum and its spatial radial variation is small. Hence, one can safely divide by μ_{eff} throughout in (5). In the same way, from the expression in (4) one can multiply throughout by κ_{eff} in (1). This leads to reduced effective parameters.

It can be shown that the reduced effective parameters in (1)-(4) are good approximations for well-known transformed density and permittivity in their reduced forms i.e.

$$\rho'_{rr} = \left(\frac{R_2(r-R_1)}{r(R_2-R_1)}\right)^2,\ \rho'_{\theta\theta} = \left(\frac{R_2}{R_2-R_1}\right)^2 \tag{6}$$

$$\varepsilon'_{rr} = \left(\frac{R_2(r-R_1)}{r(R_2-R_1)}\right)^2,\ \varepsilon'_{\theta\theta} = \left(\frac{R_2}{R_2-R_1}\right)^2 \tag{7}$$

In which case the transformed bulk modulus and permeability can be normalized to 1.

Modeling of the invisibility cloak has been carried out using the commercial finite elements package COMSOL MULTIPHYSICS. In practice, the cloak consists of a ring of inner radius $R_1=4.1$ cm and outer radius $R_2=10$ cm which has been first divided into 14 layers whose thickness varies with respect to r as $\left(\frac{R_2(r-R_1)}{r(R_2-R_1)}\right)^2$. One layer in two has been further divided in 100 identical angular sectors along the azimuthal direction θ as shown in Fig. 1(c). This alternation of homogeneous and structured layers has been optimized by solving the annex problem (3) with COMSOL, see10 for the numerical representation of periodic potential Ψ_{ij} solution of this annex problem in the context of linear surface liquid waves. This leads to the reduced effective parameters in (6) and (7). In order to quantitatively assess the cloaking efficiency of the metamaterial, following the procedure proposed in35, we numerically compute its total radar cross section (RCS)36·37 with the software CST Microwave Studio. We show in Fig. 2the electromagnetic RCS of the cloak on its own, of the cloak when its surrounds a small metal cylinder (see inset marked by dashed blue line in Fig. 2), of the small metal obstacle (see inset marked by dotted red line in Fig. 2), and of a large metal obstacle with same radius R_2 as the cloak. Moreover, in all these configurations, we consider the ground plane (5 mm thick) which is part of the cloak: this ground plane cannot be omitted in the three dimensional electromagnetic experimental setup. The RCS of the cloak (dark blue dashed curve) and of the small obstacle surrounded by the cloak are nearly superimposed in the frequency interval [2.6,7.8] GHz and much lower than the small obstacle on its own (dotted red curve). We have numerically checked that the shape and size of the small obstacle matters very little (provided the obstacle can fit within the central region of inner radius R_1 of the cloak).

We further show in Fig. 3 the electromagnetic and acoustic RCS computed with COMSOL MULTIPHYSICS using the 2D Helmholtz equation, which demonstrates cloaking for microwaves up to 7 GHz and sound waves up to 8 KHz (the lower frequency bound is irrelevant as the wave wavelength becomes larger than the cloak's height): These two-dimensional finite element simulations are valid for both sonic and transverse electric microwaves provided we make the assumption that the cloak is invariant along the vertical direction (see also Supplemental Material for RCS of linear surface water waves, which further confirms earlier results of10). We simply solve the Helmholtz equation with Neumann boundary conditions on the cloak's structural elements, which is a good model for rigid (resp. infinite conducting) inclusions in the context of pressure (resp. transverse electric) waves as announced in the introduction.

For the same 2D setup as in Fig. 3, we then show in Fig. 4 the total field scattered by a time harmonic acoustic source (resp. electromagnetic source) placed a few wavelengths away from the cloak at frequency 5 KHz (resp. 4.5 GHz), see Fig. 4 upper left panel (resp. upper right panel). The wave wavefronts are neatly reconstructed behind the cloak, as one can observe in the insets marked with the dashed parameter in both panels. The inset marked with the dashed perimeter in the middle panel is used for experimental comparisons between the pressure and microwave experiments in forward scattering in the upper panels, but both forward and backward scattering are shown for microwaves in the middle panel, with a good agreement with numerical simulation in the lower panel.

Experiments

Acoustic experiments were performed in MIT (Boston, USA) for sound waves and in Fresnel Institute (Marseille, France) for microwaves. We considered the two-dimensional Helmholtz model for the comparison between theory and acoustic experiments, therefore it came as a good surprise that our intuition was right: experiments in Fig. 4 clearly demonstrate that although the cloak's height is not large compared to the wave wavelength, it behaves as expected. An object (a large glass bottle in MIT and a small square metallic obstacle in Fresnel Institute) is rendered invisible by the structured aluminium cloak. In Fig. 4(a–b), we emphasize that slights discrepancies between phase of pressure and longitudinal magnetic field can be attributed to scattering induced by the free space experimental setups.

The experimental electromagnetic setup at Institut Fresnel is actually more in the spirit of that of Kanté et al.23 which is a nonmagnetic microwave cloak also characterized in free space, than the one of Schurig et al.4. Note however that both cloaks in4·23 heavily rely upon resonances of split ring resonators, whereas in our case cloaking is associated with some artificial, geometrical induced, anisotropy. The phase of the pressure field in the left panel of Fig. 3(a) displays strong similarities with that of11, although in this paper cloaking was achieved at ultrasonic frequencies in a shallow water experiment (the layer of water helped to guide the pressure waves, unlike for the current experiment). This suggests that although the current sound cloaking experiment was carried out in free space for this report, the cloak effectively controlled the wave path of sonic waves propagating in a layer of air passing through it.

Regarding the electromagnetic cloaking experiment, we were able to monitor both the phase and intensity of the longitudinal component of the magnetic field, in backward and forward scattering, as well as within the cloak. This experimental result clearly shows that the electromagnetic signal is much reduced inside the core of the cloak, where a metal plate is placed. Similar tests (not reported) have been performed for a cloud of metal nails and here again we observed very little electromagnetic field inside the core of the cloak (a reduction of magnetic field intensity by 13.5 to 38 dB at the cloaking frequency 3.550 GHz). This suggests some form of electromagnetic protection for the square metal plate (or another metal object) can be achieved at cloaking frequencies thanks to the cloak. We also observe that transmission is good for the electromagnetic wave propagating though the structured metal cloak. Hence, we not only achieve a good reconstruction of the wave wavefront (the phase), but also preserve its intensity.

DISCUSSION

One of the main features of the cloak is that it works through field averaging and only one phase (vacuum) need be considered in the homogenization and computational results to achieve a good prediction of experimental results. The underlying mechanism (artificial anisotropy) is so simple that it makes the cloak multiphysics: Two-dimensional numerical simulations with COMSOL MULTIPHYSICS show by a simple normalization of the wave frequency in (1) or (5) that cloaking should be at work between 5 and 16 Hz for surface liquid waves in the condition of10, (the corresponding surface water wave frequency should be $\sqrt{g\omega(1+\omega^2 d^2)\tanh(h\omega)}$ where g denotes the acceleration caused by gravity, d is the liquid capillarity length and h is the liquid depth), between 3 and 8 KHz for sound waves and between 2.6 to 7 GHz for microwaves. We note that the latter interval for cloaking frequencies is slightly shifted compared to that of [2.6,7.8] GHz found via the CST computation of RCS in Fig. 2, what can be attributed to 3D effects (the cloak's size in the vertical direction is in fact comparable to the electromagnetic wave wavelength) and the presence of a ground plane (supporting the pillars of the cloak). Therefore, our cloak cannot be considered in the strict sense as two-dimensional and our 2D homogenization and numerical models should be considered only as an approximation of full wave models. Experiments shown in Figs. 4 and 5confirm that the range of cloaking frequencies from 2.6 to 6 GHz (see also Figs. 8 and 9 for other frequencies): Although ridged horn antenna generates transverse electromagnetic waves between 0.8 and 18 GHz, our vector network analyzer only works between 50 KHz and 6 GHz.

We stress that we not only show the phase, but also the intensity of the magnetic field in the experiments. One can see that the phase of the signal is reconstructed behind the cloak in Fig. 5, but more importantly the intensity of the signal is preserved, as is also exemplified in Figs. 6and 7. The field is also measured within the cloak, and as one would hope it does seem to vanish inside the invisibility region. This is important for potential applications in protection from waves, and notably if one would like to scale up the structure in order to make a cloak for ocean waves.

The ability of the cloak to control more than one type of waves could have interesting applications in phoxonics. One could also envisage to increase the number of structural elements to improve cloaking even further. Homogenization tells us that the more structural elements, and the smaller their size compared to the wave wavelength, the better approximate cloaking, for some numerical evidence of this claim. We have thus experimentally demonstrated the first multi-physics cloaking (linear surface water waves, pressure waves and transverse electric waves) with a single cloak. The cloak is broadband, see Figs. 5,8 and 9, relatively easy to fabricate, and offers some form of protection for the object placed inside the invisibility region, see Figs. 6 and 7. We hope our experiments will foster theoretical and experimental efforts in multi-physics cloaking and exploration of invisibility versus protection for different waves. We point out that our cloak is reminiscent of labyrinthine acoustic metamaterials38, and it might inspire more elaborate designs of elasto-mechanical39 and elasto-electromagnetic40 metamaterials. Our cloak could have potential applications in telecommunications (protection against mobile phone radiations) and soundproof devices (humans are most sensitive to sound waves of frequencies between 2 and 5 kHz). The design could also be scaled down in order to achieve cloaking at optical wavelengths, what would require an analysis of cloak's dispersion, see e.g.41. Since the cloak works both in acoustic and microwave domains, it might offer new opportunities to drive and tune one field with the other one: for instance if the fluid within which pressure waves propagate is no longer air but some gas plasma or liquid electrolyte, we might be able to tune the density profile by microwave signals. On larger scales, control of sound and elastic waves could be used in anti-earthquake designs if protection is achieved via cloaking with seismic metamaterials34, that is for frequencies below 50 Hz. Finally, we note that preliminary numerical simulations suggest our cloak should also work in the context of management of thermal flux: it can be shown using homogenization methods42 that our cloak can be described by an effective conductivity tensor $k'_{rr} = \left(\frac{R_2(r-R_1)}{r(R_2-R_1)}\right)^2$, $k'_{\theta\theta} = \left(\frac{R_2}{R_2-R_1}\right)^2$ and by a product of effective mass density and specific heat averaged like the inverse of effective bulk modulus in (4). Our cloak could also serve as a prototype thermal metamaterial to mold the flow of heat or even light diffusion in the tracks of20·21·22 and43.

METHODS

Finite element method

COMSOL Multiphysics with "PDE" module has been used to compute the scattered field in Figs. 3&4, with Cartesian perfectly matched layers (PMLs) on either sides of the computational domain to avoid reflections on the computational domain boundaries. Some Neumann boundary conditions are assumed on each structural element of the cloak and the source is set in the form $-1/4\omega H_0^{(1)}(\omega|r - r_s|)$ where r_s is the position of the source (with respect to the center of the cloak, origin of the coordinates) and ω is the angular wave frequency.

Homogenization Method

The multiple scale technique30 allows us to comprehend the cloaking mechanism of the cloak. The main assumption is the decoupling between the microscopic and microscopic scales, respectively associated with fast and slow oscillations of the pressure acoustic field or longitudinal magnetic field. This method leads to a homogenized anisotropic matrix (the effective density, or the effective permittivity25, depending upon the type of wave), which is a good approximation of the transformed density, or permittivity, obtained by blowing up a point on a disc1. Details on our cloak's design via numerical solution of the annex problem (3) in the context of linear surface liquid waves can be found in10.

Experimental set-up for Microwaves

The microwaves experimental setup was positioned in an anechoic chamber (3 m long, 3 m wide and 3 m high). A ridged horn antenna [1 GHz–18 GHz] was positioned at 24 cm from the structure cloak (Fig. 6). A rectangular metallic obstacle was inside the cloak structure. The magnetic probe (Fig. 1) was positioned 5 mm above the structure cloak. The probe was a homemade magnetic loop with diameter 5 mm and the loop was positioned perpendicular to the magnetic field of the emitter antenna. In this configuration, the magnetic probe measured a single component of the three Cartesian components of the magnetic field. In this case, this is the magnetic field perpendicular component to the loop that was measured. The magnetic loop scanned a 33×71 cm^{2}surface thanks to a 3D axis positioning system36.

The antenna and the probe were connected to a vector network analyzer44 that generate and measure the electromagnetic field. The electromagnetic field was emitted by the antenna and the probe only received the field.

The vector network analyzer measured the transmission coefficient between the emitter antenna and the magnetic probe. In our case, the transmission coefficient is equivalent to the magnetic field component perpendicular to the loop. The frequency range was from 1 to 6 GHz (our vector network works between 50 KHz and 6 GHz).

The complex transmission coefficient between the antenna and the probe is displayed in terms of magnitude in dB to visualize the intensity of the field. The real part of the magnetic field is also displayed since this is related to the phase of the field.

Experimental set-up for Acoustic Waves

The acoustic wave experimental setup was positioned on an optical table with open space surrounded. A commercial loudspeaker [20 Hz–18 kHz] was placed at 20 cm from the structure cloak. The loud speaker was driven by a programmable functional generator (Tektronix AFG 3022B) and a power amplifier (Krohn-Hite Model 7500). A high sensitive microphone mounted on an x-y translation stage (Newmark NSL4) was used to record the two-dimensional acoustic pressure field distribution. The loudspeaker, acoustic cloak, and microphone were placed at the same level for characterizing how the pressure field was affected by the cloak. In the experiment, a five-period sinusoidal wave was launched by the loudspeaker. The time-dependent pre-amplified signal collected by the microphone was recorded by an oscilloscope (Agilent DSO 6104A) and downloaded to a computer for further analysis. The signal generation and acquisition were synchronized by the computer program. Therefore, we can map the 2D pressure field distribution with both amplitude and phase information. The scanning area is 25×25 cm^2.

ACKNOWLEDGEMENTS

E.G. acknowledges PhD funding from the French Fonds Unique Interministériel, with the support of the Provence Alpes **Côte d'Azur Region and** Oséo. S.G. is thankful for an ERC starting grant funding (ANAMORPHISM). S.E., S.G., R.A., P.S. acknowledge funding of PEPS INSIS (CLOAK).

REFERENCES

Pendry, J. B., Schurig, D. & Smith, D. R. Controlling electromagnetic fields. Science 312, 1780–1782 (2006).

1. Leonhardt, U. Optical conformal mapping. Science 312, 1777–1780 (2006).
2. Leonhardt, U. & Philbin, T. G. General relativity in electrical engineering. New J. Phys. 8, 247 (2006).
3. Schurig, D. et al. Metamaterial Electromagnetic Cloak at Microwave Frequencies. Science 314, 977–980 (2006).
4. Li, J. & Pendry, J. B. Hiding under the carpet: A new strategy for cloaking. Phys. Rev. Lett. 101, 203901 (2008).
5. Ergin, T., Stenger, N., Brenner, P., Pendry, J. B. & Wegener, M.Three-dimensional invisibility cloak at optical wavelengths.Science 328, 337–339 (2010).
6. Chen, X. et al. Macroscopic invisibility cloaking of visible light.Nat. Commun. 2, 176 (2011).
7. Landy, N. & Smith, D. R. A full-parameter unidirectional metamaterial cloak for microwaves. Nature Mater. 12, 25–28 (2013).
8. Cummer, S. A. & Schurig, D. One path to acoustic cloaking. New J. Phys. 9, 45 (2007).
9. Farhat, M. Enoch, S., Guenneau, S. & Movchan, A. B. Broadband cylindrical acoustic cloak for linear surface waves in a fluid. Phys. Rev. Lett. 101, 134501 (2008).
10. Zhang, S., Xia, C. & Fang, N. Broadband acoustic cloak for ultrasound waves. Phys. Rev. Lett. 106, 024301 (2011).
11. Popa, B.-I., Zigoneanu, L. & Cummer, S. A. Experimental acoustic ground cloak in air. Phys. Rev. Lett. 106, 253901 (2011).
12. Chen, H. Y. & Chan, C. T. Acoustic cloaking in three dimensions using acoustic metamaterials. Appl. Phys. Lett. 91, 183518 (2007).
13. Brun, M., Guenneau, S. & Movchan, A. B. Achieving control of in-plane elastic waves. Appl. Phys. Lett. 94, 061903 (2009).
14. Farhat, M., Guenneau, S. & Enoch, S. Ultrabroadband elastic cloaking in thin plates. Phys. Rev. Lett. 103, 024301 (2009).

15. Stenger, N., Wilhelm, M. & Wegener, M. Experiments on elastic cloaking in thin plates. Phys. Rev. Lett. 108, 014301 (2012).
16. Pendry, J. B. Negative refraction makes a perfect lens. Phys. Rev. Lett. 85, 3966 (2000).
17. Dubois, M. et al. Flat lens for pulse focusing of elastic waves in thin plates. Appl. Phys. Lett. 103, 071915 (2013).
18. Milton, G. W. & Nicorovici, N. A. P. On the cloaking effects associated with anomalous localized resonance. Proc. R. Soc. Londons, Sec. A 462, 3027–3059 (2006).
19. Guenneau, S., Amra, C. & Veynante, D. Transformation thermodynamic: cloaking and concentrating heat flux. Opt. Express 20, 8207–8218 (2012).
20. Narayana, S. & Sato, Y. Heat flux manipulation with engineered thermal materials. Phys. Rev. Lett. 108, 214303 (2012).
21. Schittny, R., Kadic, M., Guenneau, S. & Wegener, M. Experiments on transformation thermodynamics: Molding the flow of heat.Phys. Rev. Lett. 110, 195901 (2013).
22. Kanté, B., Germain, D. & de Lustrac, A. Experimental demonstration of a nonmagnetic metamaterial cloak at microwave frequencies. Phys. Rev. B 80, 201104 (2009).
23. Cai, W., Chettiar, U. K., Kildishev, A. V. & Shalaev, V. M. Optical cloaking with metamaterials. Nature Photon. 1, 224–227 (2007).
24. Farhat, M., Guenneau, S., Movchan, A. B. & Enoch, S. Achieving invisibility over a finite range of frequencies. Opt. Express 16, 5656–5661 (2008).
25. Torrent, D. & Sanchez-Dehesa, J. Anisotropic mass density by two-dimensional acoustic metamaterials. New J. Phys. 10, 023004 (2008).
26. Greenleaf, A., Kurylev, Y., Lassas, M. & Uhlmann, G. Isotropic transformation optics: approximate acoustic and quantum cloaking. New J. Physics 10, 115024 (2008).
27. Kohn, R. V., Shen, H., Vogelius, M. S. & Weinstein, M. I. Cloaking via change of variables in electric impedance tomography. Inverse Problems 24, 015016 (2008).
28. Cherkaev, A. 2000 Variational Methods for Structural Optimization (New York: Springer-Verlag).
29. Milton, G. W. 2001 The Theory of Composites (Cambridge: Cambridge University Press).
30. Fang, N. et al. Ultrasonic metamaterials with negative modulus.Nature Mater. 5, 452–456 (2006).
31. García-Chocano, V. M. et al. Acoustic cloak for airborne sound by inverse design. Appl. Phys. Lett. 99, 074102 (2011).
32. Zigoneanu, L., Popa, B.-I. & Cummer, S. A. Three-dimensional broadband omnidirectional acoustic ground cloak. Nature Mater.13, 352–355 (2014).
33. Brûlé, S., Javelaud, J., Enoch, S. & Guenneau, S. Experiments on seismic metamaterials : molding surface waves. Phys. Rev. Lett.112, 133901 (2014).
34. Urzhumov, Y. A., Kundtz, N. B., Smith, D. R. & Pendry, J. B. Cross-section comparisons of cloaks designed by transformation optical and optical conformal mapping approaches. J. Opt. 13, 024002 (2010).
35. Georget, E., Abdeddaim, R. & Sabouroux, P., A New Method to Design a Multi-Band Flexible Textile Antenna. IEEE Trans. Antennas Propag. 56, 240–248 (2014).
36. Knott, E. F., Shaeffer, J. F. & Tuley, M. T. 2004 Radar Cross Section(Raleigh: SciTech Publishing Inc, 2nd edition).
37. Frenzel, T. et al. Three-dimensional labyrinthine acoustic metamaterials. Appl. Phys. Lett. 103, 061907 (2013).
38. Bückmann, T., Thiel, M., Kadic, M., Schittny, R. & Wegener, M. An elasto-mechanical unfeelability cloak made of pentamode metamaterials. Nat. Commun. 5, 4130 (2014).
39. Shin, D., Urzhumov, Y., Lim, D., Kim, K. & Smith, D. R. A versatile smart transformation optics device with auxetic elasto-electromagnetic metamaterials. Scientific Reports 4, 4084 (2014).
40. Petiteau, D., Guenneau, S., Bellieud, M., Zerrad, M. & Amra, C.Spectral effectiveness of engineered thermal cloaks in the frequency regime. Scientific Reports 4, 7386 (2014).
41. Schittny, R., Kadic, M., Bückmann, T. & Wegener, M. Invisibility cloaking in a diffusive light scattering medium. Science 345, 427–429 (2014).
42. Farhat, M. et al. Understanding the functionality of an array of invisibility cloaks, Phys. Rev. B 84, 235105 (2011).
43. Understanding Vector Network Analysis, Anritsu Newsletter, (2013), (date ofaccess:14/10/2014) http://www.anritsu.com/en-GB/Media Room/Newsletters/files/Understanding%20vector%20network%20analysis.pdf.

CHAPTER 14 SCATTERING ENGINEERING IN CONTINUOUSLY SHAPED METASURFACE: AN APPROACH FOR ELECTROMAGNETIC ILLUSION

Yinghui Guo, Lianshan Yan, Wei Pan and Liyang Shao

INTRODUCTION

There are considerable interests in controlling and engineering the scattering electromagnetic fields from metallic objects especially for military applications[1,2,3]. The scattering property of an object is quantitatively determined by virtue of its radar cross section (RCS), which decides how far the object can be detected by a radar system. Traditionally RCS reduction methods rely on modifying the object geometry[4] and/or loading the object with absorbing materials. Recently, the rapid development of transformation optics[5,6], a powerful means to steer the light flow in almost arbitrary ways, provides a unique design paradigm for electromagnetic cloaks, which bend light around a concealed region, rendering any object inside the region "invisible". Besides, when combined with plasmonics, transformational optics can be used to realize sub-diffraction limit imaging and optical lithography[7,8,9,10], which is a tremendous advancement of optics in this century[11]. Unfortunately, the general materials required by transformational optics are in general highly inhomogeneous and anisotropic metamaterials, which significantly constrains the realization and application of practical devices and results in disadvantages of narrow operation bandwidths and high losses[12,13]. A quasi-conformal mapping method was subsequently developed[14,15], where only a modest range of isotropic refractive indices is required. Subsequently, electromagnetic cloak based on the near-zero refractive index metamaterials was also investigated[16]. Nevertheless, metamaterial-based schemes are bulky and hard to scale up[17].

The gradient phase metasurface[18,19,20], because of the reduced dimension and loss as well as powerful scattering engineering ability, is taken as a promising approach for electromagnetic cloak or electromagnetic illusion[4,17,21], a strategy used to make an object of arbitrary shape and material properties appear exactly like another object of some other shape and material makeup. In 2005, Luo *et al.* proposed the first gradient flat metasurface lens based on surface plasmon[11]. Since its ability to revise the classic Snell's law and form image beyond the diffraction limit, such lens is expected to be the next-generation optical devices[11,22] . In 2011, Capasso's group demonstrated another flat lens based on V-shaped nano-antenna array, which could also be considered as a type of metasurface[18]. In general, the behavior of light in gradient metasurface can be described by the metasurface-assisted law of refraction and reflection (MLRR)[11]. Another important gradient metasurface is based on the polarization conversion in anisotropic metasurfaces and the spin-orbit interaction (SOI) in inhomogeneous structures[11]. Luo's group proved that the traditional Fresnel's equations can be revised via the metasurface/metamirror approach and gave pioneering work on the broadband polarization conversion[11,23,24,25]. Subsequently, these artificial atoms are arranged in homogeneously to obtain nearly perfect broadband virtual shaping[4]. Theoretically, both the gradient metasurface and the surface absorber can be understood in the scenario of energy exchange between the propagating and bounding waves or the so-called "two-wave exchange"[11,26].

Citation: Guo, Y. et al. Scattering engineering in continuously shaped metasurface: An approach for electromagnetic illusion. Sci. Rep. 6, 30154; doi: 10.1038/srep30154 (2016).

Most previous metasurfaces are constructed by a serial of discrete meta-atom/molecules. Gradient phase was approximated by discrete levels of abrupt phase shift. On one hand, it is a significant challenge obtaining high-performance metasurfaces that can operate at visible frequency range with high uniformity. On the other hand, the discontinuous nature inevitably degrades the overall performance of metasurfaces due to the induced phase noise to the scattering fields, for example, decreasing the purity of generated orbit angular momentum and causing aberrations in imaging and focusing system[27,28,29]. Besides, discrete system is not electrically conductive, thus it is difficult to obtain simultaneous electric and optical functionality. Quite recently, quasi-continuous super-meta-atoms including catenary and trapezoid shaped structure are adopted in metasurfaces to overcome the shortages above[30,31,32,33,34]. Especially, the optical catenary[30], inspired from natural phenomena, is the first realization of continuous linear phase shift covering $(0, 2\pi)$ in metasurface, which invoke tremendous interests in searching the mystical interlinks between mechanical structures, religion symbols and optical phonmenona[34,35].

Inspired by the electromagnetic virtual shaping and catenary optics[4,30], here we present a metasurface constructed by true continuous super-meta-atoms, so called continuously shaped metasurface. The phase modulation is origin from the SOI in spatially inhomogeneous sinusoidal metallic strips, which redirects the reflected wave with a period-dependent angle with respect to the specular reflection direction. The polarization independent scattering engineering capability of our structure is able to change-reduce or enhance-the monostatic RCS of the underneath object dramatically.

RESULTS

Principle of Continuous Gradient Phase

It is well known that sinusoidal function is the basic oscillating form of electromagnetic wave and the Fourier expansion of an ideal triangular and rectangular waveform can be represented by the summation of sinusoidal functions. Therefore, sinusoids are chosen to realize the continuous phase shift. The geometry of the metasurface is illustrated in Fig. 1(a), where only four rows and two periods of sinusoidal metallic strips are given. As shown in Fig. 1(b), the profile of spatially inhomogeneous metallic strips is mathematically defined by a typical sinusoidal function:

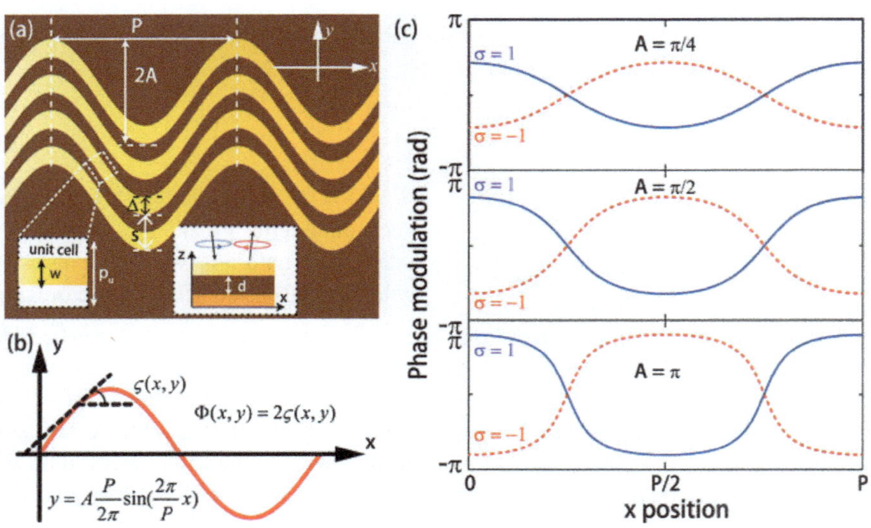

Figure 1: Principle of continuous gradient phase modulatoin based on spin-orbit interaction in space-variant half-wave plates.

(**a**) The proposed continuously shaped metasurface is constructed by sinusoidal metallic strip array, which can be considered as a serial of metallic patch with space-variant orientation angle, as shown in the inset. Space-variant half-wave plates are composed of the metasurface, dielectric spacer and a metallic ground plane, which contributes the spin-orbit interaction. (**b**) The phase shift for a rotated half waveplate under CPW. (**c**) Pancharatnam-Berry phase modulation in single period of sinusoidal metallic strip with different amplitude coefficient A.

$$y = A\frac{P}{2\pi}\sin\left(\frac{2\pi}{P}x\right) \tag{1}$$

where, A and P denote the amplitude coefficient and period of the sine curve, respectively.

In our design, a metallic ground plane and a dielectric substrate are added below the metasurface, forming a metamirror as depicted in the inset of Fig. 1(a)[23,25,36]. When the thickness of the dielectric spacer is adjusted properly, the metasurface, dielectric spacer, and the metallic ground plane would form a space-variant waveplate[4], thus subjected to the famous SOI. For half-wave plate, circular polarization would be reversed after reflection[36,37]. Spatial control of the polarization state of a beam in this manner ultimately introduces spatially varying phase distributions, known to be a manifestation of the Pancharatnam-Berry (P-B) phase[38], which is just twice the inclination angle between the curve tangent and the x axis (Fig. 1(b))[4]

$$\Phi(x, y) = 2\sigma\varsigma(x, y) = 2\sigma arc\ \tan\left(\frac{dy}{dx}\right) = 2\sigma arc\ \tan\left(A\ \cos\left(\frac{2\pi}{P}x\right)\right) \tag{2}$$

where, $\sigma = \pm 1$ denote the left-handed circular polarization (LCP) and right-handed circular polarization (RCP), respectively. According to the equation (2), the theoretical phase modulation in single period of sinusoidal metallic strip is investigated and illustrated in Fig. 1(c). As we expected, the phase profile is continuous, which may motivate the metasurface-assisted law of reflection/refraction[11,22]. Furthermore, the range of phase shift is determined by the amplitude coefficient (A) and approaches to the whole range of $(-\pi, \pi)$ with the increase of A.

Scattering Engineering via Continuously Shaped Metasurface

In order to get physical insight of scattering engineering, we subsequently consider a planer metamirror under normal illumination of circularly polarized waves (CPW) (Fig. 2(a)). Conventional mirrors, known since the dawn of civilization, obey the simple law of reflection. While, the metamirror proposed here due to the gradient phase stemming from the SOI of continuously shaped planar metasurface would force the reflected beam of opposite handedness to propagate in well-defined ways. In this way, traditional law of reflection is broken and rectified by generalized Snell's law[11,18,39], giving rise to electromagnetic illusion as elucidated in Fig. 2(b).

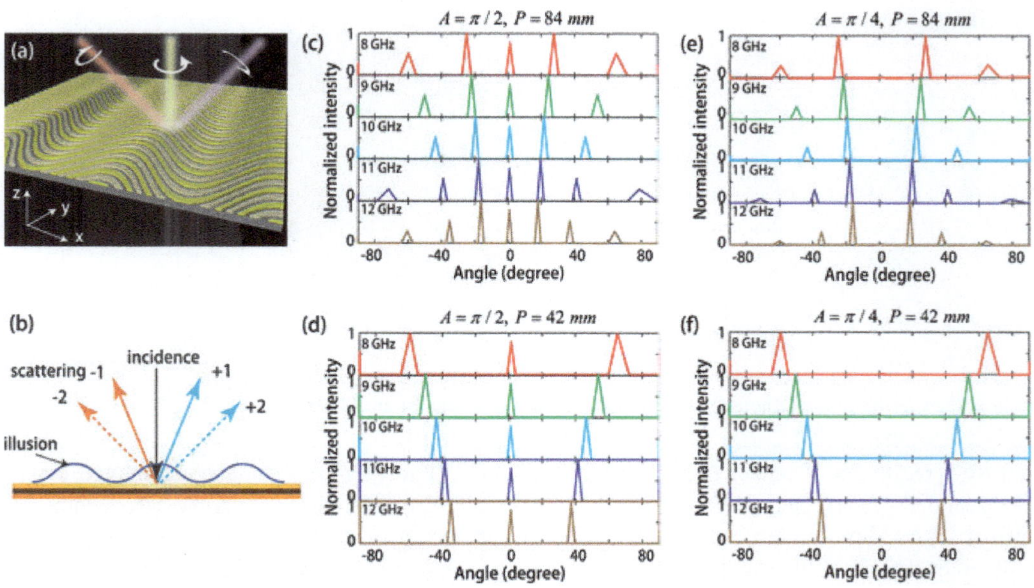

Figure 2: Sketch map of electromagnetic illusion induced by scattering engineering of continuously shaped metasurface.

(**a**) Scattering of a planer metamirror under illumination of a CPW. (**b**) Electromagnetic illusion induced by anomalous scattering. (**c–f**) Theoretically calculated scattering angular spectra for different designed parameters. (**c**) A=π/2, P=84 mm. (**d**) A=π/2, P=42 mm. (**e**) A=π/4, P=84 mm. (**f**) A=π/4, P=42 mm.

By exploiting the fast Fourier transform (FFT) command in MATLAB™for the theoretical phase shift in equation (2), the scattering properties of proposed metamirrors with an outer dimension of $20\times420\,mm^2$ are calculated under different circumstances: A=π/2, P=84 mm (Fig. 2(c)); A=π/2, P=42 mm (Fig. 2(d)); A=π/4, P=84 mm (Fig. 2(e)); A=π/4, P=42 mm (Fig. 2(f)). Evidently, due to the period of the sinusoidal metallic strip is larger than the wavelength of 8–12 GHz, multiple scattering orders emerge and the scattering angular spectra manifest several remarkable features. Initially, for each metamirror, the scattering patterns through 8–12 GHz are similar because of the frequency independent P-B phase, which may pave the avenue for broadband illusion devices. Secondly, the intensity of the scattering orders can be varied by adjusting the amplitude coefficient (A) of sinusoidal metallic strips. Especially, 0 order scattering, the main source of RCS, is totally suppressed when A is around π/4, which is incompetent for traditional grating. Ultimately, the deflection angle of m^{-th} scattering order with respect to the specular direction is frequency and period dependent, approximately satisfying:

$$\theta_m = arc\ \sin\left(\frac{m\lambda}{P}\right) \qquad (3)$$

Consequently, various kinds of scattering angular spectra can be obtained with properly adoption of A and P. Note that, although the P-B phase modulation in Fig. 1(c) is polarization sensitive, the scattering characteristics of the proposed metasurface is same regardless of the handedness of CPW and no scattering circular dichroism[40] is observed here.

Simulation and Experimental Validation of Electromagnetic Illusion

The performance of electromagnetic illusion based on proposed metasurface is characterized by the RCS metamorphosis. Full model simulation is carried out for a metamirror with a thickness of2 mm. The amplitude coefficient (A) and period (P) of the metallic winding strips is π/4 and 84 mm, respectively. The period of the sinusoidal metallic strips along y direction is s=6.8 mm, less than the incident wavelength λ to prevent the diffraction in yoz plane. The width of sinusoidal metallic strips is Δ=4 to obtain high polarization conversion efficiency around 10 GHz. As shown in Fig. 3(a), the monostatic RCS in 10–11 GHz is reduced larger than 10 dB compared with a planar metallic plate for both LCP and RCP. The maximum of the RCS reduction exceeds 25 dB around 10.6 GHz. Combination with the fact that the frequency dependent conversion efficiency of CPW in practice, it is not strange that the operation bandwidth is not broad.

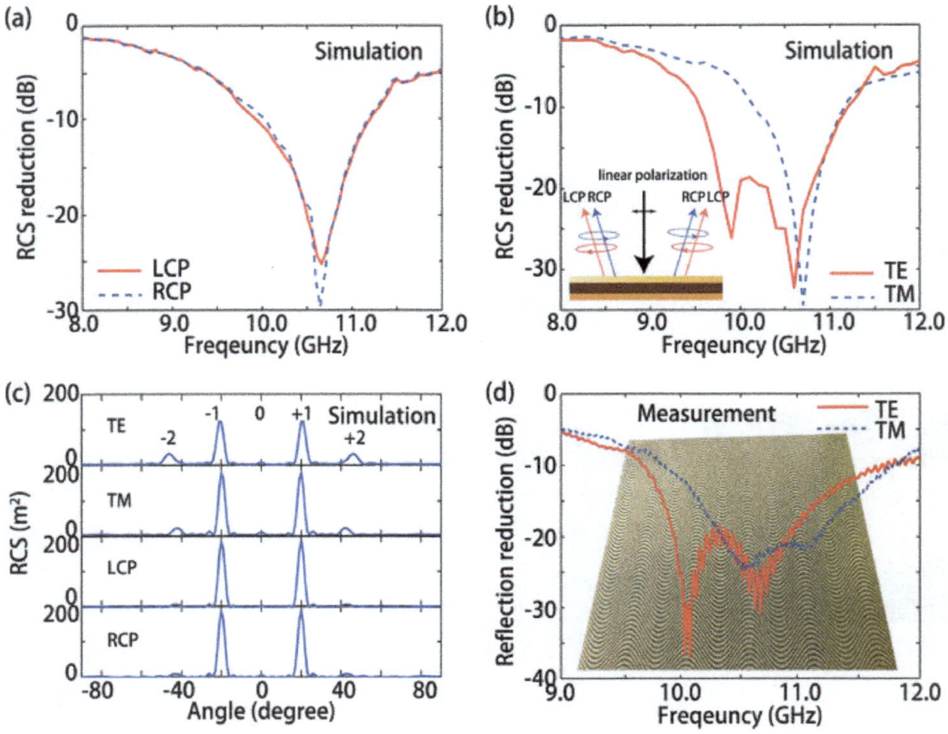

Figure 3: Simulation and experimental characterization of electromagnetic illusion in RCS reduction.

(**a**) Simulated RCS reduction of planer metamirror under illumination of LCP and RCP. (**b**) Simulated RCS reduction of planer metamirror under illumination of TE and TM. (**c**) Simulated RCS at 10.6 GHz for different polarizations as a function of angle. (**d**) Measured reflection reduction of planer metamirror under illumination of TE and TM.

A linear polarized light with an angle of polarization φ can be written as a superposition of its circular components as follows:

$$E = E_0(\hat{x}\cos\varphi + \hat{y}\sin\varphi) = \frac{E_0}{2}(\hat{r}e^{-i\varphi} + \hat{l}e^{-i\varphi}) \qquad (4)$$

Where $\hat{r} = (\hat{x} + i\hat{y})/\sqrt{2}$ and $\hat{l} = (\hat{x} - i\hat{y})/\sqrt{2}$ are the unit vectors of the RCP and LCP, respectively. Similar RCS reduction performances are achieved for the transverse electric (TE) or transverse magnetic (TM) polarization, as illustrated in Fig. 3(b). In addition, the RCS of the metamirror as a function of angle at 10.6 GHz are investigated for different polarizations. The scattering angular spectra presented in Fig. 3(c) also indicate the scattering engineering scheme is polarization independent, which is in contrast to the intuitive thought about anisotropic material. We can see the 0 order scatting is greatly suppressed and ±1 order scattering is mainly directed to ±20°, which is consistent with the theoretical result revealed in Fig. 2(e). In fact, scattering engineering can also be realized by other metasurfaces, for example, continuous triangle grating and discrete metasurfaces. However, the proposed structure in this paper has its advantages. Firstly, this continuously shaped metasurface can be scaled to visible band due to the released fabrication requirement compared with those discrete systems. Besides, the phase accuracy of continuously shaped metasurface is significantly improved owing to the nearly infinite small "pixel" size. In contrast, the "pixel" size of phase modulation is limited by the dimension of super-atoms in discrete metasurfaces. Therefore, it is a great challenge obtaining high-performance discrete metasurfaces that can operate at visible frequency range with high uniformity. Finally, although the triangle grating is continuous, the generated phase profile is discontinuous and broken at the turning point.

In order to prove the numerical results, a sample was fabricated with print circuit board (PCB) technique as shown in the inset of Fig. 3(b). Reflection measurements were taken in a microwave anechoic chamber with a network analyzer. By changing the orientation of linearly polarized horn antennas, the reflectance was measured for both the TE and TM polarizations. The results shown in Fig. 3(d) display 20 dB reflection reduction around 10.6 GHz, which are in reasonable agreement with the simulated results in Fig. (b), considering the imperfection in the fabricating and measuring process.

Finally, we would like to extend the concept of continuously shaped metasurface to RCS enhancement, dubbed as superscatterers in some literatures41·42. For example, in wireless communications where reliable communications likely rely on of direction and orientation of the mobile devices, RCS enhancement will enable tracking it with improved signal-to-noise ratio43. The configuration of RCS enhancement is presented inFig. 4(a), where the continuously shaped metasurface is warped over in a symmetrical dihedral corner reflector with a π−2α opening angle (Fig. 4(b)). Here, α is chosen as π/8 so that the flatness is improved compared to a conventional orthogonal dihedral corner reflector (with a π/2-opening angle, i.e. α=π/4). The monostatic return from such structures can be enhanced when its virtual shape behaves like an orthogonal dihedral corner reflector. The performance of RCS enhancement by proposed continuously shaped metasurface is investigated by full model simulation. The geometric parameters are same with the previous design. As shown in Fig. 4(c,d), the average monostatic RCS enhancement in 8–12 GHz is more than 10 dB for both

TE and TM polarizations compared with the cases without metasurface. Note that the RCS enhancement can be further improved by optimizing of period of the sinusoidal metallic strips to make the ±1 order scattering of metasurface is redirected back. A sample was fabricated by PCB technology and fixed on a home-made clamp to engage the opening angle is $3\pi/4$, as shown in Fig. 4(b). The measurement results under orthogonal linear polarizations (TE and TM) are respectively displayed in Fig. 4(e,f), which is consistent with the simulation results. The discrepancy between them arises mainly from the imperfect phase realization and nonuniform scattering amplitudes across the dihedral corner reflector.

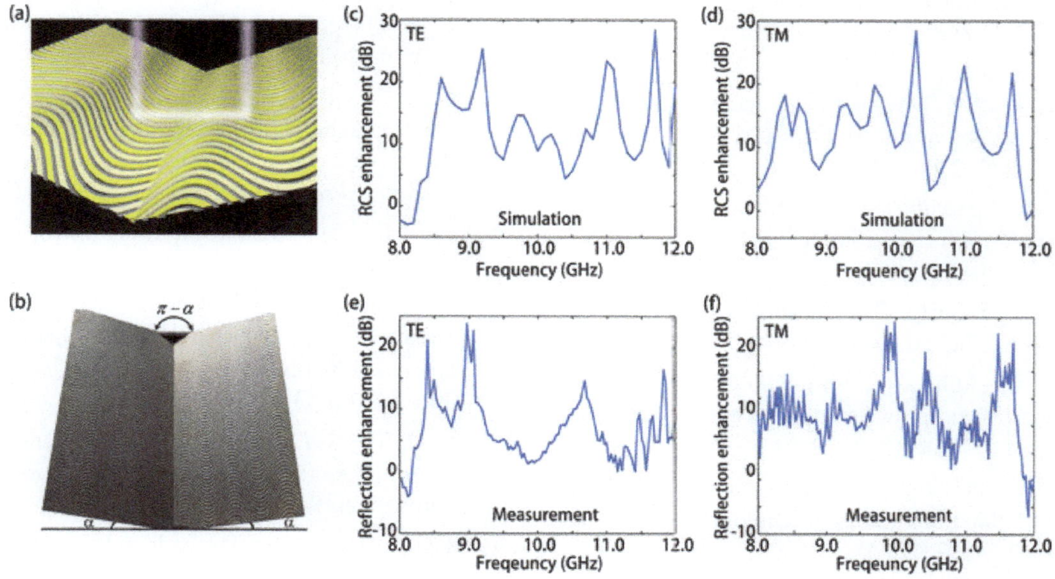

Figure 4: Simulation and experimental characterization of electromagnetic illusion in RCS enhancement.

(**a**) A prototype of electromagnetic illusion device for RCS enhancement. (**b**) Photograph of the fabricated samples. (**c,d**) Simulated RCS enhancement of dihedral corner reflector under illumination of TE and TM. (**e,f**) Measured RCS enhancement under illumination of TE and TM.

DISCUSSIONS

In summary, a new generation scheme of continuous gradient phase has been proposed and investigated, which is origin from the spin-orbit interaction in spatially inhomogeneous sinusoidal metallic stripes. Owing to the continuous phase modulation, phase noises in the case of antenna arrays can be avoided. Furthermore, by adjusting the geometric parameters of sinusoidal metallic stripes, we can engineer the gradient phase and make the scattering characteristics of the underneath object deviate from its actual status. Polarization independent electromagnetic illusion devices based on proposed continuously shaped metasurface also have been numerically and experimentally demonstrated in microwave band for significant reduction or enhancement of the scattering cross section of an object, which can be subsequently pushed to the infrared and even the optical range by downscaling the sizes of the structural dimension. Although the geometric phase shift is frequency independent, the operation bandwidth of illusion devices is limited due to the uniform polarization conversion efficiency. It is anticipated that by resorting to the versatile dispersion management strategy and multi-resonance technique presented by Luo's group[26,36,44,45,46], the operation band will be further expanded. On the other case, the pronounced Fano resonance in metasurface can also be utilized to achieve ultra-narrow band operation[47]. Finally, by utilizing dynamic control method[48,49], the performance of our device can be significantly improved.

METHODS

Numerical Analyses and Simulations

MATLAB™ is used as the basic numerical tool for the projects. The scattering patterns of the metamirror were calculated by using the fast Fourier transform command for the theoretical phase distributions in equation (2). The full model simulations were carried out by using commercial software with open boundary conditions. The amplitude coefficient (A) and period (P) of the metallic winding strips is $\pi/4$ and 84 mm, respectively. The period of the sinusoidal metallic strips along y direction is s = 6.8 mm. The width of sinusoidal metallic strips is $\Delta=4$. In the simulations, perfect electrical conductor (PEC) model was selected for the metallic patterns and the ground plane. The substrate under metallic patterns has a permittivity of 4.5 and thickness of 2 mm.

Sample Fabrication and Characterization

The sample with outer dimension (420 mm × 420 mm) was fabricated by using laser direct writing in print circuit board (PCB) technology. The performance of electromagnetic illusion was characterized by the reflection variation between the proposed metamirror and metallic reflection plane with the same configuration. The measurements were carried in a microwave anechoic chamber. Two standard linearly polarized horn antennas as transmitter and receiver, respectively, were connected to the two ports of a vector network analyzer R&S ZVA40. The incident angle was set as 5°, which is a good approximation of the normal incidence. By changing the orientation of linearly polarized horn antennas, the reflectance was measured for both the TE and TM polarizations.

REFERENCES

1. Swandic, J. R. *Bandwidth limits and other considerations for monostatic RCS reduction by virtual shaping*. (Naval Surface Warfare Center, Carderock Div., 2004).
2. Edalati, A. & Sarabandi, K. Wideband, wide angle, polarization independent RCS reduction using nonabsorptive miniaturized-element frequency selective surfaces. *IEEE Trans. Antennas Propag.* 62, 747–754 (2014).
3. Doumanis, E. *et al.* Design of engineered reflectors for radar cross section modification. *IEEE Trans. Antennas Propag.* 61, 232–239 (2013).
4. Pu, M. *et al.* Spatially and spectrally engineered spin-orbit interaction for achromatic virtual shaping. *Sci. Rep.* 5, 9822 (2015).
5. Pendry, J. B., Schurig, D. & Smith, D. R. Controlling electromagnetic fields. *Science* 312, 1780–1782 (2006).
6. Pendry, J. B., Aubry, A., Smith, D. R. & Maier, S. A. Transformation optics and subwavelength control of light. *Science* 337, 549–552 (2012).
7. Luo, X. & Ishihara, T. Surface plasmon resonant interference nanolithography technique. *Appl. Phys. Lett.* 84, 4780–4782 (2004).
8. Yao, H., Yu, G., Yan, P., Chen, X. & Luo, X. Patterining sub 100 nm isolated patterns with 436 nm lithography. In 2003 International Microprocesses and Nanotechnology Conference, Japan, (2003).
9. Luo, X. & Ishihara, T. Subwavelength photolithography based on surface-plasmon polariton resonance. *Opt. Express* 12, 3055–3065 (2004).
10. Luo, X. & Yan, L. Surface plasmon polaritons and its applications.*IEEE Photonics J.* 4, 590–595 (2012).
11. Luo, X. Principles of electromagnetic waves in metasurfaces. *Sci. China-Phys. Mech. Astron.* 58, 594201 (2015).
12. Liu, Y., Zentgraf, T., Bartal, G. & Zhang, X. Transformational plasmon optics. *Nano Lett.* 10, 1991–1997 (2010).
13. Schurig, D. *et al.* Metamaterial electromagnetic cloak at microwave frequencies. *Science* 314, 977–980 (2006).
14. Liu, R. *et al.* Broadband ground-plane cloak. *Science* 323, 366–369 (2009).
15. Valentine, J., Li, J., Zentgraf, T., Bartal, G. & Zhang, X. An optical cloak made of dielectrics. *Nat. Mater.* 8, 568–571 (2009).
16. Wu, X., Hu, C., Wang, M., Pu, M. & Luo, X. Realization of low-scattering metamaterial shell based on cylindrical wave expanding theory. *Opt. Express* 23, 10396–10403 (2015).
17. Ni, X., Wong, Z. J., Mrejen, M., Wang, Y. & Zhang, X. An ultrathin invisibility skin cloak for visible light. *Science* 349, 1310–1314 (2015).
18. Yu, N. *et al.* Light propagation with phase discontinuities: generalized laws of reflection and refraction. *Science* 334, 333–337 (2011).
19. Chen, Y. *et al.* Engineering the phase front of light with phase-change material based planar lenses. *Sci. Rep.* 5, 8660 (2015).
20. Lin, J. *et al.* Polarization-controlled tunable directional coupling of surface plasmon polaritons. *Science* 340, 331–334 (2013).
21. Lai, Y. *et al.* Illusion optics: The optical transformation of an object into another object. *Phys. Rev. Lett.* 102, 253902 (2009).
22. Luo, X., Pu, M., Ma, X. & Li, X. Taming the electromagnetic boundaries via metasurfaces: from theory and fabrication to functional devices. *Int. J. Antennas Propag.* 2015, 204127 (2015).
23. Pu, M. *et al.* Anisotropic meta-mirror for achromatic electromagnetic polarization manipulation. *Appl. Phys. Lett.* 102, 131906 (2013).
24. Huang, C. *et al.* Dual-band 90° polarization rotator using twisted split ring resonators array. *Opt. Commun.* 291, 345–348 (2013).
25. Guo, Y. *et al.* Dispersion management of anisotropic metamirror for super-octave bandwidth polarization conversion. *Sci. Rep.* 5, 8434 (2015).
26. Pu, M. *et al.* Ultrathin broadband nearly perfect absorber with symmetrical coherent illumination. *Opt. Express* 20, 2246–2254 (2012).
27. Aieta, F., Genevet, P., Kats, M. & Capasso, F. Aberrations of flat lenses and aplanatic metasurfaces. *Opt. Express* 21, 31530–31539 (2013).
28. Ma, X. *et al.* A planar chiral meta-surface for optical vortex generation and focusing. *Sci. Rep.* 5, 10365 (2015).
29. Guo, Y., Yan, L., Pan, W. & Luo, B. Generation and manipulation of orbital angular momentum by all-dielectric metasurfaces.*Plasmonics* 11, 337–344 (2015).
30. Pu, M. *et al.* Catenary optics for achromatic generation of perfect optical angular momentum. *Sci. Adv.* 1, e1500396 (2015).
31. Li, X. *et al.* Catenary nanostructures as highly efficient and compact Bessel beam generators. *Sci. Rep.* 6, 20524 (2016).
32. Li, Z., Palacios, E., Butun, S. & Aydin, K. Visible-frequency metasurfaces for broadband anomalous reflection and high-efficiency spectrum splitting. *Nano Lett.* 15, 1615–1621 (2015).
33. Wang, Y. *et al.* Quasi-continuous metasurface for ultra-broadband and polarization-controlled electromagnetic beam deflection. *Sci. Rep.* 5, 17733 (2015).
34. Sun, H. The mystical interlinks: Mechanics, religion or optics? *Sci. China-Phys. Mech. Astron.* 59, 614202 (2016).
35. Li, X. *et al.* Dynamic control of the extraordinary optical scattering in semi-continuous two-dimensional metamaterials.*Adv. Opt. Mater.* doi: 10.1002/adom.201500713 (2016).
36. Guo, Y., Yan, L., Pan, W. & Luo, B. Achromatic polarization manipulation by dispersion management

of anisotropic meta-mirror with dual-metasurface. *Opt. Express* 23, 27566–27575 (2015).

37. Wang, D. *et al.* Switchable ultrathin quarter-wave plate in terahertz using active phase-change metasurface. *Sci. Rep.* 5, 15020 (2015).
38. Bomzon, Z. 'ev., Kleiner, V. & Hasman, E. Pancharatnam-Berry phase in space-variant polarization-state manipulations with subwavelength gratings. *Opt. Lett.* 26, 1424–1426 (2001).
39. Zhao, J., Li, B., Chen, Z. & Qiu, C.-W. Manipulating acoustic wavefront by inhomogeneous impedance and steerable extraordinary reflection. *Sci. Rep.* 3, 2537 (2013).
40. Turner, M. D. *et al.* Miniature chiral beamsplitter based on gyroid photonic crystals. *Nat. Photonics* 7, 801–805 (2013).
41. Ruan, Z. & Fan, S. Superscattering of light from subwavelength nanostructures. *Phys. Rev. Lett.* 105, 13901 (2010).
42. Xu, H.-X., Wang, G.-M., Ma, K. & Cui, T. J. Superscatterer illusions without using complementary media. *Adv. Opt. Mater.* 2, 572–580 (2014).
43. Lipuma, D., Méric, S. & Gillard, R. RCS enhancement of flattened dihedral corner reflector using reflectarray approach. *Electron. Lett.* 49, 152–154 (2013).
44. Wang, Y. *et al.* Dynamic manipulation of polarization states using anisotropic meta-surface. *Opt. Commun.* 319, 14–16 (2014).
45. Ma, X. *et al.* Multi-band circular polarizer using planar spiral metamaterial structure. *Opt. Express* 20, 16050–16058 (2012).
46. Ma, X. *et al.* Dual-band asymmetry chiral metamaterial based on planar spiral structure. *Appl. Phys. Lett.* 101, 161901 (2012).
47. Pu, M. *et al.* Investigation of Fano resonance in planar metamaterial with perturbed periodicity. *Opt. Express* 21, 992–1001 (2013).
48. Ma, X. *et al.* An active metamaterial for polarization manipulating.*Adv. Opt. Mater.* 2, 945–949 (2014).
49. Pan, W. *et al.* A beam steering horn antenna using active frequency selective surface. *IEEE Trans. Antennas Propag.* 61, 6218–6223 (2013).

CHAPTER 15 SCATTERING CHARACTERISTICS OF TARGETS ABOVE A ROUGH SURFACE IN SAR IMAGES

Bo Zhao and Tie Jun Cui

INTRODUCTION

The classification to the region of interests (ROIs) or the scattering features in synthetic aperture radar (SAR) images is the final and most important stage in a SAR-image-based autotarget recognition (ATR) system. The simulation of SAR images has significant importance to the image understanding and target recognition. Compared to the inverse SAR (ISAR) images of targets in the free space such as airbornes, the simulations of SAR images to various vehicles on the ground or ships on the sea are much more complicated and difficult. In such simulations, the effects of ground or sea surface to the scattering from targets need to be considered, which is important to the real ATR system.

In the past decades, the approaches on how to take into account the coupling between the ground or sea surface and targets have been extensively studied. Johnson calculated the scattering of dielectric objects in a half space using the four-path method [1]. Guan et al. computed the scattered fields of a perfectly electrical conducting (PEC) target above a finite PEC rough surface using the method of moment (MoM) and half-space Green's function [2]. Xpatch could be used to calculate the scattering by objects—either PEC or dielectric—above a dielectric rough surface using the high-frequency (HF) methods and generate the SAR images rapidly [3–5]. However, the parameters of the ground or sea surface, including the dielectric constant, and electric conductivity could not be enumerated. On the SAR image simulations, Franceschetti et al. developed a method to produce SAR raw signals, including the stripe and spotlight mode based on the Kirchhoff approximation (KA) method [6]. Depending on this method, the ground or sea surface has been modeled as a rough surface divided by quadrilateral meshes, and the scattering from every mesh conforms to the Gaussian distribution. Xu and Jin generated a set of SAR images of wider scene with different vegetation and buildings using a mapping and projection algorithm (MPA) [7].

Under the point-target hypothesis, a real target consists of many scattering points or light spots in its SAR image. However, the light spots are not always appearing inside the SAR image, while they would change with the incident frequency and attitude angle of the target. Another obvious feature in the SAR image is that a shadow exists on the back to the radar sight. The shape of the radar shadow is similar to the shadow when illuminated by light but is not equal to that. The radar shadow is fuzzier than lighting shadow because of the coherent imaging mechanism and the resolution ratio. In the ATR system, the frequently used features include the location and intensity of the scattering centers, the contour of the target, and the shadow. However, the above features of a target alone will be different to those of the target above the ground or sea surface, while the latter will be used in the real ATR systems.

Hence the coupling between the surface and target must be considered when studying the target features in SAR images. In this paper, the frequency-domain scattered fields of targets above a rough surface are computed using the shooting and bouncing ray (SBR) technique and the truncated-wedge incremental-length diffraction coefficient (TW-ILDC) method. The SAR images of targets are then calculated by taking a two-dimensional (2D) IFFT on the scattered fields, in which the coupling effects have been investigated and interpreted in details. When taking IFFT, the resampling and window function methods are used to preprocess the raw frequency-domain data.

Citation: Bo Zhao and Tie Jun Cui, "Scattering Characteristics of Targets above a Rough Surface in SAR Images," International Journal of Antennas and Propagation, vol. 2013, Article ID 653438, 10 pages, 2013. doi:10.1155/2013/653438.

The paper is organized as follows. The SBR technique and TW-ILDC method are introduced in Section 2.1, and the simulations of SAR images are investigated in Section 2.2. Numerical examples are presented in Section 3, which include the targets in free space, the rough surface itself, and the coupling between targets and rough surface. Finally, a conclusion is made in Section 4 to illustrate the importance of the simulations in the real ATR system.

SAR-IMAGE SIMULATIONS TO TARGETS ABOVE A ROUGH SURFACE

Computation of Scattered Fields

After developed by Ling et al. [8] to calculate the scattering of cavity, the SBR technique had been used to solve the scattering problems for large electrical-size targets successfully. Xpatch is a well-known software combining SBR and other effective methods to calculate the radar cross sections (RCSs), SAR images, and ISAR images of targets.

The main idea of SBR is to use a dense grid of shooting rays as the incident field, then calculate the multiple reflections of every shooting ray by the geometrical optics (GO) method when the ray reacts to the target, and finally to compute the scattered fields using the physical optics (PO) method when the ray escapes from the target. The most time consuming part in SBR is not the electromagnetic (EM) computing but the ray tracing, which could be accelerated by means of parallel methods or GPU techniques. The disadvantage of SBR is that it does not include the diffraction and creeping waves, which would influence the accuracy of computation. The equivalent PO current on the surface of a PEC target is written as

$$\mathbf{J}_s = 2\delta\hat{n}\left(\mathbf{r}'\right) \times \mathbf{H}^i\left(\mathbf{r}'\right), \tag{1}$$

in which the coefficient δ accounts for the shadowing effects, and $\hat{n}$ is the outward unit normal vector to the illuminated object surface at $\mathbf{r}'$. We could calculate the scattering electrical field by

$$\mathbf{E}_s(\mathbf{r}) = -j\omega\mu_0 \int\int \mathbf{J}\left(\mathbf{r}'\right) G\left(\left|\mathbf{r}-\mathbf{r}'\right|\right) d^2r', \tag{2}$$

where $G(|\mathbf{r}-\mathbf{r}'|)$ is the scaler Green's function in free space

$$\mathrm{G}\left(\left|\mathbf{r}-\mathbf{r}'\right|\right) = \frac{e^{-jk_0 r}}{4\pi r} e^{-j\mathbf{k}_s\cdot\mathbf{r}'}. \tag{3}$$

To improve the accuracy of the SBR technique, the TW-ILDC method has been considered to include the diffractions from finite sized wedge faces [9, 10]. The TW-ILDC method has been confirmed to be useful in improving the accuracy, especially for the complicated targets. The detailed derivation of this method and the numerical results have been presented in [11, 12]. The total scattered field is the superposition of the SBR field and the diffraction field as follows:

$$\mathbf{E}_s = \mathbf{E}_s^{\mathrm{PO}} + \mathbf{E}_s^{\mathrm{Diffraction}}. \tag{4}$$

In general, the frequency-domain scattered field could be computed by the combined SBR and TW-ILDC methods [11].

The SAR-Image Generation

It has been proved that the SAR image could be generated from the frequency-domain scattered fields using the 2D-IFFT algorithm [13]. The incident magnetic field is obtained by $\mathbf{H}_i(\mathbf{r}) = (1/\omega\mu)\mathbf{k}_i \times \mathbf{E}_i$ from the specific incident electric field. Then the scattered field is written as

$$\mathbf{E}_s(\mathbf{r}) = -jk_0E_0 \int\int 2\hat{n}(\mathbf{r}) \times \left(\hat{k}_i \times \hat{t}\right) e^{-j\mathbf{k}_i\cdot\mathbf{r}'} G\left(\left|\mathbf{r}-\mathbf{r}'\right|\right) d^2r'. \tag{5}$$

We suppose that the $\hat{p}$ component of the scattering field is received, and define a shape function $O(\mathbf{r}')$.. Then the $\hat{p}$ component of the scattered field is written as

$$\hat{p}\cdot\mathbf{E}_s(\mathbf{r}) = -jk_0E_0 \frac{e^{-jk_0 r}}{4\pi r} \int\int\int_{-\infty}^{+\infty} O\left(\mathbf{r}'\right) e^{-2j\mathbf{k}_i\cdot\mathbf{r}'} d^3r', \tag{6}$$

from which the relationship between the scattered field and the Fourier transform of the shape function is obtained as follows:

$$\hat{p}\cdot\mathbf{E}_s(\mathbf{r}) = C\widetilde{O}\left(\mathbf{k}_s - \mathbf{k}_i\right), \tag{7}$$

in which $\widetilde{O}(\mathbf{r})$ is the Fourier transform of the shape function

$$\widetilde{O}(\mathbf{k}) = \int\int\int_{-\infty}^{+\infty} O(\mathbf{r}') e^{j\mathbf{k}\cdot\mathbf{r}'} d^3r'. \tag{8}$$

That is to say that the shape of the target could be obtained from the scattered field by using IFFT [13].

In this paper, the 2D SAR image will be considered, which is the projection of the shape function on the imaging plane, the x-y plane. Hence the relationship between the 2D SAR image and the shape function is

$$O\left(k_x, k_y\right) = \int\int_{-\infty}^{\infty} O_{\mathrm{proj}}(x, y) e^{jk_x x + jk_y y} dx\,dy. \tag{9}$$

From the above equations, it is clear that we could obtain the SAR images from the scattered fields simulated by the SBR and TW-ILDC methods. First, a set of scattered field data are simulated. Then a similar procedure is conducted to the PFA step in the spotlight SAR imaging, which takes an interpolation on the data before performing IFFT. The window function method is applied before IFFT to reduce the Gibbs' effect [14]. To obtain a high-resolution SAR image, a set of wideband and wide-azimuth data are needed. The resolution in the range and azimuth, ρ_r and , ρ_a, is to be determined as

$$\rho_r = \frac{c}{2B}, \qquad \rho_a = \frac{c}{4f_c \sin(\Delta\phi)/2}, \tag{10}$$

where c is the light velocity, B is the bandwidth of the electromagnetic wave, f_c is the central frequency, and $\Delta\phi$ is the rotation angle of SAR.

The simulated data are distributed over a sector region in the wave number domain (WND) on the $k_x - k_y$ plane. Although the distribution region is similar to a rectangle, it is still different (the azimuth is large). Hence a sector-to-rectangle interpolation, which is called as resampling in PFA, is needed before IFFT. Once the resolutions are confirmed, the resampling numbers could be computed as follows:

$$N_r = \frac{L_{\max}}{\rho_r}, \qquad N_a = \frac{L_{\max}}{\rho_a}, \tag{11}$$

where L_{max} is the maximum length of the simulation scene. On the other hand, to reduce the Gibbs' effect, we use a 2D Taylor window function on the resampled data in the k domain [14].

NUMERICAL EXAMPLES

The main purpose of this paper is to illustrate the effect of a rough surface to SAR images of targets located above the rough surface. The Gaussian distribution rough surfaces has been widely used in the scattering research, which is chosen as ground model. Because the SBR method is used to calculate scatter field of models, the rough surface is also set as PEC objects. The rough surface is described as the Gaussian spectrum by

$$W\left(k_x, k_y\right) = \frac{l_x l_y h^2}{4\pi} \exp\left[\frac{l_x^2 k_x^2 + l_y^2 k_y^2}{4}\right], \tag{12}$$

in which l_x and l_y are the correlation lengths in x- and y-directions, respectively, and h is the rms height of the rough surface. The generation progress has been shown in [15]. In the SBR method, the simulation time is related to the size and mesh density of the targets. Hence an infinite rough surface is impossible for simulations and we have to choose a finite-sized rough surface. The model should be static PEC or dielectric coated electrically large ground targets and the size of targets should be larger than the minimum resolution cell in the image. It also could be meshed before calculation. The incident parameters are defined in Figure 1(a) and the size of rough surface is set in a minimum length as shown in Figure 1(b) because it could contain the shadow area which could be an important feature in an ATR system.

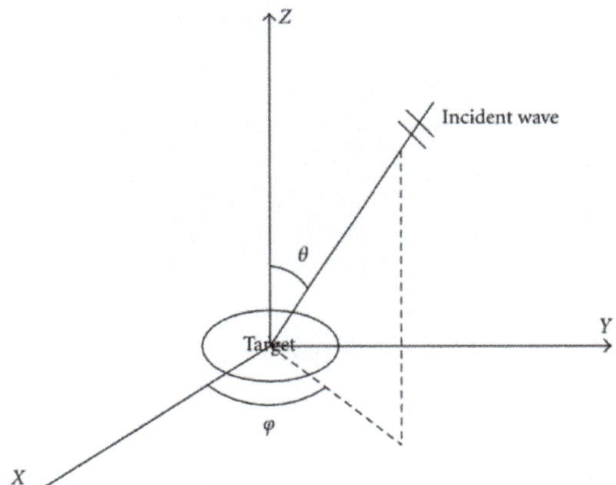

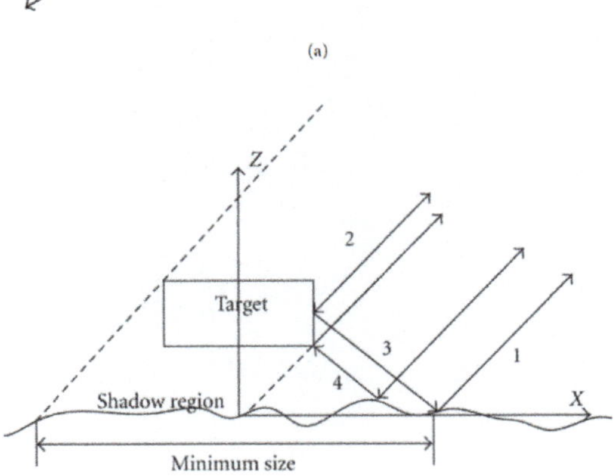

Figure 1: The incident parameters definition and the minimum size of a rough surface.

Here are the explanation of the multiple scattering routes on the rough surface. In Figure 1(b), we could see that there are several scattering routes to contribute to the back scattering, which are

1. the direct reflection by the rough surface;
2. the direct reflection by the target;
3. multiple scattering from the target to rough surface;
4. multiple scattering from the rough surface to target.

There are also other multiple scattering that could be ignored. Each route could contain numbers of reflections during SBR method [8]. In this paper, the number of reflections could be 2 or 3 according to different models and could satisfy the calculation accuracy [11, 12].

In the first example, the effect of a rough surface to targets is investigated through the simulation of a SLICY-like target above a rough surface which is shown in Figure 2. SLICY is a vehicle-size target which includes canonical shapes and cylindrical surfaces. The forward direction is looking into the trihedral corner, which we call the front of the target. The maximum dimensions of SLICY-like target shown in 2 are 81.5λ in width, 125.8λ in length, and 64.7λ in height. In the following simulations, all the rough surfaces are discretized by flat triangles whose maximum edge length is 3λ (λ is the wavelength in free space).

Figure 2: The geometrical model of a SLICY-like target above a Gaussian rough surface.

The target is located above a Gaussian rough surface with $l_x = l_y = 15\lambda$ and $h = 2\lambda$. The horizontal dimensions of the rough surface are set as $350\lambda \times 350\lambda$. The plane waves are incident from the direction of $\theta = 75^{\circ}$ and $\phi = (0^{\circ}, 15^{\circ}, 30^{\circ}, 45^{\circ}, 60^{\circ}, 75^{\circ}, 90^{\circ})$, respectively, the central frequency is 10 GHz, and the bandwidth is 1.5 GHz. Here, the incident direction implies the central direction of SAR simulation because N_a directions in the range of $\Delta\phi$ are required. From (10) we can calculate that the range resolution is 0.1 m. We also suppose that the azimuth resolution is equal to the range resolution, so that $\Delta\phi$ is 4.29546°.

Figure 3 shows the SAR image results of SLICY-like target on the Gaussian rough surface for the HH polarizations where y-axis represents range direction and x-axis represents cross-range direction. The horizontal streak in Figures 3(a), 3(b), 3(g), and 3(h) is mainly caused by the Gibbs' effect during IFFT, especially in Figures 3(a) and 3(g). An obvious feature of Gibbs' effect in SAR image is sidelobes next to the strong scattering points. Results of VV have similar characteristics, actually. To illustrate the effect of scattering from the rough surface, the corresponding SAR images of the target in free space are also simulated. Because the main strong scattering effects are brought by the first three reflections, the multiple reflections more than 4 are ignored. Limited by space, only the results of $\phi = (0^\circ, 30^\circ, 60^\circ, 90^\circ)$ are demonstrated here, which are clear enough to show the rough-surface effect.

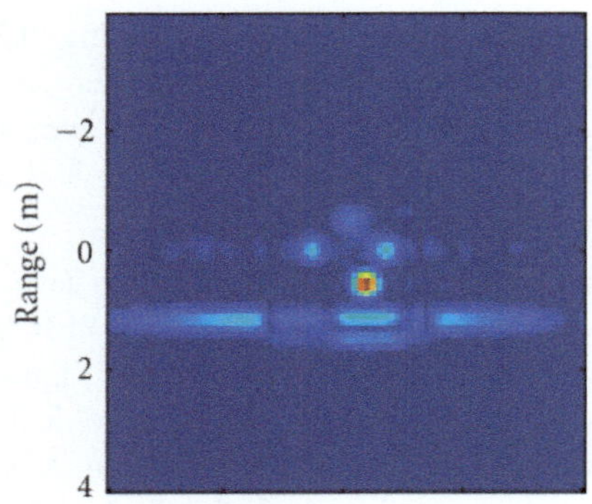

(a) Cross-range (m)

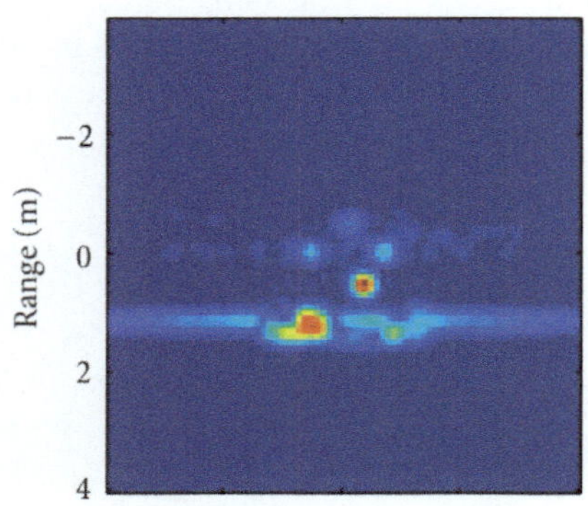

(b) Cross-range (m)

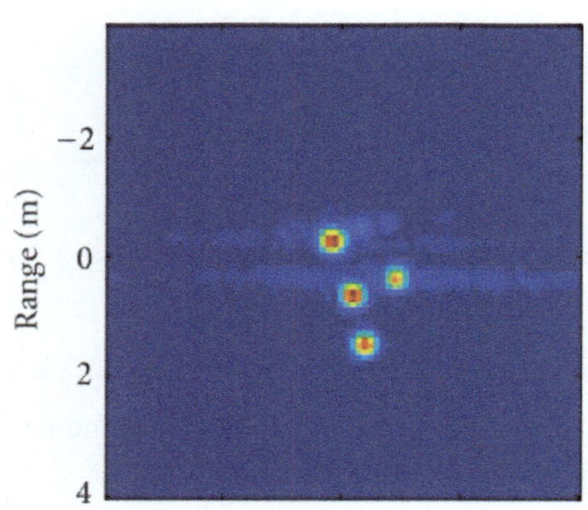

(c) Cross-range (m)

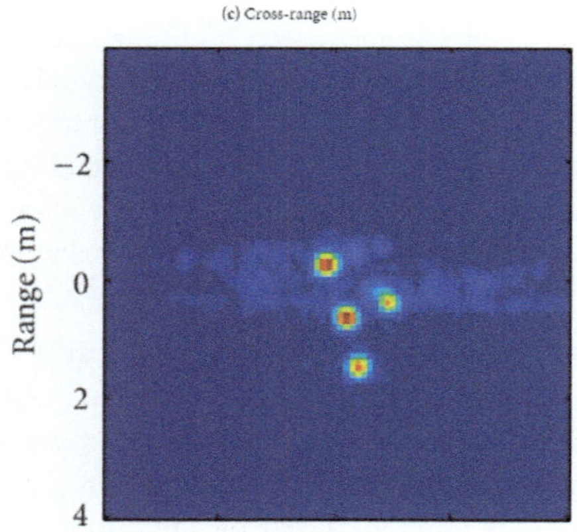

(d) Cross-range (m)

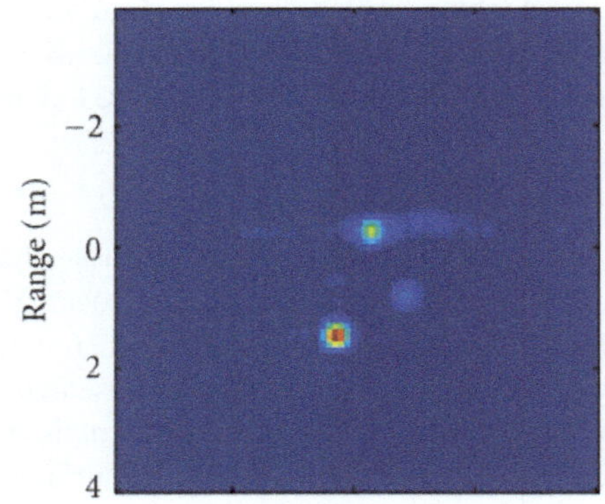

(e) Cross-range (m)

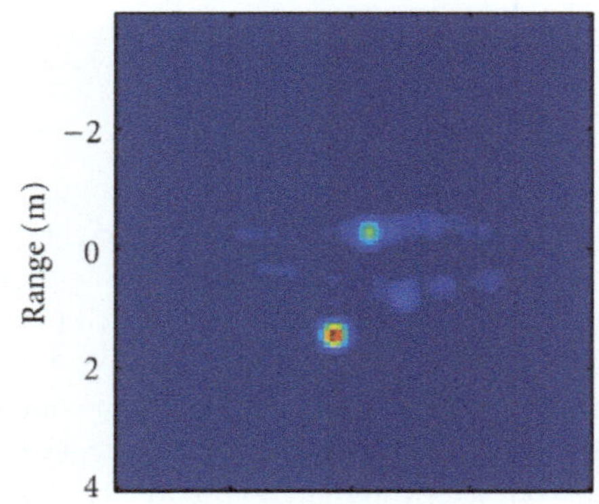

(f) Cross-range (m)

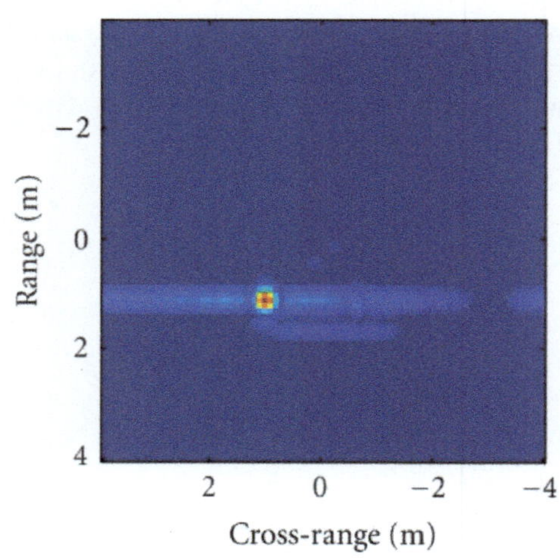

(g)

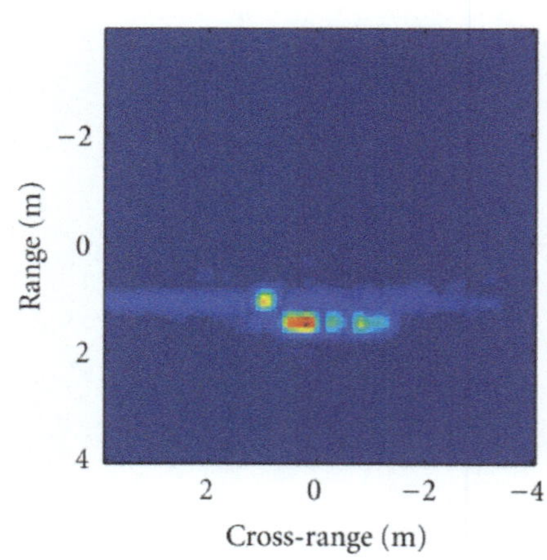

(h)

Figure 3: The 2D HH-polarization SAR images of the SLICY-like target in free space and above a Gaussian rough surface, in which the incident direction is ($(\theta = 75^\circ$ and $\phi = [0^\circ, 30^\circ, 60^\circ, 90^\circ])$), the resolution is 0.1 m, and the maximum reflection number is 3. (a) The SAR image of target in free space at the azimuth of $\phi = 0^\circ$. (b) The SAR image of target above a Gaussian rough surface at the azimuth of $\phi = 0^\circ$. (c) The SAR image of target in free space at the azimuth of $\phi = 30^\circ$. (d) The SAR image of target above a Gaussian rough surface at the azimuth of $\phi = 30^\circ$. (e) The SAR image of target in free space at the azimuth of $\phi = 60^\circ$. (f) The SAR image of target above a Gaussian rough surface at the azimuth of $\phi = 60^\circ$. (g) The SAR image of target in free space at the azimuth of $\phi = 90^\circ$. (h) The SAR image of target above a Gaussian rough surface at the azimuth of $\phi = 90^\circ$.

The strong scattering points in SAR images are generated mainly by the dihedral angles and trihedral angles on the target. When SLICY is placed above a rough surface, there is a local "dihedral angles" which supply multireflections constructed by the bottom of target and rough surface so that there are obvious differences between the results of free space and above the rough surface, as illustrated in Figures 3(a), 3(b), 3(g), and 3(h). Although the "dihedral angles" are different to those constructed by the target and a PEC plane, the construction still have similar multireflection route during ray tracing in SBR method especially enhancing the backscattering. Meanwhile, in Figures 3(c), 3(d), 3(e), and 3(f), although there are multireflection constructions at the bottom of the target, the second reflection will not return back because the direction of the reflection is different to the backscattering. Comparing to the multireflection part, the backscattering of the rough surface is so small that it could not be displayed clearly in the images.

To compare the differences of the results in free space and above the rough surface, we calculate the correlation coefficient of the two images in all of the 7 directions. The correlation coefficient is defined as

$$\text{corrcoef} = \frac{E(AB) - E(A)E(B)}{\sqrt{E(A^2) - E^2(A)}\sqrt{E(B^2) - E^2(B)}}, \quad (13)$$

where A and B are the two image matrices. The corrcoef is close to 1 when A and B are similar to each other, and if A=B then $\text{corrcoef} = 1$. The curve of the correlation coefficient is shown in Figure 4.

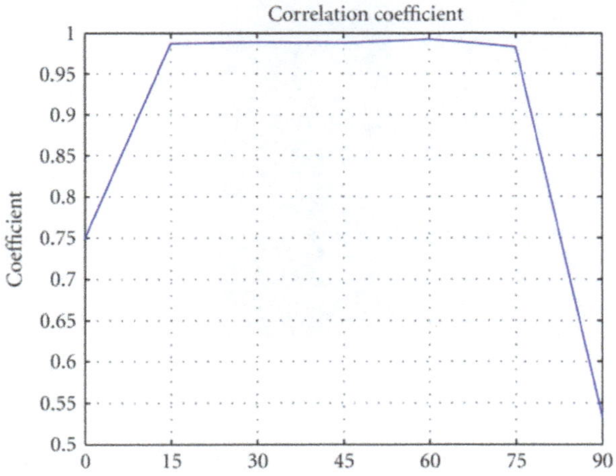

Figure 4: The correlation coefficients of the SAR images of the SLICY-like target in free space and above a rough surface at 7 azimuth directions.

From Figure 4 we can see that the correlation coefficients are close to 1 except two special directions at $\phi = 0^\circ$ and 90°. Results in Figures 3 and 4 illustrate that the rough surface always impact the SAR image results by changing the local consistence of reflectors. However, the other reflectors located on the plane of the SLICY-like target are not influenced by the rough surface so that the correlation coefficients at directions of, $\phi = (15^\circ, 30^\circ, 45^\circ, 60^\circ, 75^\circ), \theta = 75^\circ$ are close to 1.

In an SAR-based ATR system, the strong scattering points are the most important features that would be used during the matching procedure. Next we present simulation results of a complicated target, a model of tank, to illustrate the SAR images in free space and above a rough surface and the changes of amplitudes and locations of scattering points when the target is above different rough surfaces. There are also tank results that could be obtained in [4, 5]. The size of tank model is 295λ in length, 107λ in width, and 73λ in height. The three-dimensional (3D) model of the tank above a Gaussian rough surface is shown in Figure 5.

Figure 5: The 3D geometrical model of a tank above a Gaussian rough surface.

Firstly, we consider the tank model above different Gaussian rough surfaces. Generally, the coupling is weak for the parts of rough surfaces which are far from the target because small quantity of rays could reach the target. It is important to confirm the size of the Gaussian rough surface. If the size of the rough surface is small, the accuracy of the scattering effect could not be guaranteed; if the size is very large, however, huge amount of simulation time would be required. Here, we consider two sizes of rough surfaces, which are $1000\lambda \times 1000\lambda$ and $1250\lambda \times 1250\lambda$, respectively. The correlation length and height of the two rough surfaces are both $l_x = l_y = 15\lambda$ and $h = 5\lambda$. The direction of the tank gun barrel is set as 0° azimuth direction, and the three incident directions are set as $\theta = 75^\circ$ and $\phi = [30^\circ, 45^\circ, 60^\circ]$. The central frequency of the incident HH-polarization plane wave is 9.5 GHz, and the bandwidth is 0.6 GHz. The azimuth range is 3.61°. According to (10), the resolutions of the range and cross-range are $\rho_r = \rho_a = 0.25$ m.

Figure 6 shows the SAR image results of the tank model above different Gaussian rough surfaces. From top to bottom, each row means different imaging azimuth directions of the target, which are $30^\circ, 45^\circ$ and 60°, respectively. The left column of Figure 6 is the result with smaller rough surface, and the right column is the result with larger rough surface. The display ranges in the SAR images are the same, but are different from those when simulating since they need different samples to satisfy the Nyquist's sample law. The images are zoomed in to fit the size of tank model to be displayed more clearly. The grounds are generated randomly so that the noise is different.

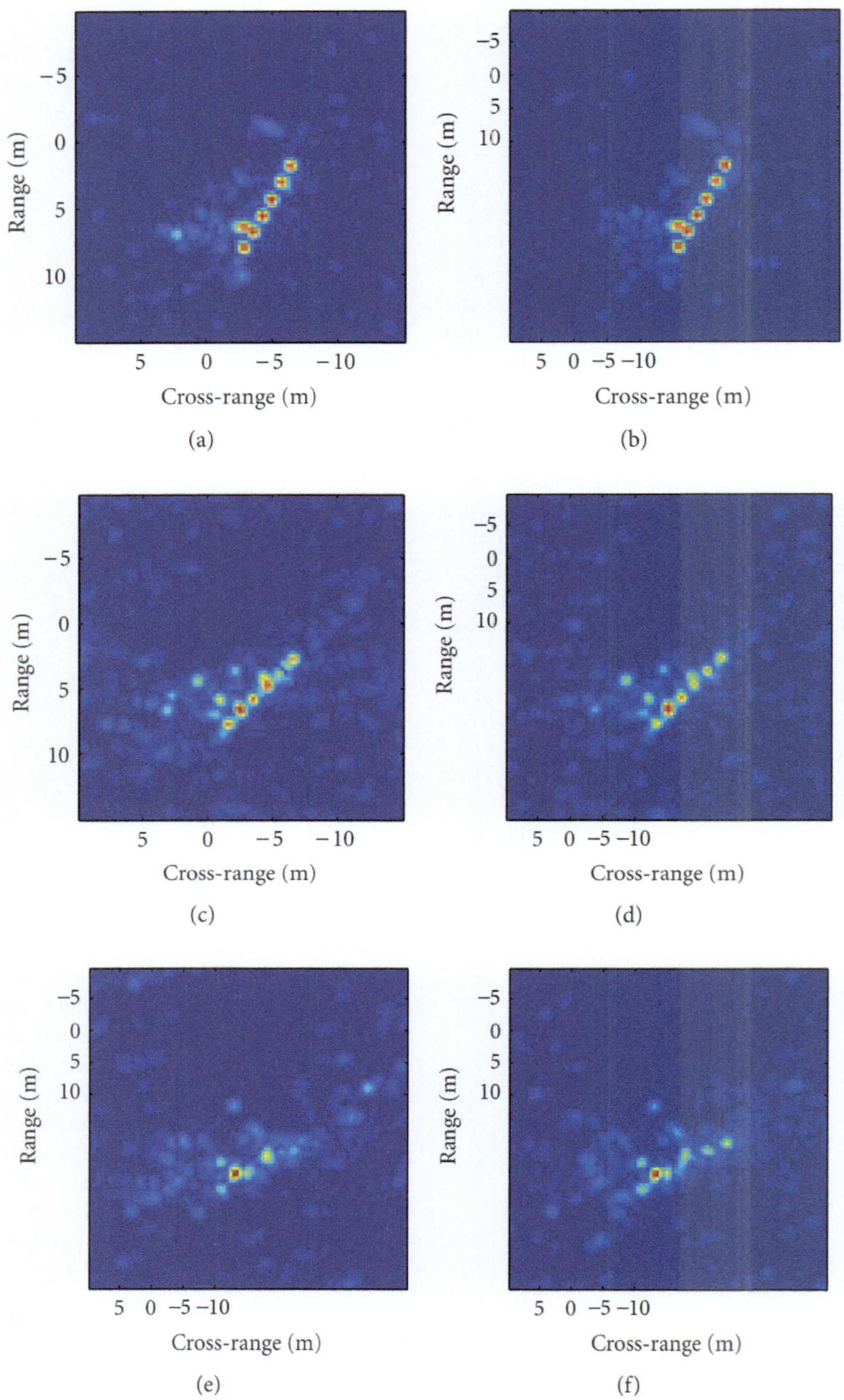

Figure 6: The 2D HH-polarization SAR images of the tank target above Gaussian rough surfaces with different sizes, in which the incident angle is $(\theta = 75^{\circ}, \phi = [30^{\circ}, 45^{\circ}, 60^{\circ}])$, the resolution is 0.25 m, and the maximum reflection number is 2. The gun barrel direction is set as $\phi = 0^{\circ}$. (a) The SAR image of the tank above the smaller rough surface at azimuth of 30°. (b) The SAR image of the tank above the larger rough surface at azimuth of 30°. (c) The SAR image of the tank above the smaller rough surface at azimuth of 45°. (d) The SAR image of the tank above the larger rough surface at azimuth of 45°. (e) The SAR image of the tank above the smaller rough surface at azimuth of 60°. (f) The SAR image of the tank above the larger rough surface at azimuth of 60°.

From Figure 6 we can see that the brightest pixels are nearly the same when the tank target is above different-size rough surfaces. In the azimuth of 30°, the brightest pixels represent the tank wheels, and the next bright pixels located in the right of the brightest ones indicate the oil drums and hatch, which are nearly the same in the two SAR images, as shown in Figures 6(a) and 6(b). The weak scattering points such as the front cover and gun barrel are different in two images because the scattering of these parts is so weak that could be easily submerged by the scattering of rough surfaces. In the azimuth of 45°, the brightest pixels represent the wheels and hatch part of the tank, and the next bright pixels are located in the two points of front cover on the right of the hatch. These features are nearly the same in the two SAR images demonstrated in Figures 6(c) and 6(d), but a tiny difference exists in the gun-barrel scattering point. It is brighter in Figure 6(c).

For the case of 60° azimuth angle, the SAR image results in such two situations have significant difference. This is because the energy is not concentrated to a few strong-scattering points but distributed in many weaker pixels. Hence the scattering by the rough surface could influent the amplitudes and locations of these pixels significantly. There are several obvious differences in the two SAR images, as illustrated in Figures 6(e) and 6(f). First of all, the point located at the back cover appears in Figure 6(f) but disappears in Figure 6(e). Secondly, the two strong scattering points generated by the hatch and toolbox are clearer in Figure 6(f). Regardless of such differences, the distributions of strong scattering points in such two images are basically similar to satisfy the requirement of an ATR system.

Figure 7 shows another SAR image results of the tank model above different rough surfaces, in which the incident angle is $(\theta = 75^{\circ}, \phi = 45^{\circ})$. The sizes of rough surfaces are $800\lambda \times 800\lambda$, $933\lambda \times 933\lambda$, $1067\lambda \times 1067\lambda$, $1200\lambda \times 1200\lambda$, and $1333\lambda \times 1333\lambda$, respectively. Other parameters are the same as those in Figure 6.

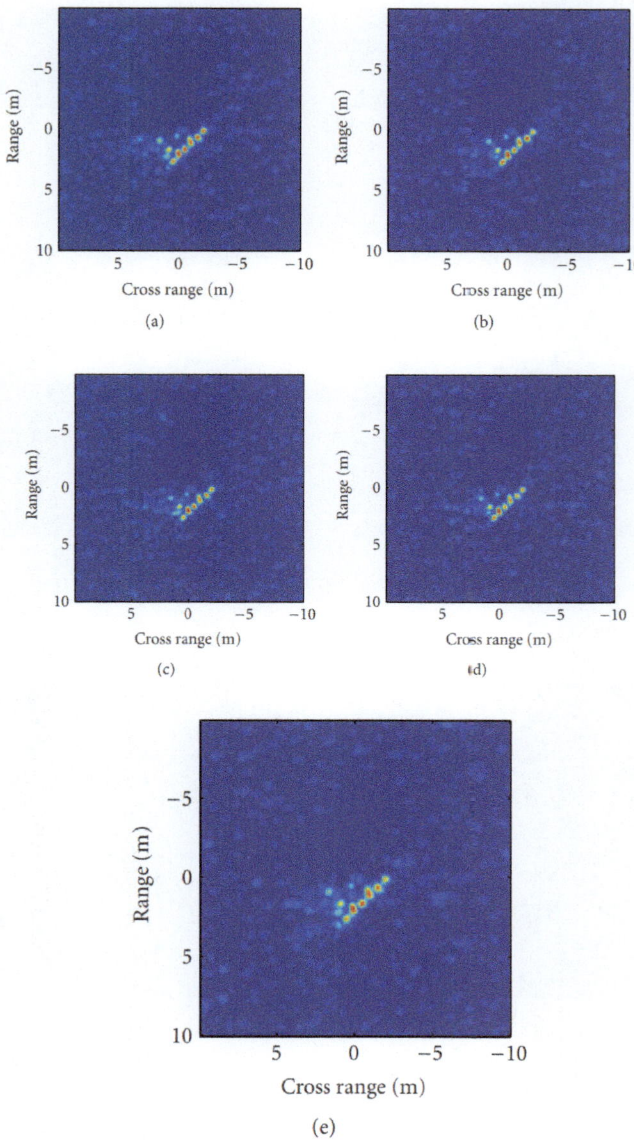

Figure 7: The 2D HH-polarization SAR images of the tank target above Gaussian rough surfaces with different sizes, in which the incident angle is $(\theta = 75^{\circ}, \phi = 45^{\circ})$, other parameters are the same as in Figure 5. (a) The SAR image of the tank above rough surface of $800\lambda \times 800\lambda$. (b) The SAR image of the tank above rough surface of $933\lambda \times 933\lambda$. (c) The SAR image of the tank above rough surface of $1067\lambda \times 1067\lambda$. (d) The SAR image of the tank above rough surface of $1200\lambda \times 1200\lambda$. (e) The SAR image of the tank above rough surface of $1333\lambda \times 1333\lambda$.

From Figure 7, we can see that the five figures are nearly the same, which means that the coupling between the tank model and rough surface is only concentrated around the tank model. Hence it would be accurate enough if the size of the rough surface is set to $1000\,\lambda \times 1000\,\lambda$.

In the above examples, the tank target and the rough surface are regarded as a combined system in the numerical simulations. To investigate the coupling effect of rough surface, several "superposed" SAR images are generated by superposing the free-space target images on the rough-surface images. The correlation coefficient is also calculated to illustrate the similarity of the two kinds of images. We present several simulations to illustrate the difference between the tank target in free space and that above a Gaussian rough surface. The size of the rough surface is $1000\,\lambda \times 1000\,\lambda$, and other parameters are the same as the earlier example. Figure 8 shows the results of "combined" and "superposed" SAR images of the tank target. The left column is the "superposed" SAR images, while the right column is the "combined" SAR images, respectively. From top to bottom, the azimuth directions are $30^\circ, 45^\circ$, and 60°, and , respectively.

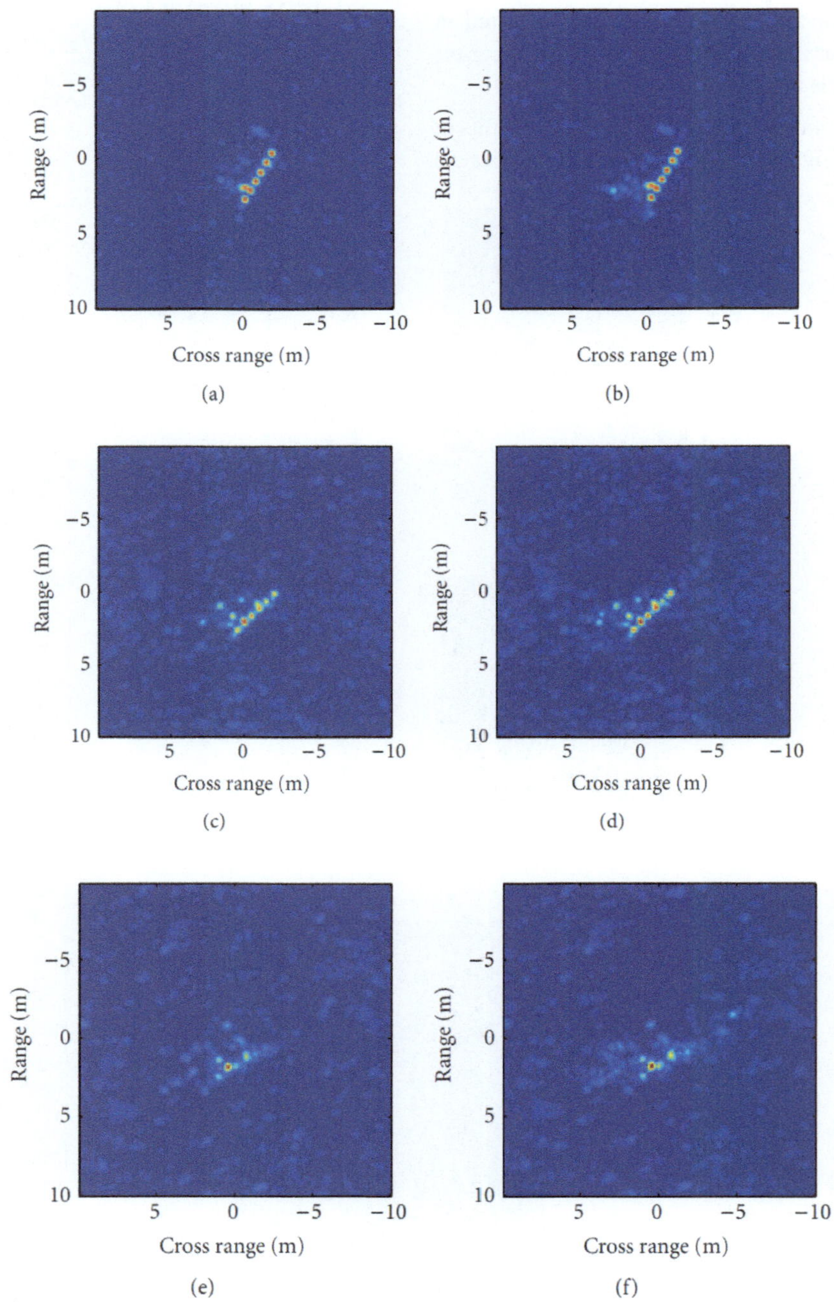

Figure 8: The 2D HH-polarization "superposed" and "combined" SAR images of the tank target and Gaussian rough surface, in which the incident angle is $(\theta = 75^\circ, \phi = [30^\circ, 45^\circ, 60^\circ])$, the resolution is 0.25 m, and the maximum reflection number is 2. The gun barrel direction is set as $\phi = 0^\circ$. (a) The SAR image of the tank in the superposed way at azimuth of 30°. (b) The SAR image of the tank in the combined way at azimuth of 30°. (c) The SAR image of the tank in the superposed way at azimuth of 45°. (d) The SAR image of the tank in the combined way at azimuth of 45°. (e) The SAR image of the tank in the superposed way at azimuth of 60°. (f) The SAR image of the tank in the combined way at azimuth of 60°.

From Figure 8 we can see that the obvious difference between the two kinds of SAR images is whether a shadow of the tank exists or not. The "combined" method is based on the ray-tracing technology, hence the occluded part of the rough surface by the tank could not be illuminated by rays. Besides, there are other differences in which we have more interest. In the azimuth of 30° shown in Figure 8(b), an extra strong scattering point exists at the right-hand of gun barrel, which disappeared in Figure 8(a). In the azimuth of 45°, the two SAR images have a difference on the scattering points of wheel locations, which implies that the rough surface and the tank confirm a local reflection structure which changes the route of reflected rays. In the azimuth of 60°, the strong scattering points are basically the same, but there is an extra point at (21, 16) in Figure 8(f). This point is caused by the multiple reflections from the tank to the rough surface, which do not exist in Figure 8(e).

CONCLUSIONS

In this paper, the scattering characteristics of targets above a Gaussian rough surface in SAR images are investigated via the HF asymptotic method. The presented numerical results demonstrate that some differences exist between the SAR images of target in free space and those above the Gaussian rough surface. This kind of phenomenon is principally caused by the local change of structure consisting of the target and the rough surface. We can use these changes of characteristics to improve the matching stage of a SAR-based ATR system.

REFERENCES

1. J. T. Johnson, "A study of the four-path model for scattering from an object above a half space,"Microwave and Optical Technology Letters, vol. 30, no. 2, pp. 130–134, 2001. · ·
2. B. Guan, J. F. Zhang, X. Y. Zhou, and T. J. Cui, "Electromagnetic scattering from objects above a rough surface using the method of moments with halfspace green›s function," IEEE Transactions on Geoscience and Remote Sensing, vol. 47, no. 10, pp. 3399–3405, 2009.
3. D. J. Andersh, S. W. Lee, J. Moore, and D. Sullivan, "Xpatch prediction improvements to support multiple ATR applications," in SPIE Conference on Radar Sensor Technology III, vol. 3395, pp. 108–119, 1998.
4. W. Coburn, C. Le, C. Kenyon, A. Sullivan, and R. Namburu, "Radar signature modeling for combat systems," in Proceedings of the 2003 User Group Conference, pp. 68–73, 2003.
5. W. Coburn, C. Le, and W. Spurgeon, "Potential discrepancies in radar signature predictions for ground vehicles," in Proceedings of the 2003 User Group Conference, pp. 45–51, July 2004.
6. G. Franceschetti, M. Migliaccio, D. Riccio, and G. Schirinzi, "SARAS: a synthetic aperture radar (SAR) raw signal simulator," IEEE Transactions on Geoscience and Remote Sensing, vol. 30, no. 1, pp. 110–123, 1992. ·
7. F. Xu and Y. Q. Jin, "Imaging simulation of polarimetric SAR for a comprehensive terrain scene using the mapping and projection algorithm," IEEE Transactions on Geoscience and Remote Sensing, vol. 44, no. 11, pp. 3219–3234, 2006. · ·
8. H. Ling, R. C. Chou, and S. W. Lee, "Shooting and bouncing rays: calculating the RCS of an arbitrarily shaped cavity," IEEE Transactions on Antennas and Propagation, vol. 37, no. 2, pp. 194–205, 1989. · ·
9. R. A. Shore and A. D. Yaghjian, "Shadow boundary incremental length diffraction coefficients applied to scattering from 3-D bodies," IEEE Transactions on Antennas and Propagation, vol. 49, no. 2, pp. 200–210, 2001. · ·
10. P. M. Johansen, "Uniform physical theory of diffraction equivalent edge currents for truncated wedge strips," IEEE Transactions on Antennas and Propagation, vol. 44, no. 7, pp. 989–995, 1996. ·
11. Z. Li, T. J. Cui, X. J. Zhong, Y. B. Tao, and H. Lin, "Electromagnetic scattering characteristics of PEC targets in terahertz regime," IEEE Transactions on Advanced Packaging, vol. 41, pp. 1–9, 2008.
12. Z. Li and T. J. Cui, "High frequency methods for simulation of high resolution imaging in terahertz regime," Journal of Infrared, Millimeter, and Terahertz Waves, vol. 31, no. 3, pp. 349–357, 2010. · ·
13. R. Bhalla and H. Ling, "Isar image formation using bistatic computed from shooting and bouncing ray technique," Journal of Electromagnetic Waves and Application, vol. 7, no. 9, pp. 1271–1287, 1993.
14. W. G. Garrara, R. S. Goodman, and R. M. Majewski, Spotlight Synthetic Aperture Radar Signal Processing Algorithms, Artech House, 1993.
15. E. I. Thorsos, "The validity of the Kirchhoff approximation for rough surface scattering using a Gaussian roughness spectrum," The Journal of the Acoustical Society of America, vol. 83, no. 1, pp. 78–92, 1988.

CHAPTER 16

COMPUTATIONAL TOOLS AND APPROACHES FOR DESIGN AND CONTROL OF COATING AND COMPOSITE COLOR, APPEARANCE, AND ELECTROMAGNETIC SIGNATURE

Erik D. Sapper and Brian R. Hinderliter

Keywords: Electromagnetic radiation, light scattering, surface scattering, bulk scattering, inclusion scattering, color, appearance, electromagnetic signature, geometric optics

INTRODUCTION

The differentiating traits of novel and high performance materials are increasingly being determined by transport properties. The transport of electromagnetic radiation through a coating is the primary cause of color, appearance, and overall electromagnetic signature for that material, and is a ubiquitous engineering consideration when designing next generation materials that have strict electromagnetic signature requirements. These materials range from consumer coatings such as architectural or automotive coatings, where color and appearance effects such as distinctness of image, gloss, color flop, and a wet look are common requirements, to more niche applications such as designed invisibility in camouflage or radar and sonar absorbing stealth coatings, into the even more exotic field of metamaterials design, where an effectively negative refractive index renders a material transparent or invisible to certain detectors, the human eye being the most common. Besides the transport of the visible wavelengths, the ultraviolet wavelengths are primary factors in photodegradation and subsequent mechanical instability of materials, and infrared radiation transport is a primary means of external heat being imparted into an object. Both UV and IR radiation transport properties are important considerations when designing materials for extreme environments, such as aerospace materials that are exposed to high altitude and space exposure, where bombarding solar photon flux is unimpeded by the presence of a gaseous planetary atmosphere. Effectively designing materials to address and satisfy all of the above concerns is the challenge of many modern day materials and coatings engineers.

The research effort presented in part in this paper attempts to combine aspects from various fields, spanning from biological transport through diffusive tissue media [1], to highly mathematical analytical solutions to many body scattering problems [2,3,4], with the hope that a unifying set of models and computational approaches may be combined into a single suite or toolbox of methods, which can address many different scattering problems at many different length scales, from nanoscale inclusions and nanoscale surface roughnesses on optically flat surfaces, to templated and highly-engineered materials like icephobic coating materials that have fractal surface roughness patterns, and multilayer systems common to automotive, aerospace, and energy harvesting industries. Indeed, the methods presented here are almost entirely scalable across many problem geometries; so that the same mathematical formalism used to address nanoscale surface roughness scattering may also be used to predict the light scattering behavior of lunar regolith and wavy ocean surfaces. The complexity hinges on the definition of a problem-specific geometric limit, and effective switching of solution regimes below, near, and above the geometric limit will ultimately determine the robustness and applicability of the solutions presented here.

Citation: Erik D. Sapper and Brian R. Hinderliter, Computational Tools and Approaches for Design and Control of Coating and Composite Color, Appearance, and Electromagnetic Signature, doi: 10.3390/coatings3020059.

Direct versus *Inverse Solutions to Electromagnetic Radiation Transport Problems*

Direct computational solutions comprise the bulk of light scattering research and literature, and aim to predict the behavior of a given, well-defined material during an explicit exposure to incident electromagnetic radiation.

Conversely, inverse computational methods attempt to indicate what material composition and geometry would produce a user-defined and desired response under a given condition of incident radiation bombardment.

It is clear that inverse methods are most valuable to the materials engineer who is tasked with designing new high performance materials, however the number and quality of inverse methods on hand is a direct consequence of the number and quality of direct solutions that have preceded the development of the inverse method of interest. Because both methods are so strongly linked, and because it is inherently more straightforward to develop direct methods, the relevancy or applicability of indirect methods usually lags behind their direct method cousins. The approaches presented here are all direct methods, though the reader is referred to the final chapters in the book by Maradudin [5] to read about two interesting inverse methods; designing surfaces that scatter light predominantly in one directional steradian, and designing surfaces that have the optical signature of a given organic compound.

One important aspect of the direct solution that is not commonly stated is its applicability to the investigation of physically intractable problems. These are problems where an exact replication in a physical environment would be difficult if not impossible, but where the solutions and modeled behavior are nonetheless important in making engineering design decisions. An example of this in practice would be the simulated single and dependent scattering regime thresholds within a cluster of closely packed specialty pigments embedded in a polymer resin; it may be difficult and cost-inefficient to replicate that material exactly in a laboratory, but the simulated material system will still offer advice regarding optimum pigment geometry and dimensions, ideal pigment volume concentration for a given effect, and predicted formulation thresholds for the onset of undesirable dependent scattering, where pigment scattering cross sections overlap due to proximity. The astute modeler and materials simulator will continue to seek out situations such as this, where the direct computational solution can assist in making quick design decisions, without having to wait for fully-realized direct or inverse solutions to be developed.

The Geometric Limit and Its Implications

Light scattering simulation modules typically represent one of the three common scattering regimes: Rayleigh scattering, Mie scattering, and geometric scattering [6,7]. These regimes are based on a dimensionless size parameter, α, which describes the size of a scattering inclusion relative to the incident wavelength,

$$\alpha = \frac{\pi D_p}{\lambda} \tag{1}$$

In the definition of size parameter α, D_p is the diameter or largest dimensional length of the scattering particle, and λ is the wavelength of the incident radiation. When $\alpha \ll 1$ for the system of interest, the scattering inclusions are much smaller than the wavelength of incident radiation and Rayleigh scattering dominates. When $\alpha \approx 1$, the scattering inclusions are of the same order of magnitude as the wavelength of incident radiation, and Mie scattering dominates in the unlikely event that the inclusions are close to being perfectly spherical.

In both of the above cases, when $\alpha \leq 1$, the most proper approach to light scattering prediction is the solution of Maxwell's equations, which govern the coupled transport of electric and magnetic fields interacting with various media, though the effect of the magnetic equations is effectively null for scattering media with magnetic permeability equal to one. This can be quite a complex computation for real, many-body systems. A finite element approach is one appropriate computational method used to address such features.

Finally, when $\alpha > 1$, the inclusion is larger than the incident radiation wavelength, and scattering events can be computed in the geometric scattering regime, where Snell's law and Fresnel approximations reduce scattering events to fairly straightforward equations that may be solved using simple arithmetic. The length scale of such features often precludes finite element approaches, and is thus most amenable to other computational approaches such as Monte Carlo ray tracing.

A challenge in developing new light scattering simulation approaches is the construction of a system that accounts for all length scales of scattering behavior, and one that can do so with minimal switching between length scales and related loss of computational efficiency. Most systems of interest for appearance and electromagnetic signature design, such as vehicle or aircraft structural components, reside on lengths scales that are 5–10 orders of magnitude above the Mie limit, so improved and multiscale approaches must be used. The hybrid geometric optics and finite element model presented in this paper represents an attempt at such a highly general yet rigorous simulation protocol for multiscale light scattering prediction in coatings, composites, and polymeric materials.

Defining a Computational Photon

In no other scientific interest does the wave-particle duality of light have such a dominating presence. A tenet of computational modeling is the complete definition of the system being modeled, ideally from physical first principles, before attempting to generate a solution. When one considers light scattering at multiple length scales, flexibility in how a photon is designed is rewarded; the most adaptive yet rigorous solutions to the omniscale scattering problem treat a photon like a particle when it is best suited to do so, and treat a photon like a wave when that is the best choice.

The designation of a problem-specific geometric limit makes the wave versus particle decision easy. When $\alpha \leq 1$, finite element methods are used to solve Maxwell's equations for an incident electromagnetic wave impinging on an object. In every other case (that is, when $\alpha > 1$) the incident radiation is treated as a photon and scattering events are simplified and reduced using the Fresnel approximation into simple photon-plane interactions that may be solved for using Snell's law and Fresnel reflection coefficient solutions. The latter method is referred to as the geometric optics approach, and its complexity lies not in the mathematical solution of the scattering event, as in wave scattering using Maxwell's equations, but in photon accounting and tracking, and thus a key aspect of those approaches are the photon queuing, collecting, and binning methods used to arrive at useful results.

SURFACE SCATTERING APPROACHES

A wealth of knowledge exists regarding surface scattering approaches. This discussion will address only the most classical approaches along with the geometric optics ray tracing approach, though the curious reader is directed to further sources in Section 2.4.

Kirchhoff Scattering, Davies Scattering, and Beckmann Scattering

Kirchhoff scattering gives the general solution to the scattering of electromagnetic radiation by a rough surface, and serves as the basis for many advanced or evolved surface scattering models [8]. The Kirchhoff solution is applicable for perfectly conducting, homogenous surfaces where the mutual interaction of surface irregularities, like shadowing and multiple surface scattering, is neglected [9,10]. In modern coatings and composites science surfaces are rarely perfectly conducting and homogenous, and the increased emphasis on electromagnetic signature design for real, in-service materials has necessitated the inclusion of surface irregularity interactions. The set of approaches developed and presented in this work allow for heterogeneous and dielectric surfaces, and account for shadowing and multiple scattering events on the surface.

The Davies scattering solution was one of the first to make explicit use of statistically rough surfaces, that is, surfaces whose peaks and valleys are described by a continuous Gaussian process statistical distribution and correlation function [11]. Beckmann provided a similar treatment describing the surface as a Markov chain, where the surface is described by a stochastic matrix [12]. Both approaches have contributed greatly to the study of light scatter by rough surfaces, and have served as the basis for many evolved models and statistical mathematical approaches. There is a usually-light-hearted divide amongst surface scattering modelers as to whether real random surfaces follow Gaussian or fractal height distributions [5,13]; the surfaces generated here follow the historical assumption and use Gaussian distributions to represent random surfaces.

Geometric Optics Ray Tracing and Nanoscale Surface Roughness

The geometric optics approach models surface scattering behavior by reducing any real surface to a discrete or a statistical distribution of microfacets that are perfectly flat on the photon-local length scale [14,15,16,17,18]. It is assumed that the size dimensions of the microfacets will be large relative to the incident radiation wavelength. Given a distribution of microfacets overlaid onto a surface, ray tracing and tracking approaches are used in computational scripts to compute the scattering behavior for each incident photon or ray at each microfacet being bombarded by incident radiation, accomplished through the straightforward solution of Snell's law and calculation of the Fresnel reflection coefficient for that interaction. The individual results of the microfacet scattering events are summed to yield the global scattering event result. The user-defined spatial and wavelength binning collection methods will then determine how the final result is interpreted, whether it's a color or gloss value prediction, or something of a larger scale such as a complete bidirectional reflectance distribution function, or BRDF [19,20].

The geometric optics approximation incorporates multiple scattering events into its solutions, and can therefore predict diffuse scattering behavior when a surface becomes very rough and multiple scattering begins to occur. The geometric optics approximation is thus able to predict forward, diffuse, and retro-reflectance.

The geometric optics approximation is computationally less expensive than other exact solutions of electromagnetic scattering theory, since the matrices in electromagnetic scattering integrals are not needed to be solved. The geometric optics approximation requires about one hundredth the CPU memory and one fifth the time to solve, as compared to the exact solution and is a suitable approximation for many rough surfaces having varying surface feature parameters [16].

First, random rough surface profiles with Gaussian peak and valley distributions are generated by using the Spectrum method outlined by Thorsos [10] and surface derivatives, tangents, and slope values are calculated for each node or point on the surface.

The angle of incident radiation is specified, and a number of first reflection points are chosen. The number of first reflection points may be less than, equal to, or greater than the number of surface points, and can be randomly or equidistantly distributed along the surface or surface profile.

Next, shadowing tests are performed at each randomly chosen first reflection point until a point is selected that does not appear shadowed or masked by the incident photon. Employing the Fresnel approximation that all local scattering events occur on a locally perfectly flat plane, the angle of the scattered ray is calculated using Snell's Law. To account for absorption or extinction, the energy of the scattered photon is calculated as the energy of the incident photon multiplied by the Fresnel coefficient for a surface material with a given index of refraction, extinction coefficient, and local angle of incidence.

The presence or absence of a second reflection point is determined by comparing the local topology around the first reflection point to the outgoing photon scattering angle, similar to the shadowing test performed earlier. If a new reflection point is located for the scattered photon, that photon becomes an incident ray and the scattering process (including any possible successive reflections) is continued until the photon exits the simulation space. A new first reflection point is then selected and the ray-tracing process continues until all first reflection points and initial incident photons have been accounted for.

After all first reflection points have been addressed, the scattering energy at each scattering angle is divided by the total amount of energy incident on the surface, which is the defined as the differential reflection coefficient. The BRDF is then calculated by dividing the differential reflection coefficient by the cosine of the outgoing scattering angle and the size of the scattering region $d\Omega_s$, in radians, and multiplying by π.

Rough surfaces used in scatter simulation may be constructed computationally from statistical parameters such as root mean squared (rms) roughness and autocorrelation length (τ), shown inFigure 1 (Left), or may be input directly from empirical measurement, for example, an atomic force microscope height profile scan may be used to generate a surface based on the properties of a specific real material, as in Figure 1 (Right).

When using surface statistics to generate rough surfaces, the total number of surface points is arbitrary and is chosen by the user, in this case 40,000 points distributed equally across the surface of the material. The number of first incoming ray strike points is equal to the number of rays fired at each surface, and is here set at 1600. To obtain more statistically meaningful results for the surfaces described by the surface roughness and autocorrelation length parameters, five separate surface realizations are performed for each defined rms and autocorrelation length surface of interest; the results from the five surfaces will be averaged to yield a Monte Carlo value of average scattering properties for each surface type.

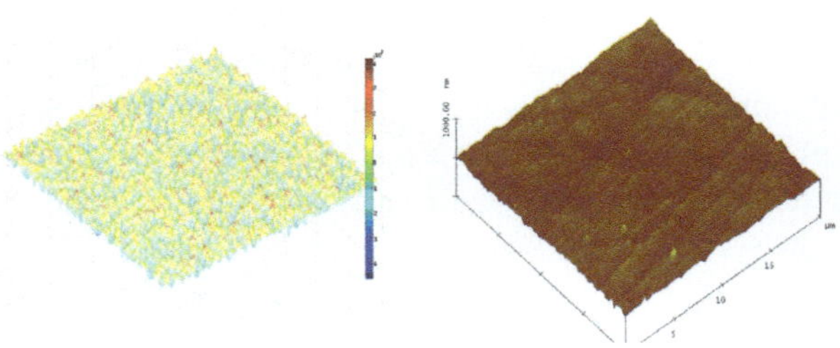

Figure 1. (**Left**) A computer-generated surface with nanoscale surface roughness constructed with a Gaussian distribution of peaks and valleys. (**Right**) A 20 × 20 micrometer height profile scan of a diamond-polished epoxy surface, obtained using an AFM, which can be input directly into a scattering simulation module.

Surface Scattering Simulation Results

Because geometric optics ray tracing approaches are so general and adaptive to many different problem types, the outputs from such a simulation are usually limited only by the creativity and needs of the user. The produced outputs are largely a matter of collecting photons exiting the simulation space into bins that can then be summed into a result that has meaningful value. For example, one can the construction of an outgoing photon collection bin that replicate a gloss meter may be used, which would be situated above the simulation space in a position located with mirror symmetry about the surface normal. If injected photons are globally incident at 20° to the surface normal, the computational collector bin will detect and count outgoing photons that leave the surface at −20° to the surface normal. The lateral width of such a bin will be related to the spot size of the gloss meter being simulated.

Figure 2 displays the mean hemispherical reflectance of a simulated poly(methyl methacrylate), or PMMA, surface having a fixed autocorrelation length of 1μm, being bombarded by incident radiation with wavelength of 590 nanometers. The simulation runs two parameterized loops, effectively increasing the rms surface roughness from 0–2μm while also increasing the angle of incident radiation from 0 (head-on) to 85 (near grazing). Because the autocorrelation length is fixed, the effect of increasing rms roughness is equivalent to increasing the slope of the surface microfacets; the spacing of the microfacet peaks and valleys remains unchanged throughout the simulation.

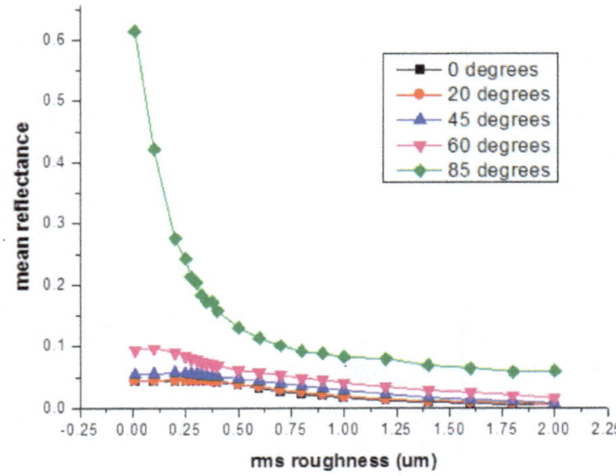

Figure 2. Simulated mean hemispherical reflectance for a PMMA surface with varying surface roughness and fixed autocorrelation length (1 μm). The 590 nanometer incident radiation impinges on the surface at angles ranging from 0° to 85° relative to the global surface normal.

Figure 2 indicates that, regardless of surface topography, the average surface reflectance decreases as the incident radiation approaches a head-on orientation. This is intuitive, especially when considering the corollary statement; as incident angle increases toward grazing, more light is reflected off the surface. All light not reflected off the surface will enter the material, and so a head-on bombardment of light will impart the largest flux of photons into the material.

Figure 2 also indicates that, regardless of incident radiation angle of attack, the amount of surface reflected photons decreases as the surface becomes rougher and the facets that comprise the surface are more steeply oriented. Having a surface with very high roughness essentially "traps" most incoming photons as they repeatedly scatter off of side walls and are eventually transmitted into the material. The effect of rough surface trapping onset is most dramatic in the case of near-grazing incident angle of 85°.

One can use the results from a geometric optics simulation to assess not only scattering properties, but also characteristics of the surface, such as the amount or proportion of the surface effectively caught in shadow due to surface roughness. Figure 3 shows the proportion of shadowed surface in the same material geometry and incident radiation parameters used in Figure 2, where a PMMA surface is bombarded by 590 nanometer wavelength incident radiation.

Figure 3 indicates that as long as the incident radiation is head-on, at 0°, none of the surface is ever caught in shadow. This is not a surprising result, but serves as a good check on the validity and soundness of the chosen geometric optics code. The relationship between increasing angle of incidence and increasing proportion of surface caught in shadow is interesting in that it is non-linear, and the onset of quick increases in shadowing proportion occurs at increasingly smooth surface values as angle of incidence increases. That is to say, very little surface roughness is needed to shadow the majority of the surface for an incoming photon with a near-grazing 85 degree angle of attack. This is also an intuitive result, but one that is difficult to quantify without using geometric optics simulation.

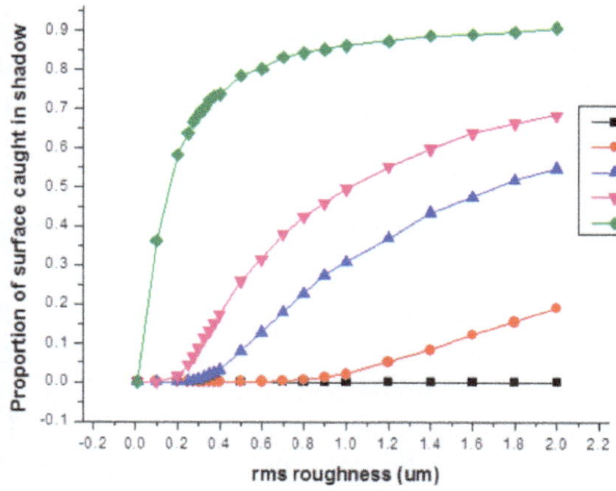

Figure 3. Proportion of surface caught in shadow for a PMMA surface with varying surface roughness and fixed autocorrelation length (1 μm). The 590 nanometer wavelength incident radiation impinges on the surface at incident angles ranging from 0° to 85° relative to the global surface normal.

The shadowing results in Figure 3 have been compared to the shadowing function derived by Brockelman and Hagfors [21], as well as to the shadowing function corresponding to Beckmann scattering theory [8]. Brockelman and Hagfors derive a shadowing function, R(θ), which accounts for facets of the surface that are tilted but are not in fact in shadow due to the parallel alignment of the incident radiation. These favorably oriented surface features will serve to reduce the amount of surface effectively in caught in shadow. This is in contrast to the Beckmann shadowing function, S(θ), which is independent of the tilt of any particular surface features, and which may overestimate the effect of shadowing by rough surface features.

Figure 4 shows the comparison between the shadowing results presented here and the shadowing functions developed by Brockelman and Hagfors (S(θ)) and Beckmann (R(θ)). The results are plotted as the proportion of the surface caught in shadow, corresponding to 1-S(θ) and 1-R(θ) in the historical models, as a function of the unitless value L/h_o, the ratio of autocorrelation length to rms surface roughness. Results from the same model are presented using the same color, and results from the same incident angle of radiation are indicated by identical marker geometries. No data for 85° incident angle was available for the Brockelman and Hagfors simulation.

Figure 4 indicates that the shadowing results presented here fall between the shadowing functions put forth by Brockelman and Hagfors and by Beckmann. At each value of incident angle, the Beckmann shadowing function overestimates shadowing and the Brockelman and Hagfors shadowing function underestimates shadowing, as compared to the new results from the current model. It is believed that the individual treatment of individual local microfacet geometry and tilt during the ray tracing algorithm here leads to a more robust method for shadowing prediction and quantification. The concept is the same as that provided by Brockelman and Hagfors, though increased computing power has improved the accuracy of the results.

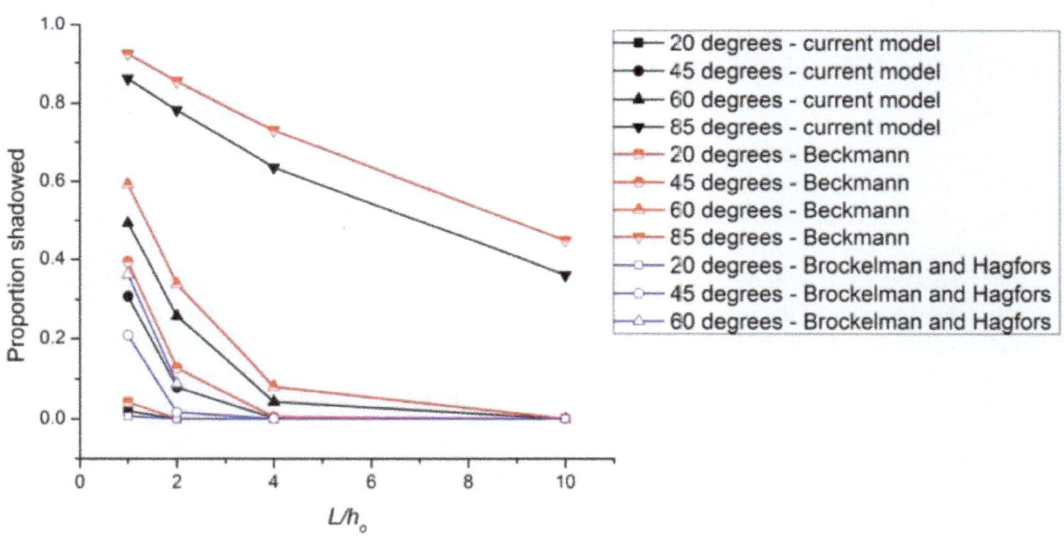

Figure 4. Comparison of the surface shadowing results inFigure 3 to those produced by the shadowing functions derived by Brockelman and Hagfors and by Beckmann. Data from both historical models is taken from reference [21].

These results barely represent a tiny fraction of the output possible with a fully-realized surface scattering modeling approach. A forthcoming paper by the authors will investigate further results in more depth [22].

Further Reading

The reader interested in learning the basics of surface scattering approaches has much seminal literature to choose from. For starters, one is directed to the classic book by Beckmann and Spinoza [8], as well as the primer by Bennett and Mattson [23]. Tsang, Kong, and Ding [24] provide excellent examples of numerical approaches to light scattering problems, and Stover [25] presents the science from the perspective of the quality control and inspection of silicon wafer production, where nearly imperceptible surface irregularities may indicate a bad production run. The collection edited by Maradudin [5] is a well put-together overview of the problem and theory of surface scattering, composed with an eye towards surfaces having nanoscale surface roughness. Finally, the doctoral thesis and related papers by Bergström [26] are concise presentations of surface scattering solutions given by geometric optics approaches and bidirectional reflectance distribution function formalism, and Bergström's codes are wonderful starting places for those looking to run simulations before writing their own code.

BULK SCATTERING APPROACHES

No different than the surface scattering approaches, the field of bulk, inclusion, or volume scattering runs wide and deep. The approaches presented here represent the two historical approaches that the coatings scientist is most likely to come across; Mie theory and Kubelka-Munk two-flux theory. Section 3.5 points the reader towards more definitive references for further reading.

Mie Theory Scattering

Mie theory scattering solves for the scattering behavior of particles that are on the same order of magnitude as incident radiation (*i.e.*, $\alpha \approx 1$). The computation is quick, and many codes and solutions are available in the literature [7,27,28]. A caveat with Mie scattering treatment is that the scatterer should be perfectly spherical in geometry. In practice this relegates Mie scattering treatment to the solution of spherical pigments or air bubbles in materials and extremely large ensembles of pseudo-spherical scatterers, where the deviations from perfectly spherical geometry are nullified by the law of averages during the computation. A discretized Mie formalism has been presented in the literature, which modifies the treatment for axisymmetric and irregularly-shaped particles [29,30].

Kubelka-Munk Theory: A Two-Flux Scattering Model

Kubelka-Munk theory is a popular light scattering prediction method for computing the diffuse reflectance of thin layers of materials containing pigments in a surrounding binder or medium such as an organic coating [31,32,33,34,35]. The description of "thin" here does not apply to the incident wavelength, but rather the diameter of the overall object being studied, *i.e.*, the thickness of the organic paint. Using Kubelka-Munk theory, the reflectance of a paint layer that is adequately hiding its substrate is given by Equation 2.

$$R = 1 + \frac{K}{S} - \sqrt{\frac{K^2}{S} + 2\frac{K}{S}} \quad (2)$$

Where, $K = \Sigma K_i c_i$, $S = \Sigma S_i c_i$.

In Equation 2, K_i and S_i are the fraction of light absorbed and the fraction of light scattered, respectively, per pigment i having unit weighting fraction c_i. Values for K and S are generally measured from a series of prepared coating samples with known density, thickness, and material composition.

Although improvements have been made to the Kubelka-Munk model to account for surface roughness and film nonuniformity [36], the theory remains a so-called two-flux theory [37], as all transport behavior is simplified to an "in-flux" and an "out-flux." In this formalism, material behavior or scattering characteristics besides hiding power and bulk film reflectance are hard to assess, such as the effects of pigment or inclusion geometry, distribution, and orientation. Still, however, it is possible to utilize aspects of Kubelka-Munk theory for more complex systems where color matching ability is desired, as presented in this paper in Section 5.2.

Finite Element Solutions to Maxwell's Equations

No contribution has been more meaningful to the study of electromagnetic radiation than that of James Clerk Maxwell, a Scottish mathematician and physicist who is credited with formulating classical electromagnetic theory with his set of partial differential equations now referred to as Maxwell's equations. Maxwell first presented his equations in a four-part paper [38], "On Physical Lines of Force," in 1861 and 1862. Many first principles simplifications of light scattering phenomenon, such as Snell's law and Fresnel reflectance functions, can be derived from Maxwell's equations, and the most accurate and precise solution to any scattering problem is still the coupled solution to Maxwell's equations, a task that is readily performed on today's modern computers and in commercially-available software packages such as COMSOL [39], provided the geometry of the problem does not reside on multiple length scales and the mesh size around relevant features allows for the accurate computation of the relevant equations. If the size scale of the feature of interest is much smaller than the size scale of the entire system, then a multiscale approach such as the one presented here must be employed.

Figure 5 shows representative output from a COMSOL solution of Maxwell's equations for a 300 nanometer diameter silicon carbide (SiC) fiber embedded in PMMA, being impinged on by incident radiation having a wavelength of 590 nanometers. The computation was performed in the radiofrequency module, using the frequency domain electromagnetic waves solver mode. A standard default COMSOL convergence criterion of $1e^{-6}$ was used. The total, time-averaged energy density in joules per cubic meter is provided as a contour map with an associated legend, while the time-averaged power flow in watts per square meter is given by arrows of varying length depending on power flow magnitude. General validity was confirmed by ensuring a constant flux along the boundaries in a pure scattering system with no extinction/absorption; no new flux was created and no flux was destroyed. The problem geometry for SiC fibers in this case accommodates a 2D, rather than a 3D, solution, and the view shown in Figure 4 is a cross section of the system geometry with scattered field power flux represented by a contour plot overlaid with an arrow plot. The simulation space that contains the fiber and polymer matrix is surrounded by an empty concentric ring called a perfectly matched layer, which ensures that no unwanted internal reflections will occur on the edge of the simulation space. Degrees of freedom solved for range from 10,000 to 30,000, depending on geometry complexity and fineness of the user-defined variable mesh along material and component interfaces within the geometry. In the example shown here, 13,000 degrees of freedom are solved for.

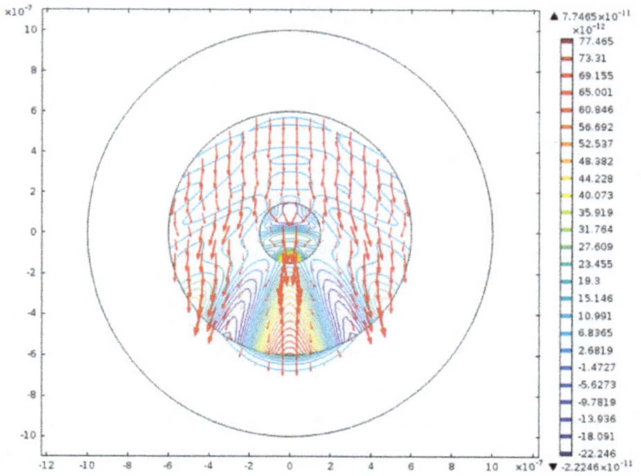

Figure 5. Finite element solution to Maxwell's equations for a 300 nanometer diameter SiC fiber embedded in PMMA matrix. This representative output shows the scattered field power flux as both contour and arrow plots. The length scale of the *x* and *y* axes is given in meters.

The output from COMSOL or a similar solver package for a single scattering event around a particular inclusion can be extracted and selectively integrated around the simulation space boundaries to produce a set of outgoing photons or photon portions, which may be carried over into a subsequent ray tracing simulation, if so desired. For instance, a single scattering event on a SiC fiber in a polymer resin may generate five or more scattered photons that can continue on in a ray tracing simulation. If the scattering event is assumed to be purely elastic and thus no absorption occurs then the sum of the outgoing photon weights will equal the weight of the initial incoming photon. One should note, however, that purely elastic collisions are not necessary in the approach; carbon black and other blackbody absorbers that are commonly used in the coatings industry may readily be used in this system. For example, in one extension of this approach, tracking the lossy component during an inelastic scattering event can be used to assess heat accumulation within a coating.

Inclusions to Consider in a Finite Element Scattering Module

The user-friendly design of modern finite element simulation packages means that many different types of inclusions may easily be simulated, ranging from two-dimensional fiber cross sections as in the above example, to completely asymmetric and irregular three-dimensional pigment geometries. Inclusions of great interest to the coatings community are pigments, be they the ubiquitous titanium dioxide, or more complex multilayered pearlescent flakes. Figure 6 shows the simulated scattered field response for a 300 nanometer long-axis diameter mica flake coated with titanium dioxide, being impinged on by 590 nanometer incident radiation at 0°, 20° and 70° angles of incidence. When parameterized across many wavelengths of incident radiation, such results may be summed into predictions and visual estimators of pearlescent appearance and color flop.

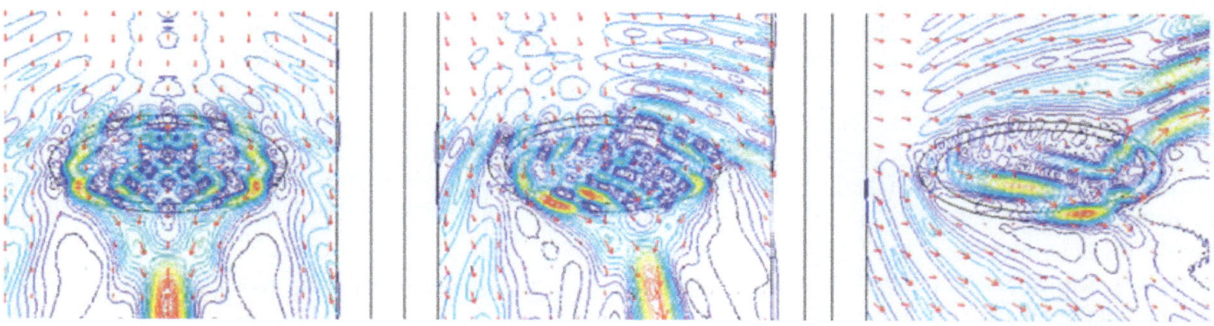

Figure 6. Simulated scatter response for a 300 nanometer mica flake coated with titanium dioxide, being impinged on by 590 nm incident radiation at 0° (**left**), 20° (**center**) and 70° (**right**) angles of incidence.

Air voids are a common nuisance in many coatings and composite applications, and these may be simulated by any number of gaseous void geometries included in the simulation space. A useful application of air void scattering is to include the presence of voids near fibers or pigments to simulate changes in electromagnetic signature before and after fiber debonding or onset of the critical pigment volume concentration, when there is no longer sufficient polymer binder to completely wet the embedded pigment. Both situations are undesirable, and both may be easily addressed by finite element simulations. Such incorporation may lead to improved non-destructive evaluation and assessment methods.

Finally, heterogeneous phases in copolymer blends may also be investigated using finite element approaches, given the refractive index of the separate phases are nonequivalent. Coupled with molecular and coarse grained simulation tools like dissipative particle dynamics approaches for phase separation of material blends, this approach may lead to another new class of non-destructive interrogation and assessment methods.

Further Reading

For more information on bulk, volume, or inclusion scattering, one is firstly directed to the seminal book by Van de Hulst [6], as well as the similar yet updated treatment given by Bohren and Huffman [7]. Both sources use as their foundation the classical electrodynamics math put forth in such classics texts as Jackson [40], Schwartz [41], and Shadowitz [41].

HYBRID OR MULTISCALE APPROACHES TO OMNISCALE ELECTROMAGNETIC RADIATION TRANSPORT

After introducing surface and inclusion scattering, it naturally follows that one consider combining the two approaches into a comprehensive modeling approach for light scattering simulation. The hybrid and multiscale scattering approach presented here uses the simplification that surface scattering will be performed by geometric optics, as shown in the above cases, and that any and all inclusion scattering will be treated using the complete solution to Maxwell's equations. This is not as computationally expensive as it initially sounds; each inclusion scattering profile only needs to be computed once, while the results from such a one-off simulation are held in a numerical matrix that is regularly called upon by the overall geometric optics framework. Thus, all transport between scattering events within the material will occur through geometric optics ray tracing processes. This means that incident radiation will strike the surface of a material while acting as a particle, be partially absorbed and travel to an inclusion as a particle, and interact with an inclusion as an incident wave which produces scattered photons behaving as particles, each of which will further propagate through the material as particles. This approach makes full use of the flexible nature of wave-particle duality that is inherent to light scattering problems.

Discrete versus Statistical Material Approaches

The hybrid model for nanocomposite scattering presented here represents a marriage of the surface geometric optics ray tracing model with a finite element bulk transport model that accommodates many sorts of inclusions and various scattering phenomena. There are two approaches to this hybrid, total-geometry model; those addressing discrete (or deterministic) materials, and those addressing statistical (or stochastic) materials.

In the discrete material model the geometry of the system is strictly defined, including the surface topography and the precise placement and size of the inclusions in the bulk matrix. This model is useful for applications where a specific surface texture is of interest (patterned or etched ridges, grooves, or other templated surfaces with directionality in either one or two dimensions), or where the placement of inclusions in the bulk material is well defined, such as fiber-containing composite materials where the fibers are directionally or anisotropically aligned.

The second type of model is the statistical or stochastic material model, where global and local statistical parameters are applied to the model either at the generation of the geometry, or on-the-fly and in real time, when individual scattering events occur in the bulk material. The statistical model relies on inputs such as volume percent loading of an inclusion such as a pigment or an air void, and employs a random number generator determination scheme at defined steps during the simulation to determine whether an inclusion resides at a strikeable set of coordinates for any particular incident photon. In the statistical model every photon that enters will be tracked until it dies of extinction or exits the material simulation space, while in the discrete model some photons may continue out of the simulation space in lateral directions and will be lost or missing from the summed scattering results, due to the well-defined geometry of the simulation space. The width of the statistical scatter model simulation space is unrestricted because the compositional input is volumetric and statistical, not discrete and spatially fixed.

Hybrid Scattering in an Ideal, Discrete Material

This section introduces a typical set of results from a very simple hybrid scattering simulation of a discrete material, in this case a 300 nm diameter SiC fiber embedded in a slab of PMMA that is bounded on the bottom by an aluminum substrate and on the top by a layer of ambient air.

Figure 7 shows a pictorial summary of the scattering events involved in this discrete material simulation. An incident beam of photons is fired towards the polymer surface at an incident angle of 20° to the global surface normal. In this example, the surface at the polymer/air interface is ideally and perfectly flat, so scattering is determined by the straightforward solution of Snell's laws and Fresnel reflectance functions, which predicts that of 100 incident photons, 4 are reflected specularly and leave the system after exactly one surface scattering event. Thus, of the initial 100 photons in the impinging beam, 96 are transmitted into the polymer slab.

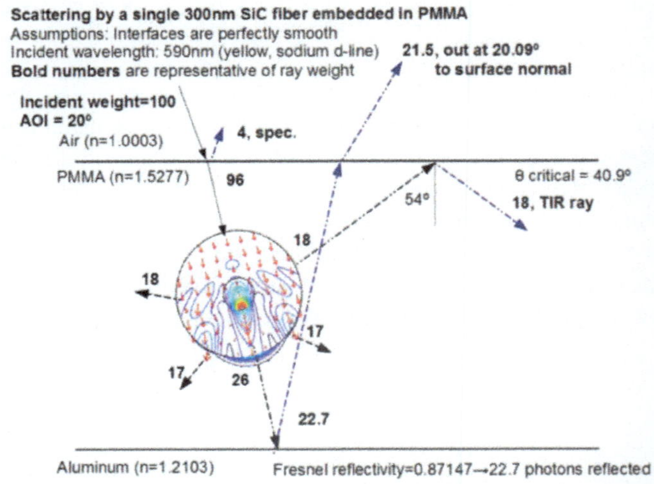

Figure 7. Pictorial summary of a hybrid light scattering simulation of a simple discrete material, a 300 nm diameter SiC fiber embedded in PMMA.

The transmitted photon beam with 96% of the original incident weight (*i.e.*, with a relative weight of 96) now travels in a straight line through the material until it strikes a domain occupied by a scatterer. At this point the model calls the finite element results for that particular scattering inclusion, and the incident rays from that scattering event, five in this case, with relative weights 17, 17, 18, 18, and 26, are allowed to transport further through the material.

Of the five scattered photons only two are tracked here in order to demonstrate the approach. One of the side-scattered photons, with relative weight of 18, travels towards the polymer/air interface, where it undergoes total internal reflection and is scattered back into the polymer material.

The forward-scattered photon, with relative weight of 26, travels towards the aluminum substrate, where some photon weight is absorbed by the metal, and the balance of the incident weight is reflected towards the polymer/air interface. It arrives at the polymer/air interface uninterrupted and exits with an angle of twenty degrees relative to the surface normal, and travels to an arbitrary collection bin where it is counted and summed into a result.

This formalism can be used to study or investigate the scatter processes or phenomena by nanoscale inclusions that are generally considered to be invisible, translucent, or clear. In reality, they are invisible to the human eye, but not to instrumental detectors, and certainly not to mathematical treatment! However, these scattering events are definitely "invisible" to closed form ray tracing where photons treated as particles wouldn't effectively "see" the inclusions because their scattering cross sections are smaller than the geometric limit of the radiation system of interest. The approach demonstrated here is capable of capturing the nuances involved in electromagnetic radiation transport through nanocomposite and electromagnetic signature-designed materials.

EXTENSION TO COLOR, COLOR MATCHING, AND METAMERISM PREDICTION AND VISUALIZATION

Oftentimes, light scattering simulations are only useful if they can ultimately provide visual results. Fortunately, this is readily accomplished by parameterizing the simulations across the visible spectrum, and summing those results into values for tristimulus values, CIE L*a*b*, or RGB. Literature in computational color science, such as the book by Westland and Ripamonti [42] and the classic text by Billmeyer and Saltzman, now maintained by Berns [31], offer many numerical methods for summing and translating between different color spaces.

The Color of Purely Diffuse Scattering Systems

In the case of purely diffuse scattering systems, such as dyed coatings and composites, quantum mechanical or semi-empirical molecular orbital theory calculations are used to compute the spectral absorbance of the individual material components, or of small clusters of components if dispersion is heterogeneous.

The semi-empirical molecular orbital theory package VAMP [43] uses the same linear combination of atomic orbitals-self consistent field (LCAO-SCF) theory that more computationally expensive *ab initio* programs use. However, some of the complex and more computation intensive integrals are removed and replaced using either simple approximations, empirical relationships, or known parameters where appropriate. The following example of implementation uses the NDDO Hamiltonian approximation set [44] with AM1 parameterized elements [45]. This approach is appropriate for molecules containing carbon, hydrogen, oxygen, and nitrogen. Modified approximation sets and parameterized elements should be used with other molecular systems, or a more dedicated computational approach should be taken in order to accommodate *ab initio*calculation.

For quinacridone violet in poly(methyl methacrylate), PMMA, the spectral absorbance of each of the components is calculated in VAMP, along with a user-defined mixed cell containing both components, shown in Figure 8.

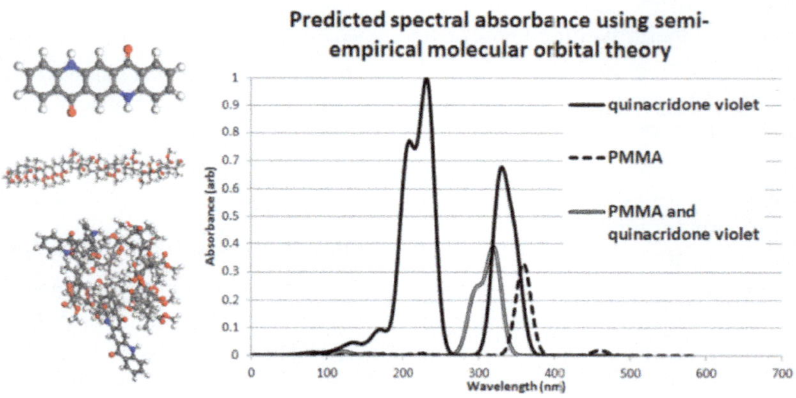

Figure 8. Semi-empirical molecular orbital theory predicts the optimized geometry of various components in a colored system, as well as the spectral absorbance spectra.

For a statistical material prediction, the weighted absorbance spectrum for the complete material may be computed from the individual component weights and their respective absorbance spectra. For discrete material predictions, a finite element approach to averaging may be used, where a grid of material is defined, wherein each grid element may have a different concentration of dye or other chromophore. The local grid properties are summed into the gross length scale absorbance values, as seen in Figure 9.

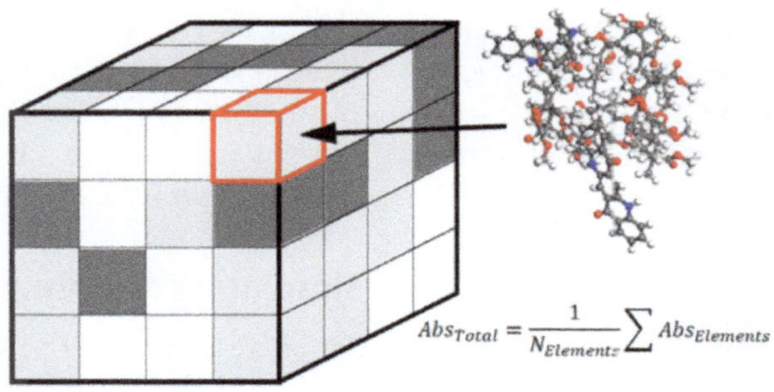

Figure 9. In systems with heterogeneous distribution of chromophores, a finite element approach to averaging may be used, whereby molecular modeling computes the individual absorbance of separate grid elements, and the bulk property exhibits the summed absorbance of its components.

If the total material has significant absorbance in the visible wavelength range, color prediction and visualization codes presented elsewhere [42] may be used to compute XYZ, RGB, and visual appearance color patches for qualitative and aesthetic assessment.

If the resultant total material absorbance is desired for color mixing and matching applications, the predicted absorbance response may be converted to Kubelka-Munk *K*/*S* values, using the formalism presented below.

It should be noted that many pigments and dyes contain, and indeed rely upon, inorganic components to impart their chromophoric behavior. For many such systems the *ab initio* and semi-empirical methods readily available may not suffice, and it may be easier or more appropriate to find values for spectral reflectance to be used as starting points in the open color science literature, such as the technical report by Mohammadi *et al.* [46].

Kubelka-Munk Theory and Application to Multiscale Light Scattering Simulations

The Kubelka-Munk set of color mixing theories [31,32,33,34,35,47,48,49,50] are useful in situations where the material being modeled scatters as well as absorbs light, and where only general flux information is desired. Although light in a scattering as well as absorbing material undergoes complex-subtractive mixing [31], the employment of Kubelka-Munk theory greatly simplifies matters. In this case, it is assumed that all light will be designated as either travelling up or down in a direction perpendicular to the plane of the sample. In other words, there is an "in-flux" and an "out flux," and the most common form of the Kubelka-Munk theory is generally referred to as a "two-flux" theory. A second important assumption in the theory is that the colorant layer must be completely diffuse, so that no directional information is incorporated into the model. Because of this, metallic, pearlescent, and many other directional special effects pigment types cannot be considered with Kubelka-Munk theory. Another negative aspect of the approach is that the set of parameters used apply to only one wavelength at a time, so a spectrophotometer must be used to generate data, and care must be taken in deciding what wavelengths to perform the math on.

For opaque systems like heavily pigmented or dyed coatings, a Kubelka-Munk value of *K*/*S* ("K over S") is determined from the reflectance at a given wavelength, and represents the ratio of absorption (*K*) to scattering (*S*).

$$\left(\frac{K}{S}\right)_{\lambda} = \frac{(1-R_{\lambda,i})^2}{2R_{\lambda,i}} \quad (3)$$

The *K*/*S* ratio of a mixture of components in a single material is simply an additive combination of each colorant component's unit absorptivity, k_λ, and unit scattering, s_λ, scaled by the effective concentration, and added to the absorption and scattering of the substrate, noted with subscript *t*, as in Equation 4.

$$\left(\frac{K}{S}\right)_{\lambda,\mathrm{mix}} = \frac{K_{\lambda,\mathrm{mix}}}{S_{\lambda,\mathrm{mix}}} = \frac{k_{\lambda,t} + c_1 k_{\lambda,1} + c_2 k_{\lambda,2} + \cdots}{s_{\lambda,t} + c_1 s_{\lambda,1} + c_2 s_{\lambda,2} + \cdots} \quad (4)$$

If one assumes no scattering and only absorption, the mixing equation is simplified and only the component *K*/*S* value needs to be known and Equation 5 may be used.

$$\left(\frac{K}{S}\right)_{\lambda,\mathrm{mix}} = \left(\frac{k}{s}\right)_{\lambda,t} + c_1\left(\frac{k}{s}\right)_{\lambda,1} + c_2\left(\frac{k}{s}\right)_{\lambda,2} + \cdots \quad (5)$$

Surface roughness and refractive index changes at the surface or surfaces are included through the Saunderson correction, Equation 6 [51], where K_1 and K_2 are the Fresnel reflection coefficients for light incident on the material from the outside and striking the surface from the inside, respectively.

$$R_{\lambda,i} = \frac{R_{\lambda,m} - K_1}{1 - K_1 - K_2 + K_2 R_{\lambda,m}} \quad (6)$$

For measurements taken with a specular excluded geometry, the K_1 value is removed from the numerator [31]. The Kubelka-Munk (*K*/*S*) value is roughly linear with respect to colorant concentration and can, with slight or no modification, be used when the appearance of a coating must be related to the type and amount of colorants in them, provided the colorants are typical dyes and pigments [52].

For the quinacridone violet (QV) and PMMA system noted above, the material is only weakly absorbing in a narrow absorbance band centered on 320 nanometers wavelength, with a relative unit value of 0.4. Thus, the (*K*/*S*) for this system is given by:

$$\left(\frac{K}{S}\right)_{QV/PMMA,320\ \mathrm{nm}} = \frac{(1-(1-0.4))^2}{2(1-0.4)} = 0.133 \quad (7)$$

This simulated material is now ready to be added to a virtual color mixing program as a component or colorant. Due to the discrete nature of the atomistic simulation used to compute the absorbance, care should be taken to adjust the virtual concentration of dye within the surrounding polymer matrix, and an effort should be made to create standard curves using known physical formulations to benchmark the simulations against.

Color Prediction of Scattering Systems with Diffuse and Directional Components

For systems containing both diffuse and directional components, a hybrid approach to color prediction must be taken. The diffuse components should be treated separately from the directional components. The directional components may undergo the usual ray tracing and finite element inclusion scattering approach, and results can be combined to produce the final predicted material response.

A typical example of such an approach would be an organic coating containing an organic dye, such as quinacridone violet, and a pigment such as titanium dioxide. The QV in polymer matrix will be computed using the atomistic and Kubelka-Munk approaches, and the rough surface and titanium dioxide pigments will be treated using the ray tracing and finite element approach used for directional materials. Results can be summed at the material tristimulus value construction, as the tristimulus value and, subsequently, RGB and CIELAB values, are not length scale or directionally-dependent.

Simulating Metamerism

An interesting aspect of color visualization is the prediction of metamerism. These methods replicate the effects of color blindness (observer metamerism) and the effect of changing the incident radiation source (source metamerism). Using tristimulus values, the predicted spectral response of a material generated by coupled surface and inclusion scattering methods can be combined with known light source spectral power distributions to predict the resultant color of the material seen by a neutral observer. Figure 10 shows the predicted source metamerism effects of an epoxy loaded with quinacridone violet dye, illumined under different common light sources C, D65, A, and F2 incandescent bulb. Clearly, the illuminant will render the same surface many colors, ranging from orange to purple. These methods enable the coatings scientist to predict color and appearance, even when a large selection of laboratory light sources is physically unavailable.

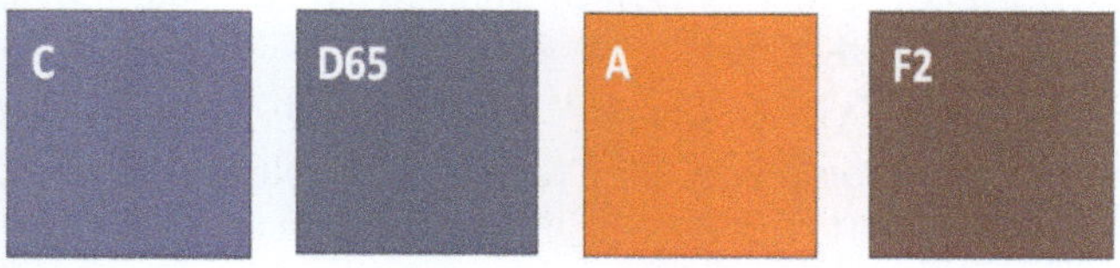

Figure 10. Simulating the effects of source metamerism; an epoxy loaded with quinacridone violet illuminated under various lighting conditions.

CONCLUSIONS

An ongoing challenge in any simulation-driven area of research is the continual incorporation of new methods with historically proven methods, as well as the repackaging of those methods to serve new audiences of users. The work presented here combines two well-established light scattering simulation methods, one for surfaces and one for bulk inclusions, into a combined approach for whole-material, omniscale light scattering, which is flexible and adaptive to many material types, but which is also rigorous in its treatment of the radiative transfer mathematics. Forthcoming papers from the authors will present the results of scaled up hybrid light scattering simulations; address nuances in photon queue and accounting structures; and present experimental validation of the models using various materials of industrial importance.

ACKNOWLEDGMENTS

The authors wish to acknowledge the Army Research Laboratory for funding this effort, under ARL grant #14518, as well as that grant's principal investigator, Victoria Johnston Gelling, at North Dakota State University, for her support of the work.

REFERENCES

1. Martelli, F.; Del Bianco, S.; Ismaelli, A.; Zaccanti, G. *Light Propagation through Biological Tissue and Other Diffusive Media*; SPIE Press: Bellingham, WA, USA, 2010.
2. Tishkovets, V.P. Incoherent and coherent backscattering of light by a layer of densely packed random medium. *J. Quant. Spectrosc. Radiat. Transfer* **2007**, *108*, 454–463.
3. Tishkovets, V.P. Light scattering by closely packed clusters: Shielding of particles by each other in the near field. *J. Quant. Spectrosc. Radiat. Transfer* **2008**, *109*, 2665–2672.
4. Tishkovets, V.P.; Jockers, K. Multiple scattering of light by densely packed random media of spherical particles: Dense media vector radiative transfer equation. *J. Quant. Spectrosc. Radiat. Transfer* **2006**, *101*, 52–74.
5. *Light Scattering and Nanoscale Surface Roughness*; Maradudin, A.A., Ed.; Nanostructure Science and Technology Series XXIV; Springer: New York, NY, USA, 2007.
6. Van de Hulst, H.C. *Light Scattering by Small Particles*; John Wiley & Sons: New York, NY, USA, 1957.
7. Bohren, C.F.; Huffman, D.R. *Absorption and Scattering of Light by Small Particles*; John Wiley & Sons: New York, NY, USA, 1983; p. 530.
8. Beckmann, P.; Spizzichino, A. *The Scattering of Electromagnetic Waves from Rough Surfaces*; Oxford: New York, NY, USA, 1963; p. 503.
9. Chen, M.F.; Fung, A.K. A numerical study of the regions of validity of the Kirchhoff and small-perturbation rough surface scattering models. *Radio Sci.* **1988**, *23*, 163–170.
10. Thorsos, E.I. The validity of the Kirchhoff approximation for rough surface scattering using a Gaussian roughness spectrum. *J. Acoust. Soc. Am.* **1988**, *83*, 78–92.
11. Davies, H. The reflection of electromagnetic waves from a rough surface. *Proc. IEE* **1954**, *101*, 209–214.
12. Beckmann, P. A new approach to the problem of reflection from a rough surface. *Acta Techn. CSAV* **1957**, *2*, 311–355.
13. Thomas, T.R. *Rough Surfaces*, 2nd ed; Imperial College Press: London, UK, 1999.
14. Dorsey, J.; Rushmeier, H.; Sillion, F. *Digital Modeling of Material Appearance*; Elsevier: New York, NY, USA, 2008.
15. Tang, K.; Buckius, R.O. The geometric optics approximation for reflection from two-dimensional random rough surfaces. *Int. J. Heat Mass Transfer* **1998**, *41*, 2037–2047.
16. Tang, K.; Dimenna, R.A.; Buckius, R.O. Regions of validity of the geometric optics approximation for angular scattering from very rough surfaces. *Int. J. Heat Mass Transfer***1997**, *40*, 49–59.
17. Warnick, K.F.; Arnold, D.V. Generalization of the geometrical-optics scattering limit for a rough conduting surface. *Opt. Soc. Am.* **1998**, *15*, 2355–2361.
18. Bergstrom, D.; Powell, D.; Kaplan, A.F.H. A ray-tracing analysis of the absorption of light by smooth and rough metal surfaces. *J. Appl. Phys.* **2007**, *101*, 113504.
19. Nicodemus, F.E. Directional reflectance and emissivity of an opaque surface. *Appl. Opt.* **1965**,*4*, 767–773.
20. Parvianen, H.; Muinonen, K. Bidirectional reflectance of rough particualte media: Ray-tracing solution. *J. Quant. Spectrosc. Radiat. Trans.* **2009**, *110*, 1418–1440.
21. Brockelman, R.A.; Hagfors, T. Note on the effect of shadowing on the backscattering of waves from a random rough surface. *IEEE Trans. Ant. Prop.* **1966**, *14*, 621–626.
22. Sapper, E.; Hinderliter, B. A Hybrid Computational Approach for Simulation the Scattering of Electromagnetic Radiation in Coatings and Composite

Materials. CoatingsTech Conference, Rosemont, IL, USA, 2013.

23. Bennett, J.M.; Mattsson, L. *Introduction to Surface Roughness and Scattering*; Optical Society of America: Washington, D.C., WA, USA, 1989; p. 110.
24. Tsang, L.; Kong, J.A.; Ding, K.-H.; Ao, C.O. *Scattering of Electromagnetic Waves: Numerical Simulations*; John Wiley & Sons: New York, NY, USA, 2001.
25. Stover, J.C. *Optical Scattering: Measurement and Analysis*; SPIE Press: Bellingham, WA, USA, 1995.
26. Bergstrom, D. The Absorption of Laser Light by Rough Metal Surfaces. Ph.D. Thesis, Luleå University of Technology, Luleå, Sweden, 2008.
27. Mishenko, M.I.; Travis, L.D. Gustav Mie and the evolving discipline of electromagnetic scattering by particles. *Bull. Am. Meteorol. Soc.* **2008**, *89*, 1853–1861.
28. Wiscombe, W.J. Improved Mie scattering algorithms. *Appl. Opt.* **1980**, *19*, 1505–1509.
29. Rother, T.; Schmidt, K. The discretized Mie-formalism—A novel algorithm to treat scattering on axisymmetric particles. *J. Electromag. Wav. Appl.* **1996**, *10*, 273–297.
30. Rother, T.; Schmidt, K. The discretized Mie-formalism for plane wave scattering on dielectric objects with non-separable geometries. *J. Quant. Spectrosc. Radiat. Transfer* **1996**, *55*, 615–625.
31. Berns, R.S. *Billmeyer and Saltzman's Principles of Color Technology*, 3rd ed; John Wiley & Sons: New York, NY, USA, 2000.
32. Phillips, D.G.; Billmeyer, F.W. Predicting reflectance and color of paint films by Kubelka-Munk analysis. *J. Coat. Technol.* **1976**, *48*, 30–36.
33. Kubelka, P. New contributions to the optics of intensely light-scattering material, part I. *J. Opt. Soc. Am.* **1948**, *38*, 448.
34. Kubelka, P. New contributions to the optics of intensely light-scattering material, part II. *J. Opt. Soc. Am.* **1954**, *44*, 330.
35. Kubelka, P.; Munk, F. Ein beitrag zur optik der farbanstriche. *Z. Tech. Physik.* **1931**, *12*, 330.
36. Bierwagen, G.P. Estimation of film thickness nonuniformity effects on coating optical properties. *Col. Res. Appl.* **1992**, *17*, 284–292.
37. Billmeyer, F.W.; Chassaigne, P.G.; Dubois, J.F. Determining pigment optical properties for use in the mie and many-flux theories. *Col. Res. Appl.* **1980**, *5*, 108–112.
38. Maxwell, J.C. On physical lines of force. *Lond. Edinb. Dublin Philos. Mag. J. Sci.* **1861**, *21*, 161–175, 281–291, 338–348.
39. COMSOL, *COMSOL Multiphysics RF Module User's Guide*; COMSOL AB: Burlington, MA, USA, 2006.
40. Jackson, J. *Classical Electrodynamics*; John Wiley & Sons: New York, NY, USA, 1998.
41. Shadowitz, A. The Electromagnetic Field. In *Dover Books on Physics*; Dover: New York, NY, USA, 2010.
42. Westland, S.; Ripamonti, C. *Computaitonal Colour Science Using MATLAB*; John Wiley & Sons: West Sussex, UK, 2004.
43. Leach, A. *Molecular Modeling: Principles and Applications*, 2nd ed; Prentice Hall: New York, NY, USA, 2001.
44. Pople, J.A.; Santry, D.P.; Segal, G.A. Approximate self-consistent molecular orbital theory. I. Invariant procedures. *J. Chem. Phys.* **1965**, *43*, S129.
45. Dewar, M.J.S.; Zoebisch, E.G.; Healy, E.F.; Stewart, J.J.P. Development and use of quantum mechanical molecular models. 76. AM1: A new general purpose quantum mechanical molecular model. *J. Am. Chem. Soc.* **1985**, *107*, 3902–3909.
46. Mahnaz, M.; Mahdi, N.; Lawrence, T.; Roy, B. Pigment selection using Kubelka-Munk turbid media theory and non-negative least square technique. Available online: https://ritdml.rit.edu/handle/1850/4353 (accessed on 3 April 2013).
47. Latimer, P.; Noh, S.J. Light propagation in moderately dense particle systems: A reexamination of the Kubelka-Monk theory. *Appl. Opt.* **1987**, *26*, 514–523.
48. Molenaar, R.; Bosch, J.J.T.; Zijp, J.R. Determination of Kubelka-Munk scattering and absorption coefficients by diffuse illumination. *Appl. Opt.* **1999**, *38*, 2068–2077.
49. Murphy, A.B. Modified Kubelka-Munk model for calculation of the reflectance of coatins with optically-rough surfaces. *J. Phys. D* **2006**, *39*, 3571–3581.
50. Vargas, W.E.; Niklasson, G.A. Applicability conditions of the Kubelka-Munk theory. *Appl. Opt.* **1997**, *36*, 5580–5586.
51. Saunderson, J.L. Calculation of the color of pigmented plastics. *J. Opt. Soc. Am.* **1942**, *32*, 727–736.
52. HunterLab. The Kubelka-Monk theory and K/S. Available online: http://www.hunterlab.com/appnotes/an07_06.pdf (accessed on 3 April 2013)

INDEX

N

O

P

R

S

T

V